Atomic Masses and Fundamental Constants 6

Previously Published Conferences on Atomic Masses

Nuclear Masses and Their Determinations, H. Hintenberger (ed.), Pergamon Press, London (1957).

Proceedings of the International Conference on Nuclidic Masses, H. E. Duckworth (ed.), University of Toronto Press, Toronto (1960).

Nuclidic Masses, Proceedings of the Second International Conference on Nuclidic Masses, W. H. Johnson, Jr. (ed.), Springer-Verlag, Vienna and New York (1964).

Proceedings of the Third International Conference on Atomic Masses, R. C. Barber (ed.), University of Manitoba Press, Winnipeg (1968).

Atomic Masses and Fundamental Constants 4, J. H. Sanders and A. H. Wapstra (eds.), Plenum Press, London and New York (1972)

Atomic Masses and Fundamental Constants 5, J. H. Sanders and A. H. Wapstra (eds.), Plenum Press, New York and London (1975)

Atomic Masses and Fundamental Constants 6

Edited by

Jerry A. Nolen, Jr.

and

Walter Benenson

Michigan State University
East Lansing, Michigan

Plenum Press · New York and London

Library of Congress Cataloging in Publication Data

International Conference on Atomic Masses, 6th, East Lansing, Mich., 1979.
    Atomic masses and fundamental constants, 6.

    Includes index.
    1. Atomic mass—Congresses. 2. Atomic mass—Measurement—Congresses. I. Nolen,
Jerry A. II. Benenson, Walter. III. Title. IV. Title: Fundamental constants.
QC172.I56 1979   541.2'42   80-12163
ISBN 0-306-40441-9

Proceedings of the Sixth International Conference on Atomic Masses,
held in East Lansing, Michigan, September 18—21, 1979.

© 1980 Plenum Press, New York
A Division of Plenum Publishing Corporation
227 West 17th Street, New York, N.Y. 10011

Printed in the United States of America

PREFACE

The Sixth International Conference on Atomic Masses was held in East Lansing, Michigan, Sept. 18-21, 1979. The conference was initiated, organized, and sponsored by the Commission on Atomic Masses and Fundamental Constants of the International Union of Pure and Applied Physics.

The members of the conference committee are listed below:

W. Benenson, Chairman	Michigan State University
R.C. Barber	University of Manitoba
E.R. Cohen	Rockwell International
V.I. Goldanskii	Institute of Chemical Physics, Moscow
J.C. Hardy	Chalk River, Canada
W.H. Johnson	University of Minnesota
E. Kashy	Michigan State University
R. Klapisch	Orsay, France
J.A. Nolen, Jr.	Michigan State University
R.G.H. Robertson	Michigan State University
E. Roeckl	G.S.I., Darmstadt
B.N. Taylor	National Bureau of Standards
O. Schult	IKF, Julich
A.H. Wapstra	IFO, Amsterdam
N. Zeldes	Racah Institute, Jerusalem

The conference was a little different from the preceding one (in Paris, 1975) in that the fundamental constant aspects were limited to those directly related to atomic masses. The gap is to be filled by the second International Conference on Precision Measurement and Fundamental Constants which is now scheduled for June 1981 in Gaithersburg, Maryland. Only one of the seven sessions in this conference was devoted to fundamental constant determinations.

The conference was very strongly supported by the Department of Energy, the National Science Foundation, and the International Union of Pure and Applied Physics. In addition, the organizers are grateful for contributions from Oak Ridge Technical Enterprises and Copper and Brass Sales of Grand Rapids, Michigan.

The summary papers in the last section of these precedings show
that the fields represented are indeed in a very productive period.
A gratifying development is the very high theoretical interest in
the results, as is indicated by the two entire sessions devoted to
the theoretical treatment of atomic masses. The conference was
also enhanced by the presence of several vigorous and active
attendees from the original "zeroth conference" in Mainz, 1956.
We hope that the continuity of scientific interest that this
indicates will persist, and judging from the quality and quantity
of the papers presented at this conference we have no doubts that
it will.

J.A. Nolen, Jr.
W. Benenson
Michigan State University

CONTENTS

Session II Theory of Atomic Masses, First Section
Chairman: W.M. Myers

Session III Fundamental Constants
Chairman: E.R. Cohen

Session IV   Direct Mass Measurements
Chairman: H.E. Duckworth

Session V   Theory of Atomic Masses, Second Section
Chairman: N. Zeldes

Session VI   Studies of Isotopes with On-Line Mass Separators
Chairman: D.E. Alburger

Session VII   Summaries
Chairman: K. Way

STUDIES OF ISOSPIN QUINTETS AND NEUTRON-DEFICIENT INDIUM
ISOTOPES WITH THE ON-LINE MASS ANALYSIS SYSTEM RAMA*

Joseph Cerny, J. Äystö[+], M. D. Cable, P. E. Haustein[++],
D. M. Moltz, R. D. von Dincklage[+++], R. F. Parry and
J. M. Wouters

Department of Chemistry and Lawrence Berkeley Laboratory
University of California, Berkeley, California 94720

I. INTRODUCTION

The characterization of the nuclidic mass in those regions
which are far removed from the line of beta stability is of
fundamental importance for testing and improving mass theories.
An increasing component of our knowledge on nuclear properties
of light exotic nuclei is derived from the analysis of mass-
separated radioactivity, complementing the much more extensive
investigations by this technique of heavy nuclei off the stability
line.[1] Such studies provide new insights into the limits of nuclear
stability, the behavior of nuclear shapes and the onset of new
radioactive decay modes.

The success of the isobaric multiplet mass equation in relating
mass-excesses of members of light isospin quartets has permitted it
to play a significant role in predicting ground state masses of
highly proton-rich 1f2p shell nuclides. In addition, the extensive
studies of complete isospin quartets[2] plus more recent research on
isospin quintets provide valuable nuclear structure information as
well as deepening our basic understanding of charge-dependent
effects in the interactions between nucleons.

This report in part gives an overview of an experimental
program that was developed to study the most proton rich members of
high isospin multiplets through their radioactive decay. In
experiments on light nuclei far from stability, the capability for
on-line mass analysis of nuclides of many chemical elements with
half-lives as short as 50 ms is clearly of importance. In order
to accomplish this, an instrument known as RAMA, for Recoil Atom

1

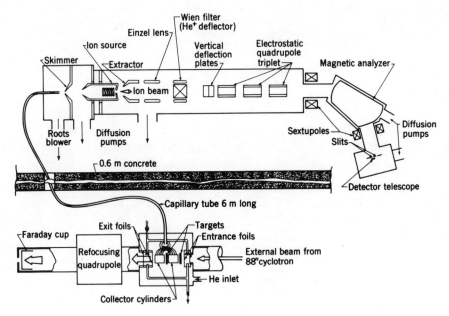

RAMA - 88 SCHEMATIC

Fig. 1   Schematic view of the on-line mass separator RAMA.

Mass Analyzer, has been constructed and will be briefly described below.[3]

Another part of our program in studying nuclei far from stability is the initiation of a series of experiments to determine total decay energies of isotopes in the vicinity of the doubly-magic nucleus $^{100}_{50}$Sn. Mass-excess determinations for nuclides in this region should highlight the influence of the closed shells on the decay energies and establish whether such proton-rich magic nuclei follow the same mass systematics as do those nearer to or at stability. In the following we give preliminary results of experiments determining the total decay energies, and hence mass-excesses, of $^{103}_{49}$In and $^{105}_{49}$In.

II.   DESCRIPTION OF THE RAMA SYSTEM

Figure 1 presents the experimental lay-out of the on-line mass analysis system RAMA at the Lawrence Berkeley Laboratory 88-inch cyclotron. RAMA employs a helium jet to transport activity from the target area to a Sidenius-type hollow cathode ion source which is coupled to a mass separator. A special multiple target and multiple capillary system was constructed to provide optimal yield. Reaction recoils are thermalized inside a cylinder and are collected by a set of capillaries spaced evenly over the distance of a maximum recoil range. A 6 m long stainless steel capillary is then fed by the multiple capillary system, transporting the activity to the skimmer-ion source region. Ethylene glycol is employed as an additive in the helium to build up high molecular weight aerosols; nuclides attached to these aerosols possess excellent transport and skimming properties. After skimming, the reaction products are ionized in the ion source which is operating at temperatures up to 2000 °C. Singly charged ions are extracted at 18 kV and mass analyzed as shown in Fig. 1. Overall efficiencies for RAMA currently range from 0.1-0.5% for such elements as Na, Mg, Si, Ca, In, Te, Cs, Ho, Er and At. The shortest half-lives observed to date have been ∿100 ms.

Detection systems for beta-delayed proton spectroscopy as well as for beta-, gamma-, and X-ray spectroscopy have been developed. Nuclides with short half-lives (∿100-500 ms) are observed by electrostatically deflecting the mass analyzed ion beam from one to another of a pair of vertically positioned detector systems placed about the RAMA focal plane. This technique gives half-life information as well as particle identification and energy measurements. For the beta-delayed proton studies, solid-state counter telescopes were employed which were protected from the ion beam by a 50 μg/cm^2 carbon foil. Investigation of the longer-lived, beta-

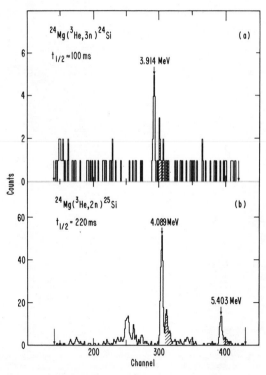

Fig. 2   Spectra of beta-delayed protons from a) ^{24}Si and b) ^{25}Si obtained at 70 and 41 MeV beam energy, respectively. Shaded areas in both spectra are due to a pile-up effect.

gamma emitting indium isotopes was accomplished by collecting the
mass analyzed ions on common magnetic computer tape; this tape can
be moved at a speed of 75 cm/s from air into and out of the evacuated focal plane region via two differentially-pumped vacuum chambers.
Singles and coincident beta-gamma spectra were obtained by high
geometry viewing of the collected activity.

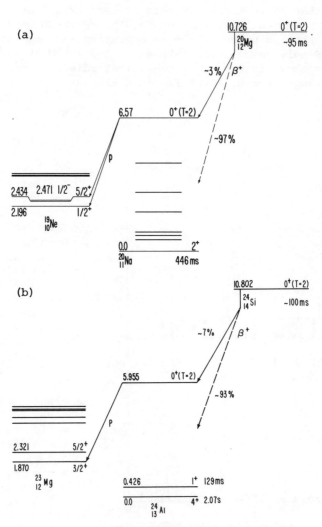

Fig. 3(a)   Proposed decay scheme of ^{20}Mg.
       (b)   Proposed decay scheme of ^{24}Si.
The ^{24}Al-^{24}Si mass difference was taken
from the quadratic IMME prediction.

III.   ISOSPIN QUINTETS

The masses of the 2T+1 members of an isospin multiplet are given in first order by the isobaric multiplet mass equation (IMME) as

$$M(A,T,T_z) = a(A,T) + b(A,T) \cdot T_z + c(A,T) \cdot T_z^2 .$$

This equation results from the assumptions that the wavefunctions of analog states in an isospin multiplet are identical and that charge-dependent forces (Coulomb plus nuclear) are of two-body character and may be treated as perturbations. Although this quadratic form of the IMME fits the vast majority of the data on isospin quartets, a persistent deviation has been reported in the mass 9 quartet.[2] In addition the mass 8 isospin quintet also shows a deviation from the simple IMME,[4] so that tests at higher masses have become imperative to establish comparable systematics for quintets. Deviations from the quadratic form are generally represented by additional terms $d(A,T) \cdot T_z^3$ and $e(A,T) \cdot T_z^4$, in which the d and e coefficients can be derived from second order perturbation theory.

At present, experimental data on isospin quintets are almost exclusively limited to the A = 4n series. Although mass excesses of the T_z = + 2, + 1 and 0 members are well known from A = 8 to A = 40, progress with regard to mass determinations of the T_z = -1 and -2 members has been much slower. Probably the most general approach for locating the analog states in the T_z = -1 nuclei is through investigation of decays of the T_z = -2 nuclei,[5] while the (^{4}He, ^{8}He) four-neutron transfer reaction is commonly employed to determine the ground state mass of the T_z = -2 member.

One of the initial scientific motivations for the construction of RAMA was to exploit its properties to permit

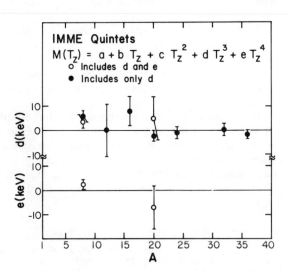

IMME Quintets
$M(T_z) = a + b\, T_z + c\, T_z^2 + d\, T_z^3 + e\, T_z^4$
○ Includes d and e
● Includes only d

Fig. 4   A summary of the d and e coefficents of the IMME required to fit 0^+, T = 2 multiplets in which four or more members are known.

observation of the $T = 2$ states in the $T_z = -1$ nuclei via the beta-delayed proton decay of the mass-separated $T_z = -2$ nuclei. So far we have observed the decays of ^{20}Mg [6] and ^{24}Si,[7] completing the mass 20 isospin quintet and determining the mass-excess of the fourth member, ^{24}Al, of the mass 24 quintet. ^{20}Mg is the lightest nucleon-stable member of the series of beta-delayed proton emitters, since ^{12}O and ^{16}Ne are known[8] to have unbound ground states.

As an example, the beta-delayed proton spectrum arising from the decay of mass-separated ^{24}Si is shown in Fig. 2(a). The single peak evident in the spectrum occurs at a laboratory energy of 3.914±0.009 MeV and results from the isospin-forbidden proton decay of the lowest $T = 2$ state in ^{24}Al. The decay of ^{25}Si provides a convenient proton-energy calibration as is shown in Fig. 2(b). Similar beta-delayed proton spectra were also obtained for ^{20}Mg and ^{21}Mg. Decay schemes for ^{20}Mg and ^{24}Si are shown in Fig. 3(a) and 3(b), respectively. The mass excesses of the lowest $T = 2$ states in ^{20}Na and ^{24}Al, as calculated from the observed proton energies, are 13.42±0.05 and 5.903±0.009 MeV, respectively. For both these multiplets, an excellent fit to the mass-excesses is obtained by using only the quadratic form of the isobaric multiplet mass equation.

A summary of the present experimental situation regarding fits via the IMME for studies of isospin quintets in which four or more members are known[4-9] is presented in Fig. 4. Here the d coefficient has been determined for each multiplet as well as the d and e coefficients for the complete quintets at masses 8 and 20.

The only deviations from the quadratic form of the IMME that have been observed arise in the mass 8 [4] and mass 16 [8] quintets with the latter being less statistically significant. In the case of the mass 8 quintet, the nonzero d and e coefficients have been attributed to the strong Coulombic repulsion associated with its particle-unbound members, in addition to the effect of isospin mixing in the $T_z = 0$ member of this multiplet. [All members of the complete mass 20 quintet are stable toward isospin-allowed particle decay.] Generally, however, the results on isospin quintets together with the numerous measurements on isospin quartets support the validity of the simple quadratic mass equation and provide no evidence for substantial higher-order, charge-dependent effects in the nuclear interaction.

Extension of these experiments to detect other unknown $T_z = -2$ beta-delayed proton emitters, such as ^{28}S and ^{36}Ca, is planned. Our greatest interest centers on the A = 36 quintet, since a determination of the $T = 2$ state in ^{36}K would complete the heaviest quintet possible with established techniques and stable targets.

IV.   NEUTRON-DEFICIENT INDIUM ISOTOPES

Total decay energies of the indium isotopes of interest were determined by utilizing RAMA for mass separation together with β^+ - γ coincidence techniques for data collection. Activity collected on magnetic tape on the focal plane was rapidly transported to a detection station. Isotopic identification and end point determinations were accomplished by gating a beta-telescope with known γ-rays of the daughter nucleus.

IV.1.   Detector System

A telescope designed to measure Q_β values up to 20 MeV was used for β-particle detection. It consisted of a 10 mm diameter and 1 mm thick NE 102 plastic scintillator as a ΔE detector [for γ-ray rejection] and a large cylindrical, 11.4 cm diameter and 11.4 cm long, NE 102 plastic scintillator as an E detector. Typical coincidence timing between the ΔE and E counters was 5 ns full width at half-maximum. Gamma-ray detection was accomplished using a large volume (15%) Ge(Li) counter with a resolution of 2.1 keV at 1332 keV. Due to the low yields of the nuclei of interest, a high geometry was employed between the β- and γ-counters; the coincidence resolution between them was 20 ns.

The recorded positron spectra were distorted by energy-dependent effects, such as the finite energy resolution, back-scattering at the scintillation surface and pile-up by Compton recoil electrons arising from the annihilation radiation. However, in the rather narrow energy range of 2 to 5 MeV used in these experiments, the last two effects were of considerably less importance than the first in determining positron end-point energies.[10] As a result the measured β^+-spectra were corrected only for the resolution. Conversion electron measurements gave an energy resolution of

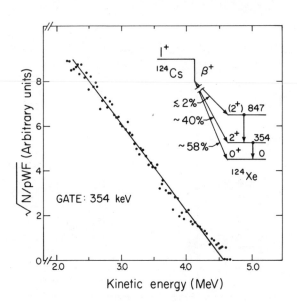

Fig. 5 Fermi Kurie plot and partial decay scheme for ^{124}Cs. Beta-branching ratios were obtained from the γ-spectrum in coincidence with positrons. Ground branching was taken from Ref. 12.

200 keV (FWHM) at $\sim$1 MeV and the response function of the E detector
was then assumed to be a Gaussian curve whose width varied with
energy as $\sqrt{E}$.[10] The overall energy calibration using this response
function and employing known positron activities was found to provide
a good linear fit.

IV.2. Decay Energy Measurements

Energy calibration of the beta-telescope was based on the beta
end points of the strongest allowed decay branches of ^{123}Cs(3.410±
0.120 MeV--ref. 11,12); ^{124}Cs(4.574±0.150 MeV--ref. 11,12); and
^{112}Sb(4.750±0.050 MeV--ref.13) produced in ^{20}Ne + Ag and ^{20}Ne + Mo
reactions at 150 MeV bombarding energy, respectively. As an example
of these calibrations, the Fermi-Kurie plot of the response-function-
corrected beta spectrum of ^{124}Cs is given in Fig. 5. The weak β⁺-
decay component feeding the second 2⁺ state does not appear to affect
the excellent straight line fit.

Indium isotopes were produced by bombarding 2 mg/cm^2 ^{102}Pd
targets with a beam of 200 MeV ^{14}N ions; this relatively high beam
energy was found to produce optimum yields of the isotopes of
interest in our particular multiple target setup. Decay energies
were measured for $^{103-106}$In. A typical result is shown in Fig. 6,
which presents a
similar Fermi-Kurie
analysis of the posi-
tron spectrum of ^{105}In
that is in coincidence
with the 131 keV γ
transition in the
^{105}Cd daughter. About
80% of the beta decay
was found to feed this
131 keV level, while
most of the remaining
decay strength feeds
levels at 196 and 260
keV which directly de-
excite to the ground
state.[12,14] As a
result, the Fermi-
Kurie plot is quite
linear in the energy
interval from 2.5 to
4.0 MeV.

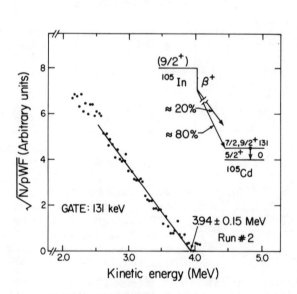

Fig. 6 Fermi-Kurie plot and partial decay
scheme for ^{105}In. Beta-branching ratios
were determined from the γ-spectrum
in coincidence with positrons.

The positron end
point of the decay of

^{103}In was obtained by gating the beta counter with the 188 keV, $7/2^+ \rightarrow 5/2^+$, ground state transition in the daughter ^{103}Cd.[15] This 188 keV state is strongly fed in the beta-decay ($\sim$80%) and only one additional transition--direct de-excitation of a level at 202 keV to the ground state--was observed. The resulting end point of 4.35±0.35 MeV is in reasonable agreement with the Louvain on-line isotope separator result.[15] [End point energies for ^{104,106}In were also determined; however, because of the unsatisfactory under-standing of isomerism in these isotopes and substantial disagreement with previous results, additional studies appear to be necessary.] A summary of our measurements of the ^{103,105}In decay energies is given in Table I.

Table I. SUMMARY OF Q_{EC} DETERMINATIONS

NUCLIDE	GATE [keV]	LEVEL [keV]	DECAY ENERGY [MeV]	
			THIS WORK	LITERATURE
^{103}In	188	188	5.56±0.35	5.8±0.5[a]
^{105}In	131	131	5.12±0.13[b]	-

a) ref. 15, b) average of two measurements.

The mass excesses of ^{103}In and ^{105}In are presented in Table II.[16,17] Deviations of the experimental values from a semi-empirical shell model formula by Liran and Zeldes,[16a] from a Garvey-Kelson transverse equation,[16b] as well as from a droplet model calculation by Myers[16c] are given for comparison.

Table II. MASS EXCESSES OF ^{103}In and ^{105}In
COMPARED TO DIFFERENT MASS PREDICTIONS

NUCLIDE	MASS EXCESS [MeV]	Δ = ME(EXP)-ME(THEOR) [MeV]		
		L-Z[a]	G-K[b]	Myers[c]
^{103}In	-75.04±0.38[d]	-1.00	-0.61	0.87
^{105}In	-79.22±0.13[d]	0.24	0.55	1.21

a) ref. 16a; b) ref. 16b; c) ref. 16c; d) mass-excesses of the ^{103,105}Cd daughters were taken from ref. 17.

In the main, these predicted masses agree fairly well with the experimental measurements--the observed deviations are of the same order of magnitude as were observed for the neutron-rich $^{120-129}$In isotopes.[18] An interesting disagreement of -1.0 MeV of the ^{103}In mass excess from the Liran-Zeldes calculations[16a] is inconsistent with the good agreement of their approach in predicting the other experimentally known In mass excesses. Further systematic studies are required to understand both this behavior and whether it might have any possible relationship with a nearby double shell closure.

Footnotes and References

*   This work was supported by the Nuclear Physics and Nuclear Sciences Divisions of the U.S. Department of Energy under contract No. W-7405-ENG-48.

+   On leave from the University of Jyväskylä, Finland.

++ Permanent address: Brookhaven National Laborabory, Upton, N.Y.

+++ Present address: University of Göttingen, West Germany.

1. Proc. Int. Conf. on Nuclei Far from Stability, Cargese, 1976, 3rd. Rep. CERN 76-13 and P. G. Hansen, to be published in Ann. Rev. Particle and Nuclear Science.
2. W. Benenson and E. Kashy, to be published in Revs. Mod. Phys. (1979).
3. J. Cerny, D. M. Moltz, H. C. Evans, D. J. Vieira, R. F. Parry, J. M. Wouters, R. A. Gough and M. S. Zisman, Proc. of the Isotope Separator On-Line Workshop, ed. R. E. Chrien, BNL-50847 (1977) p. 57.
4. R. G. H. Robertson, W. Benenson, E. Kashy, and D. Mueller, Phys. Rev. C13 (1976) 1018; R. E. Tribble, R. A. Kenefick and R. L. Spross, Phys. Rev. C13 (1976) 50.
5. E. G. Hagberg, P. G. Hansen, J. C. Hardy, A. Huck, B. Jonson, S. Mattsson, H. L. Ravn, P. Tidemand-Peterson, and G. Walter, Phys. Rev. Lett. 39 (1977) 792.
6. D. M. Moltz, J. Äystö, M. D. Cable, R. D. von Dincklage, R. F. Parry, J. M. Wouters, and J. Cerny, Phys. Rev. Lett. 42 (1979) 43.
7. J. Äystö, D. M. Moltz, M. D. Cable, R. D. von Dincklage, R. F. Parry, J. M. Wouters and J. Cerny, Phys. Lett. 82B (1979) 43.
8. G. J. Kekelis, M. S. Zisman, D. K. Scott, R. Jahn, D. J. Vieira, J. Cerny and F. Ajzenberg-Selove, Phys. Rev. C17 (1978) 1929.
9. R. E. Tribble, J. D. Cossairt, and R. A. Kenefick, Phys. Rev. C15 (1977) 2028.
10. E. Beck, Nucl. Instr. Meth. 76 (1969) 77.

11. M. Epherre, G. Audi, C. Thibault, R. Klapisch, G. Huber, F. Touchard and H. Wollnik, Phys. Rev. C19 (1979) 1504.

12. C. M. Lederer and V. S. Shirley, editors, Table of Isotopes, 7th edition, John Wiley and Sons, New York, 1978.

13. M. Singh, J. W. Sunier, R. M. DeVries and G. E. Thompson, Nucl. Phys. A193 (1972) 449.

14. J. Rivier and R. Moret, Rad. Acta 22 (1975) 27.

15. G. Lhersonneau, G. Dumont, K. Cornelis, M. Huyse and J. Verplancke, Phys. Rev. C18 (1978) 2688.

16. (a) S. Liran and N. Zeldes, At. Data and Nucl. Data Tables 17 (1976) 431; (b) J. Jänecke, ibid. 17 (1976) 455; (c) W. D. Myers, ibid. 17 (1976) 411.

17. A. H. Wapstra and K. Bos, At. Data and Nucl. Data Tables 19 (1977) 175.

18. K. Aleklett, E. Lund, and G. Rudstam, Phys. Rev. C18 (1978) 462.

MASS MEASUREMENTS OF PROTON-RICH NUCLEI AND EXOTIC SPECTROSCOPY

Robert E. Tribble*

Cyclotron Institute
Texas A&M University
College Station, Texas 77843

INTRODUCTION

In this report we review the progress of recent work aimed at extending our knowledge of proton-rich nuclei. The work focuses on Q-value measurements via the (^{4}He,^{8}He) reaction, a reaction that populates very proton-rich nuclei by the removal of four neutrons from the most neutron-deficient stable targets. The (^{4}He,^{8}He) reaction has quite negative Q-values (typically -60 MeV) and low cross section (a few nb/sr), necessitating incident beams with both high energy and high intensity. We first review techniques for performing Q-value measurements with reactions of such low cross section. We then discuss the recent results, including new isobaric quintet tests of the isobaric multiplet mass equation (IMME).

In addition to the mass measurements, we have been investigating the reaction mechanism responsible for this exotic four neutron transfer. Initial results can be described by calculations that assume a one-step transfer of a four-neutron cluster. From this evidence, we shall suggest some possible new uses of the (^{4}He,^{8}He) reaction as a spectroscopic tool.

EXPERIMENTAL PROCEDURE

Since general techniques for performing precision Q-value measurements have already been considered in detail by several authors,[1] the discussion below will be limited to some of the special features inherent in our measurements. The experiments were performed by observing ^{8}He's in the focal plane of an Enge

13

split-pole magnetic spectrograph, following reactions induced by alpha beams accelerated to energies between 80 and 128 MeV by the Texas A&M University 224 cm cyclotron. By combining the defining aperture and faraday cup into one unit, we performed our most recent measurements at θ_{lab} = 3° with a 2.1 msr solid angle corresponding to a 3° horizontal acceptance. Typical beam currents on target range from 0.6 to 3 eμA.

The standard detector system consists of a 10-cm single-wire gas proportional counter backed by a 50 mm x 10 mm x 600 μm Si solid state detector. Typically, the gas counter is operated at a pressure of 1 atm, and the entrance is collimated to 47 mm x 9.5 mm to match the solid state detector. Four analog signals -- corresponding to the gas counter left (ΔE_L) and right (ΔE_R) energies, E_{Si}, and time of flight (TOF) -- are fed to an on-line computer for processing, and are also event-mode recorded for further analysis off-line. Particle position is determined by charge division and particle identification is obtained from three constraints: (1) $(dE/dx)_{gas}$ = $(\Delta E_R + \Delta E_L)$, (2) E_{Si}, and (3) TOF relative to the cyclotron r.f. A typical plot of E_{Si} versus TOF is shown in Fig. 1. For ^{8}He events too energetic to stop in the Si detector, a degrader foil is inserted between the gas counter and the Si detector. Using this technique, ^{8}He discrimination to less than 100 pb/sr-MeV has been obtained for targets of A ≈ 60.

The incident beam energy is one of several parameters that must be determined to perform a Q-value measurement. To measure alpha beam energies in the range of 100 MeV, we use the momentum-matching technique[2] with analog d or H_2^+ beams accelerated to the

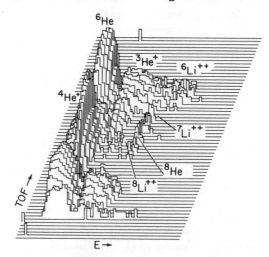

Fig. 1   E vs TOF profile with the Z-dimension (number of counts) logarithmic.

same magnetic rigidity as the alpha beam. The analog beams can
be transported without any change in the beam optics. For
example, we determined the energy of a 123-MeV alpha beam to 20
keV by observing simultaneously reaction products from
^{28}Si(p,p)^{28}Si elastic scattering and ^{28}Si(p,d)^{27}Si(g.s.). The
proton beam was obtained by accelerating H_2^+. To determine other
parameters, such as target thicknesses, scattering angle and
focal-plane calibration, we utilize standard techniques.

The Mass of ^{8}He

In order to determine a mass from a Q-value measurement, it
is necessary to know the masses of the projectile, ejectile, and
target. Any uncertainty in these masses directly limits the mass
determination, independent of the Q-value uncertainty. In the
case of the (^{4}He,^{8}He) reaction, the ^{8}He ejectile is itself an
exotic nucleus whose mass has only recently been determined to
good precision. In Fig. 2, a history of ^{8}He mass measurements,
which includes our most recent results, is presented. Our
measurement was performed by determining the Q-value of the
^{64}Ni(^{4}He,^{8}He)^{60}Ni reaction. The incident beam energy was chosen
at 80 MeV in order to observe the ^{64}Ni(^{4}He,^{6}He)^{62}Ni reaction to
the ground and first excited states of ^{62}Ni simultaneously with
the ^{8}He events populating the ground state of ^{60}Ni. The two ^{6}He
peaks provided the focal plane calibration, thereby reducing the
sensitivity of the measurement to beam energy, scattering angle,
and target thickness uncertainties. Two focal plane detectors,
one for the ^{6}He and ^{8}He events and a second for inelastic scattering, continuously monitored magnetic field fluctuations in
either the spectrograph or the beam transport system.[7]

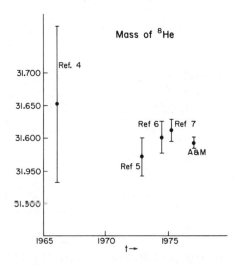

Fig. 2   History of ^{8}He measurements.

Measurements were performed at both θ_{lab} = 5° and 7° with
typical ^{6}He and ^{8}He spectra, obtained at the 7° scattering angle,
as shown in Fig. 3.   Results of the measurements and a list of
experimental uncertainties are given in Table 1.   The overall
uncertainty in our measurement was 8 keV, an improvement of a
factor of two over any previous single result.   By combining the
best three measurements in quadrature, the ^{8}He mass is found to
be 31.595±0.007 MeV, a result that does not represent a large
uncertainty for other masses obtained by the (^{4}He,^{8}He) reaction.

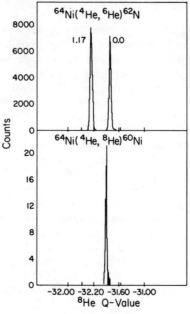

Fig. 3 ^{6}He and ^{8}He spectra obtained simultaneously at θ_{lab} = 7°.

Table 1.   Experimental Uncertainties and Results for the ^{8}He
Mass Measurement

Source	Uncertainty	Results		Q-value[b]
		θ_{lab}	$(d\sigma/d\Omega)_{lab}$	(MeV)
Beam energy	2.5			
Reaction angle	<1	5	36±8 nb/sr	-31.799(5)
Target thickness	3	7	44±8 nb/sr	-31.797(5)
Centroid	3.7			
Calibration[a]	4.2	With all uncertainties		
$\sigma(\theta)$	<1	Q = -31.798(8) MeV		
Other masses	2	Mass Excess = 31.591(8) MeV		

[a]Q-value plus ^{6}He centroid uncertainties.
[b]Centroid, beam energy, and calibration (centroid only) uncertainties.

$T_z = -2$ Nuclei and the IMME

Within the last couple of years, mass measurements have been completed for most of the $T_z = -2$ nuclides with even N and Z below A = 40. Our measurements include masses for ^{8}C, ^{20}Mg, ^{24}Si, and ^{36}Ca. In addition, KeKelis et al.[8] have determined the masses of ^{12}O and ^{16}Ne. Results of these measurements, including Q-values, masses, and cross sections, are reviewed in Table 2. A ^{8}He spectrum from a very recent measurement of ^{24}Si is shown in Fig. 4.

In all cases, the masses of the $T_z = -2$ nuclei can be used to test the IMME. In three of the quintets -- A = 8, 20, and 24 -- all five members of the T = 2 multiplets have been determined. The ($T_z = -1$) member has not been located in the other quintets. The quadratic IMME, which relates the members of an isobaric multiplet by the equation $M(A,T,T_z) = a(A,T) + b(A,T)T_z + c(A,T)T_z^2$, a, b, and c being constant across the multiplet, has been tested in over 20 isobaric quartets.[13] It has been remarkably

Table 2. Results of Recent $T_z = -2$ Mass Measurements

Isotope	Q-Value (MeV)	Mass Excess (MeV)	$(\frac{d\sigma}{d\Omega})_{lab}$ (nb/sr)	θ_ℓ	Ref.
^{8}C	-64.167(24)	35.097(24)	9(3)	5°	9
^{12}O	-66.02(12)	32.10(12)	2(1)	8°	8
^{16}Ne	-60.15(8)	23.92(8)	5(3)	8°	8
^{20}Mg	-60.67(3)	17.57(3)	3.0(15)	5°	10
^{24}Si	-61.48(4)	10.80(4)	5(3)	3°	11
^{36}Ca	-57.58(4)	6.44(4)	1.3(6)	5°	12

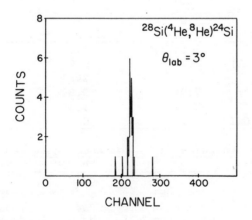

Fig. 4 ^{8}He spectrum populating the ground state of ^{24}Si.

successful in that only the four members of the A = 9 quartet
cannot be fit with the three parameter equation. Results of the
IMME tests for the isobaric quintets are summarized in Table 3.
We have included predictions for higher order coefficients to the
mass formula (it can be extended to $M = a + bT_z + cT_z^2 + dT_z^3 + eT_z^4$). For completed quintets, both d and e coefficients can be
predicted. For those quintets with only four members known, only
a d coefficient is given. Also included in the table is the χ^2
for a three parameter fit. The only mass quintet that shows a
signficant deviation from the quadratic IMME is A = 8. The
deviation in A = 8 can be traced in part to isospin mixing in the
$T_z = 0$ system. Also three of the members of this quintet are
unbound to isospin allowed particle decay, and, hence, their
physical energies are likely shifted from the corresponding bound
state energies. These two effects, which have been discussed in
detail elsewhere[15,16] could lead to nonzero d and e coefficients
in the quintet.

The quadratic IMME can reproduce masses within an isobaric
multiplet to high precision, but it fails to provide any information about the strength of a nuclear charge dependent interaction
(CDI). For this, comparisons of experimental and calculated
Coulomb energies will likely be superior. In Table 4, we compare
the experimental Coulomb energies for the T = 2 quintet in A = 36
to two predictions, one that includes a nuclear CDI and one that
does not. The calculations in this case are actually based on
experimental Coulomb energies for T = 1/2 and T = 1 nuclei in the
$d_{3/2}$ shell.[17] There is remarkable agreement between the experimental results and the predictions when the CDI is included. The
effect seems to be anamolously large in this mass region, however.[18]
While these results are enticing, a more systematic approach will
be needed to pin down the size and character of a nuclear CDI.

Table 3.   IMME Fits for Mass Quintets

A	d (keV)	e (keV)	χ^2 (Quadratic Fit)	Ref.
8	3.6(27)	2.6(21)	7.7	9
12	0(11)	...	0.0	8
16	8(5)	...	2.8	8
20	5(9)	-7(9)	0.98	14
24	-4(4)	2(2)	1.35	11
36	-1.6(18)	...	0.78	12

Table 4.   Coulomb Energies (keV) for the A = 36 Isobaric Quintet

Nuclide	No CDI	With CDI	Exp
^{36}Cl	6368	6210	6224(2)
^{36}Ar	6681	6628	6629(2)
^{36}K	6994	7047	7033(6)*
^{36}Ca	7307	7465	7469(40)*

*Based on the quadratic IMME prediction for the ^{36}K T = 2 state.

$T_z = -1$ Nuclei ^{50}Fe and ^{54}Ni

^{8}He spectra from the ^{54}Fe(^{4}He,^{8}He)^{50}Fe and ^{58}Ni(^{4}He,^{8}He)^{54}Ni
reactions are shown in Fig. 5. The laboratory cross sections
corresponding to the yield to both the ^{50}Fe and ^{54}Ni ground
states is ~0.5 nb/sr at a scattering angle θ_{lab} = 5° and incident
energy of 110 MeV. Comparison of the two spectra points out the
need for very pure targets in these measurements. The background
apparent in the ^{50}Fe spectrum is due to impurities in a target
that was 97% enriched in ^{54}Fe. The Ni target was isotopically
enriched to 99.9% ^{58}Ni, and no such comparable background level
is observed in that spectrum.

The results of the measurements are summarized in Table 5
where the masses are compared to predictions from a recent
Coulomb energy calculation,[18] and the symmetric Garvey-Kelson
mass formula.[19] In both cases, the two predictions are nearly
identical and in reasonably good accord with the measurements.
In a recent review of $f_{7/2}$ Coulomb displacement energies, Brown
and Sherr[20] have pointed out that the T = 1, 0^{+} configurations
show strong evidence of a nuclear CDI. In Table 5, we include
their predictions for the difference between experimental and
theoretical Coulomb energies obtained with and without provision
for a nuclear CDI. Once again, the results are enticing but point
out the need for more precise measurements of both ^{50}Fe and ^{54}Ni.

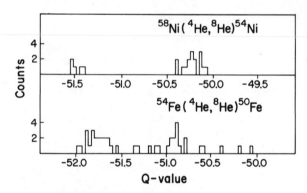

Fig. 5 ^{8}He spectra from ^{54}Fe and ^{58}Ni targets. The ^{54}Fe target
had 3% impurities from other Fe isotopes which produced
the background.

Table 5. Properties of ^{50}Fe and ^{54}Ni

	Mass Excess (MeV)			$(\Delta E_{exp}^{Coul} - \Delta E_{calc}^{Coul})$ (keV)	
Nuclide	Exp	Garvey-Kelson	Coul. Energy	No CDI	With CDI
^{50}Fe	-34.48(6)	-34.50	-34.472(13)	41	-14
^{54}Ni	-39.21(5)	-39.27	-39.296(13)	105	34

The (^{4}He,^{8}He) Reaction as a Four-Neutron Cluster Transfer

Very little is known about the transfer mechanism of any
exotic multi-nucleon reaction. We have begun a series of measurements to develop a better understanding of the (^{4}He,^{8}He) reaction
mechanism and to ascertain the feasibility of using the reaction
for unusual spectroscopy studies. The initial work focussed on
measuring the angular distribution of the ^{64}Ni(^{4}He,^{8}He)^{60}Ni
reaction. Data have been obtained for c.m. angles between 4° and
60° at an incident energy of 80 MeV. The measurement was facilitated by using a thick target (2.9 mg/cm^2) and obtaining on
target-beam currents of 3 eµA. Data were obtained simultaneously
for the ^{64}Ni(^{4}He,^{6}He)^{62}Ni (E_x = 0.0 and 1.17 MeV) reaction.

As the theoretical analysis has already been discussed in
detail,[21] we shall provide only a few important points below.
Spectroscopic factors were calculated by assuming a four neutron
cluster transfer with a {22} spatial symmetry and cluster quantum
numbers (S = 0, ℓ = 0, T = 2). In the distorted wave (DW) calculation, the number of nodes in the cluster wave function was
determined by matching the number of oscillator quanta of the
transferred particles to that of the cluster. One node was
required in the internal structure of the cluster to satisfy the
Pauli principle, thus leaving five nodes in the cluster wave
function assuming an L = 0 transfer. The calculation was found
to be sensitive to radial cutoffs in the range from 0-5 fm. To
reduce the sensitivity to the interior, the cluster wave function
was multipled by a Wood-Saxon shape damping factor of the form
{1 - (1 + exp [(r-R)/a])$^{-1}$} and was then renormalized. Similar
calculations were carried out for the ^{64}Ni(^{4}He,^{6}He)^{62}Ni reaction.
In this case, the two-nucleon cluster wave function was found to
be insensitive to the damping factor. The DW calculations are
compared to the experimental results in Fig. 6. The fits for
both the two and four neutron transfers are quite acceptable.
Spectroscopic factors were calculated for the two neutron transfer

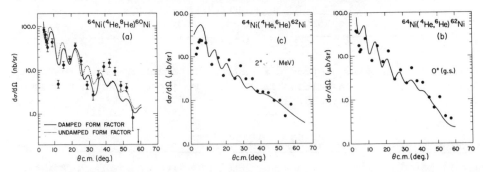

Fig. 6   DWBA fits to (^{4}He,^{6}He) and (^{4}He,^{8}He) angular distributions
         from a ^{64}Ni target.

reactions and the $(^4\text{He},^8\text{He})$ reaction for ^{64}Ni and ^{58}Ni targets.
Results of these calculations are reported in Table 6, where both
single shell model configuration and realistic wavefunction
results are given. It is clear from the table that collective
effects are expected to be important for both the two and four
neutron transfer. The collective spectroscopic amplitudes
coupled with the DW results predict the relative 0^+ and 2^+ cross
sections in ^{62}Ni to within 50%. Results for the four neutron
transfers from ^{64}Ni and ^{58}Ni predict a 40:1 cross section ratio
(most of this comes from the spectroscopic factors). Experi-
mentally, we find a ratio of $(100\pm50):1$. Thus in this case, both
the shape of the angular distribution and the relative cross
sections are adequately explained by a simple one-step four
neutron cluster transfer mechanism.

Table 6.   Spectroscopic amplitudes.

	J	SSM Configuration	Nodes	S(SSM)	S(Collective)
^{62}Ni + (2n)	0^+	$(f_{5/2}^4)$	3	0.19	0.64^a
	0^+				0.62^b
	2^+	$(f_{5/2}^4)$	2	0.11	0.58^a
	2^+				0.73^b
^{60}Ni + (4n)	0^+	$(f_{5/2}^4)$	5	0.0014	0.049^a
^{54}Ni + (4n)	0^+	$(p_{3/2}^2 f_{7/2}^2)$	5	0.0097	0.011^a

[a] BCS model calculation. The U and V factor are from Ref. 22.
[b] Shell model calculation. Wave functions are the D3 set in Ref. 23.

To pursue the problem of the reaction mechanism further, we
have begun a test of the reaction selectivity by observing the
$^{44}\text{Ca}(^4\text{He},^8\text{He})^{40}\text{Ca}$ reaction. Spectra spanning several regions of
excitation are shown in Fig. 7. Some general observations can be
made immediately. First, we note that states with dominant
$[(fp)^2(sd)^{-2}]$ configurations, including the 0^+ T = 1 and T = 2
states, are relatively weak. The low-lying 3^- state, mostly
$[(fp)^1(sd)^{-1}]$, is quite strong, likely indicating a collective
enhancement for this transition. The low-lying 0^+ and 2^+ states
are thought to be mostly $\{[(fp)^4]^{T=0} \times [(sd)^{-4}]^{T=0}\}$ configurations.[24]
A one-step transfer to these states would only proceed via
$[(fp)^6(sd)^{-2}]$ admixtures in the ^{44}Ca ground state wavefunction
and would be expected to be small, as the data indicates. To be
more quantitative requires calculating spectroscopic factors and
DW predictions. Since the transferred particles come from both
the fp and sd shell, the calculation is not as simple as for the
Ni targets. We are using ^{40}Ca wavefunctions of Gerace and Green[24]

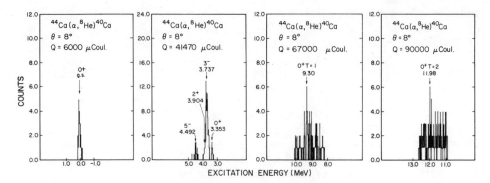

Fig. 7 ^{8}He spectra showing the final state selectivity in ^{40}Ca from ^{44}Ca(^{4}He,^{8}He)^{40}Ca at θ_{lab} = 8°.

and a shell model ^{44}Ca wavefunction[25] that includes particle-hole admixtures.[26] The calculations are proceeding assuming a double two-neutron transfer much like the microscopic description of the (p,t) reaction.

Once the calculations have been completed, we hope to be able to better assess the feasibility of using this reaction as a tool for exotic spectroscopy. As an example, the measurements in ^{40}Ca could be extended to search for the $\{[(fp)^4]^{T=2} \times [(sd)^{-4}]^{T=2}\}$ strength. This configuration can couple to T = 0-4 and should dominate the lowest 0^+ T = 3 and T = 4 levels. This is only one of many examples where the four neutron reaction might be used to search for new nuclear structures inaccessible by more conventional reactions.

ACKNOWLEDGMENTS

This work could not have been completed without the assistance of several different individuals over the past four years. In particular, I wish to acknowledge the contributions of Drs. J. D. Cossairt, R. A. Kenefick, and D. P. May.

REFERENCES

*Supported in part by the National Science Foundation, the R. A. Welch Foundation, and the A. P. Sloan Foundation.
1.   See, for example, J. A. Nolen, Jr., G. Hamilton, E. Kashy, and I. D. Proctor, Nucl. Inst. and Meth. <u>115</u>, 189 (1974).
2.   G. F. Trentelman and E. Kashy, Nucl. Inst. and Meth. <u>82</u>, 304 (1970).
3.   J. Cerny, S. W. Cooper, G. W. Butler, R. H. Phel, F. S. Goulding, D. A. Landis, and C. Detraz, Phys. Rev. Lett. <u>16</u>,

469 (1966).

4. J. Cerny, N. A. Jelley, D. L. Hendrie, C. F. Maguire, J. Mahoney, D. K. Scott, and R. B. Weisenmiller, Phys. Rev. C 10, 2654 (1974).

5. J. Jänecke, F. D. Becchetti, L. T. Chua, and A. M. Vander-Molen, Phys. Rev. C 11, 2114 (1975).

6. R. Kouzes and W. H. Moore, Phys. Rev. C 12, 1511 (1975).

7. R. E. Tribble, J. D. Cossairt, D. P. May, and R. A. Kenefick, Phys. Rev. C 16, 1835 (1977).

8. G. J. KeKelis, M. S. Zisman, D. K. Scott, R. Jahn, D. J. Vieria, J. Cerny, and F. Ajzenberg-Selove, Phys. Rev. C 17, 1929 (1978).

9. R. E. Tribble, R. A. Kenefick, and R. L. Spross, Phys. Rev. C 13, 50 (1976); mass revised by R. G. H. Robertson, E. Kashy, W. Benenson, and A. Ledebuhr, Phys. Rev. C 17, 4 (1978).

10. R. E. Tribble, J. D. Cossairt, and R. A. Kenefick, Phys. Lett. 61B, 353 (1976).

11. R. E. Tribble, A. F. Zeller, and D. M. Tanner, preliminary result, to be published.

12. R. E. Tribble, J. D. Cossairt, and R. A. Kenefick, Phys. Rev. C 15, 2028 (1977).

13. W. Benenson and E. Kashy, Rev. Mod. Phys. 51, 527 (1979).

14. D. M. Moltz, J. Aystö, M. D. Cable, R. D. von Dincklage, R. F. Parry, J. M. Wouters, and J. Cerny, Phys. Rev. Lett. 42, 43 (1979).

15. R. E. Tribble, R. A. Kenefick, and R. L. Spross, Phys. Rev. C 13, 50 (1976).

16. R. G. H. Robertson, W. Benenson, E. Kashy, and P. Mueller, Phys. Rev. C 13, 1018 (1976).

17. R. Sherr and I. Talmi, Phys. Lett. 56B, 212 (1975).

18. R. Sherr, Phys. Rev. C 16, 1159 (1977).

19. D. Mueller, E. Kashy, W. Benenson, and H. Nann, Phys. Rev. C 12, 51 (1975); E. Kashy, private communication.

20. B. A. Brown and R. Sherr, Nucl. Phys. A 322, 61 (1979).

21. R. E. Tribble, J. D. Cossairt, K.-I. Kubo, and D. P. May, Phys. Rev. Lett. 40, 13 (1978).

22. B. Bayman and N. F. Hintz, Phys. Rev. 172, 1113 (1968).

23. D. H. Kong-A-Siou and H. Nann, Phys. Rev. C 11, 1681 (1975).

24. W. J. Gerace and A. M. Green, Nucl. Phys. A 93, 110 (1967).

25. J. B. McGrory, private communication.

26. H. T. Fortune, private communication.

MASSES OF MEDIUM WEIGHT NUCLEI BY TRANSFER REACTIONS

R. C. Pardo

Cyclotron Laboratory[*]
Michigan State University
E. Lansing, Michigan 48824

I. TESTS OF THE IMME

In the last decade a large experimental program has been
in progress at MSU with the aim of testing the isobaric mass
multiplet equation (IMME) first proposed by Wigner.[1] This relation
holds that the masses of the members of an isospin multiplet of
mass A and isospin T can be expressed in a simple quadratic form:

$$M(A,T_z) = a + bT_z + cT_z^2,$$

where T_z is the isospin projection. There are numerous derivations
of this formula in the literature,[2,3] which employ perturbation
methods to expand the nuclear Hamiltonian. In the above form, the
lowest isospin multiplet which can be used to test the IMME is
$T = 3/2$.

Prior to this work the masses of all members of 21 $T = 3/2$
multiplets have been determined. The results have been recently
reviewed by Benenson.[4] Only one quartet has shown any significant
deviation from the quadratic relationship given above. That
quartet, the $A = 9$ quartet, is currently undergoing investigation
at MSU and is reported on by Kashy[5] at this conference.

In addition to directly testing the IMME, the completed
quartets give information which is directly calculable in the
context of the shell model. The b-coefficient is the analog to the
Coulomb energy in $T = 1/2$ nuclei. That is, the energy differences
between the $T_z = \pm 1/2$ members of the quartet is just b. The b-

[*]Present address: Argonne National Laboratory, Argonne, IL. 60439.

and c-coefficients together relate the masses of the $T_z = \pm 3/2$ and $T_z = \pm 1/2$ levels such that

$$\Delta m = b - 2(T_{z_>} - T_{z_<})c$$

where Δm is the splitting between $T_z = 3/2$, $T_z = 1/2$ or $T_z = -3/2$, $T_z = -1/2$ members of the quartet.

One of the benefits of transfer reactions used to measure ground state masses is the fact that a number of excited states are populated. Each of the states observed in (^{3}He,^{6}He) reactions are members of an isospin quartet. Currently only nine excited-state quartets have been completed; and only one mass system, A = 33, has three complete quartets. By testing the IMME at higher excitation energies, one should be increasingly sensitive to the most probable sources of a d-coefficient. These sources are isospin mixing of T = 1/2 levels with the T = 3/2 members of a quartet and expansion of the wave function due to Coulomb repulsion.

We have continued the experimental study of the IMME by emphasizing the study of excited state quartets in these nuclei and by improving the mass determinations of some quartets already completed. Specifically, I shall discuss the results of our work on the A = 21, 29, and 37 quartets.

One of the benefits of transfer reactions used to measure ground state masses is the fact that a number of excited states are populated. Each of these excited states is a member of an isospin multiplet. We used the (^{3}He,^{6}He) reaction at 70 MeV bombarding energy on targets of ^{24}Mg, ^{32}S, and ^{40}Ca in order to determine the excitation energy of T = 3/2 states in ^{21}Mg, ^{29}S, and ^{37}Ca, respectively, as well as the ground state masses of ^{29}S and ^{37}Ca.

The experimental method employed used the MSU Enge split-pole spectrograph with a 2-wire charge division proportional counter in the focal plane of the spectrograph backed by a stopping scintillator from which light and time-of-flight (TOF) information were obtained. The resulting ΔE, TOF, light, and position information in various two-dimensional slices gave an essentially unique identification of the ^{6}He ejectiles and allowed the use of this low cross-section reaction for precision measurements. This detection system and the techniques were described by Kashy et al.[6] A significant improvement in light resolution resulted from the use of an adiabatic light pipe to couple the scintillator to the photomultiplier tube. The energy resolution obtained was as good as 25 keV in the ^{24}Mg(^{3}He,^{6}He)^{21}Mg experiment but more typically a resolution of 30-40 keV was obtained. A spectrum of this reaction is shown in Fig. 1a.

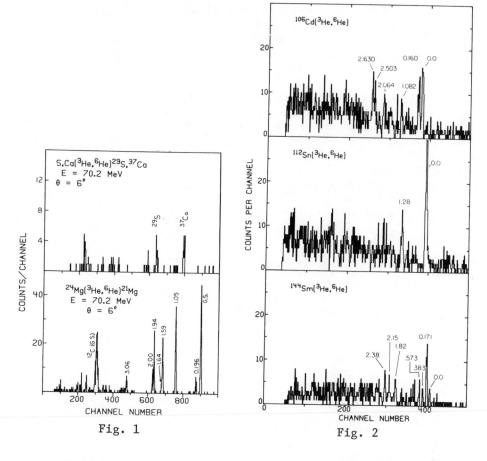

Fig. 1

Fig. 2

Experimental difficulties with sulfur and calcium targets have contributed to significant uncertainties in mass excess determinations for these isotopes in the past. In Fig. 1b the ^{6}He spectrum from a target made up of approximately 25 µg/cm^2 layers of alternately deposited ^{32}S and ^{40}Ca is shown. This work and a redetermination of the mass of ^{37}Ca has resulted in significant improvements in the uncertainties of the IMME coefficients for the A = 29 and A = 37 quartets. From these experiments, the mass excess of ^{29}S and ^{37}Ca was determined to be −3.105 ± .016 MeV and −13.135 ± .012 MeV, respectively.

We have searched for the analogs to the excited states observed in ^{21}Mg and ^{29}S by using the (p,t) and (p,^{3}He) reactions at 45 MeV on ^{23}Na and ^{31}P. In addition to the focal plane detector described above, photographic plates were used in this study in order to separate the T = 3/2 states of interest from the large number of T = 1/2 states populated at these excitation energies.

Identification of the T = 3/2 states is based on cross-section
ratios observed in the (p,t) and (p,^{3}He) reactions for assumed
multiplet members, on the relatively narrow widths of the T = 3/2
states when compared to the nearby T = 1/2 states, and on the
requirement that the excitation energy be approximately predicted
by the IMME. The data analysis is still in progress, but our early
results allow the completion of a third quartet for A = 21 which
corresponds to the second excited state in the T_z = ±3/2 nuclei.
Relevant information on the complete isospin quartets resulting
from our work for the A = 21, 29 and 37 nuclei is compiled in
Table I.

Table I. Status of complete isospin quadruplets in A = 21, 29,
and 37 nuclei.

J^π	Excitation Energy (MeV)				d-term
	^{21}Mg	^{21}Na	^{21}Ne	^{21}F	(keV)
$5/2^+$	0.0	8.970(5)	8.856(6)	0.0	5(4)
$1/2^+$	0.196(5)[a]	9.219(5)	9.139(6)	0.279(2)	2(5)
(1/2, 3/2)$^-$	1.054(5)[a]	10.049(12)[a]	9.963(6)	1.101(2)	-3(7)
	^{29}S	^{29}P	^{29}Si	^{29}Al	
$5/2^+$	0.0	8.382(5)	8.331(1)[7]	0.0	12(9)
$1/2^+$	1.29 (20)[a]	9.743(?)[7]	9.619(2)[7]	1.398(<1)[7]	49(14)
	^{37}Ca	^{37}K	^{37}Ar	^{37}Cl	
$3/2^+$	0.0	5.047(2)	4.993(6)	0.0	-3(4)
$1/2^+$	1.613(17)[a]	6.670(20)	6.654(10)	1.7266(1)	-3(12)

[a]Results of work reported on in this talk. Other data taken from
reference cited or Reference 4 and references therein when no
citation given.

For the A = 29 quartets, the situation is presently in a
state of flux. The recent tabulations of Endt and Van der Leun[7]
assign the ^{29}Si member of the J^π = 5/2$^+$ quartet an excitation
energy of 8.331 MeV, 40 keV higher than previously accepted. Also,
in ^{29}P a level observed in ^{29}Si(p,p') at 9.735 MeV is assigned to
the J^π = 1/2$^+$ quartet. These assignments, combined with our results
give d-terms of 12.3 ± 8.6 keV and 48.8 ± 13.7 keV. No d-terms as
large as these have ever been observed. The s-state 1/2$^+$ quartet
should be the most sensitive to such possible d-term sources and
Coulomb wave function expansion, but it would be surprising if the
effect would be this large in A = 29 and not observable in the
A = 27 quartet.

The ^{31}P(p,t)^{29}P and ^{31}P(p,^{3}He)^{29}Si data which we have taken should
be able to address this problem.

II.   THE MEASUREMENT OF THE MASSES OF MEDIUM WEIGHT PROTON–RICH
NUCLEI

The second area of active research has been the measurement of masses of proton-rich medium-weight nuclei using the
(^{3}He,^{6}He) reaction. In the main, the masses of nuclei far from
stability are based on the determination of the beta decay endpoint
energy. The mass excess for a nucleus far from stability is
determined by combining the endpoint energy of a decaying species
with that of its daughters until either a stable nucleus or a
nucleus whose mass is known by some other technique is reached.
Because the endpoint energy in beta decay is difficult to determine
accurately for these nuclei, the mass excess uncertainty becomes
quite large for nuclei far removed from stability. So, in addition
to the direct tests of mass relations which our results allow, they
provide valuable mass outposts in their region from which improved
measurements of the masses of other nuclei can result.

This study used the experimental system previously
discussed. The MSU cyclotron provided approximately 1 μa beams
of 70 MeV ^{3}He which were used to irradiate targets of ^{70}Ge, ^{90}Zr,
^{92}Mo, ^{106}Cd, ^{112}Sn, and ^{144}Sm. Calibration reactions with similar
Q–values were chosen so that the spectrograph field remained
unchanged throughout the experiment and targets as heavy as possible
were chosen to reduce errors due to scattering angle uncertainties.
The calibration reactions used were ^{60,62}Ni(^{3}He,^{6}He) and
^{26}Mg(^{3}He,^{6}He). The resolution obtained in these experiments was
approximately 60 keV except where limited by target thickness
considerations. Figure 2 shows some of the spectra obtained in
this work.

The results of this study have allowed the determination
of the previously unmeasured masses of ^{109}Sn and ^{89}Mo. In addition
significant improvements in precision have been obtained for most
of the other nuclei studied. These results are summarized in
Table II. The new mass determinations can be compared to the
predictions of various mass formulae. This is done in Table III.

The ability to use the (^{3}He,^{6}He) reaction for the study
of heavier nuclei was pleasantly surprising. It has been thought
that the cross section for this reaction would be too small for it
to be a useful tool. In Fig. 3 I have plotted the maximum
observed differential cross section to any state as a function of
mass number. This shows that the cross section for the (^{3}He,^{6}He)
reaction has begun to level off for A $\gtrsim$ 100 and should provide
many useful results for the careful experimenter and valuable tests

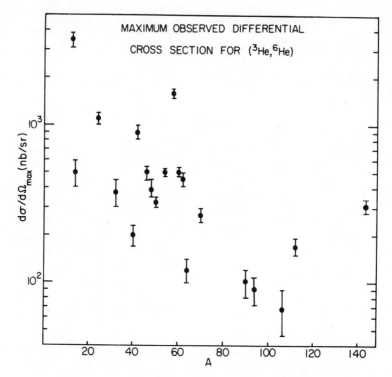

Fig. 3.   The maximum differential cross section for the most
          strongly excited state observed at any angle.

of mass relations in the deformed region. This also allows one to
hope that other exotic reactions will maintain enough of their
already meager cross section to allow useful results to be
obtained.

          We have attempted to observe $^{144}Sm(^3He,^8Li)^{139}Pm$ reaction
at   70-MeV ^3He energy. Although we were limited in time, 5 ^8Li
events were observed. This implies a cross section of ≈ 10 nb/sr
in 5 MeV energy span for this reaction. I hope this information
will be an enticement for others to continue this effort.

III.   THE MASSES OF ^{146}Gd AND ^{147}Gd AND SHELL CLOSURE AT Z = 64,
       N = 82

          Finally, I would like to describe our recent use of heavy
ion transfer reactions to obtain accurate mass measurements of
exotic nuclei. Heavy ions have just recently been available at
MSU. For cyclotrons a major problem with the measurement of low
cross section heavy ion reactions has been source lifetime and

TABLE II. Mass excess and Q-values.

Nucleus	# of Measurements	Measured Q-Value (MeV)	Mass Excess (MeV)	Previous Mass Excess (MeV)
^{67}Ge [a]	2	-10.572 ± .030	-62.65 ±.03	-62.45 ± .05
^{87}Zr	6	-12.083 ± .008	-79.344 ±.009	-79.43 ± .08
^{89}Mo	3	-14.465 ± .015	-75.008 ±.015	New
^{103}Cd	3	9.173 ± .017	-80.620 ±.018	-80.60 ± .14
^{109}Sn	4	-8.686 ± .009	-82.634 ±.011	New
^{144}Sm	5	-8.693 ± .012	-75.933 ±.014	-75.91 ± .06

[a]Lowest energy state observed is assumed to be 18-keV level. Uncertainty quoted reflects this.

machine stability. Some recent tests[8] with various cathode materials at MSU has resulted in significant improvements in these areas.

Heavy ion transfer reactions have been used for many purposes, but only rarely have they been used for the measurement of nuclear masses. A quick glance at the Chart of the Nuclides will confirm that even such mundane reactions as two-proton stripping can reach nuclei which have either unmeasured or poorly determined mass excesses. Multinucleon transfer to neutron rich nuclei reaches similarly interesting species.

Of course heavy ions have certain limitations. The most prominent is the large energy loss and energy straggling which occurs for heavy ions at lower energies, where many of these reactions will be most important. This problem exhibits itself strongly in mass determinations by demanding the use of thin targets. The low cross sections ($\lesssim$50 μb/sr) compound this problem by making target deterioration a serious concern as well.

Recently, Bhatia et al.[9] have used the ^{48}Ca$(^{18}$O$,^{15}$B$)^{51}$V reaction to measure the mass of ^{15}B. The mass of ^{21}O has been determined by Ball et al.[10] using the ^{208}Pb$(^{18}$O$,^{21}$O$)^{205}$Pb reaction and by Naulin et al.[11] using the ^{18}O$(^{18}$O$,^{15}$O$)^{21}$O reaction. These

Table III. Comparison of results to predictions.

Source	^{109}Sn Mass Excess	^{89}Mo Mass Excess
Experimental	-82.634 ± .011	-75.008 ± .015
Groote, HIlf, Takahashi	-83.65	-75.72
Seeger and Howard	-82.52	-75.2
Liran and Zeldes	-82.52	-74.65
Bauer	-84.38	
Beiner, Lombard, and Mas	-81.9	-74.5
Janecke, Garvey and Kelson	-82.92	-74.68
Comay and Kelson	-82.9	-74.90
Janecke and Cynou	-82.87	-74.48
Wapstra and Bos	-82.620	-75.220
Myers		-75.72

These experiments were conducted with 91— 105–MeV ^{18}O ions on a
neutron-rich target. In general, the more interesting reactions
would be the three proton stripping reaction on a proton-rich
target, or the three-proton pickup reaction on a neutron-rich
target. We have looked for the three-proton transfer reaction
($^{12}C,^{9}Li$) at 75 MeV on targets of ^{58}Ni and ^{144}Sm and have observed
no cross section at the 1 nb/sr level in 5 MeV of excitation energy.
Hopefully the search for these reactions at higher energies will be
more successful.

The specific study which I would like to report on here
is the measurement of the masses of ^{146}Gd and ^{147}Gd. This region
has received considerable attention because of the discovery by
Kleinheinz et al.[12] that ^{146}Gd was the second isotope known to have
a first excited state with $J^{\pi} = 3^{-}$. Only ^{208}Pb was previously
known to have this property. This property of only the most magic
of the magic nuclei raised the question of the size of the energy
gap at Z = 64, N = 82.

The nuclei whose masses are needed in order to address the
question of the energy gap between the $1g_{7/2}-2d_{5/2}$ orbitals and the
$1h_{11/2}-3s_{1/2}-2d_{3/2}$ orbitals are ^{145}Gd, ^{146}Gd, ^{147}Gd, ^{147}Tb and
^{145}Eu. At the time of Kleinheinz's work, only ^{145}Eu and ^{147}Gd
masses has been experimentally determined.

We have used the $^{144}Sm(^{12}C, ^{10}Be)^{146}Gd$
$^{144}Sm(^{12}C, ^{9}Be)^{147}Gd$ reactions to determine the mass excess of ^{146}Gd
and ^{147}Gd simultaneously. The reactions were produced with 75-MeV
^{12}C ions on targets of approximately 100 µgm/cm^2 enriched ^{144}Sm.
The $^{10}Be^{4+}$ and $^{9}Be^{4+}$ ejectiles were simultaneously detected in the
focal plane of the MSU Enge split-pole spectrograph. The detector
used was that described earlier in this talk. Resolution obtained
in this experiment was as good as 90 keV for short runs. Due to
the combined effects of ion source instabilities and target
effects more typical resolution results were 100— 120 keV. The
Q-values for the ^{10}Be and ^{9}Be reactions are such that the position
of the focus for $^{9}Be^{4+}$ ions populating the ground state of ^{147}Gd
are separated by approximately 5 inches from the focus point for
$^{9}Be^{4+}$ ions populating the ground state of ^{146}Gd. This allowed
the simultaneous determination of the mass excesses of ^{146}Gd and
^{147}Gd by simply replaying the event recorded data. Figure 4 shows
typical spectra obtained.

The calibration reactions used were $^{92}Mo(^{12}C, ^{10}Be^{4+})^{94}Ru$
and $^{92}Mo(^{12}C, ^{9}Be^{4+})^{95}Ru$. These reactions were chosen because of
their similar atomic mass number and the fact that the Q-value
for these reactions placed the peaks of interest on our detector
without any change in the spectrograph magnetic field.

The major source of error in this experiment came from
the uncertainty in target thicknesses. Because of the choice of
calibration reactions uncertainties in beam energy, scattering
angle, and focal plane calibration have either been eliminated
or reduced substantially. In order to minimize the effect of
target thickness uncertainties we used a total of four targets.
In computing the contributions due to target thickness the
uncertainties were treated as uncorrelated. This should be valid
since any systematic effects should be accounted for in the
calibration targets which were measured in the same way.

For ^{146}Gd we determined the mass excess to be
−76.096 ± .025 MeV while for ^{147}Gd we found the mass excess to
be −75.48 ± .03 MeV. The result for ^{146}Gd is in good agreement
with the value recently determined by Alford et al.[13] using the
($^{3}He,n$) reaction. Our value for ^{147}Gd is 270 keV more negative
from that based on the results of a beta endpoint measurement by
Adam and Toth.[14]

With the results of this experiment and the recent ^{145}Gd
β^+ endpoint measurement of Firestone et al.,[15] the masses of four
of the five nuclei needed in order to experimentally address the
energy gap in the single particle level scheme around ^{146}Gd have
been determined. Only ^{147}Tb remains to be studied along with its
metastable state, ^{147m}Tb.

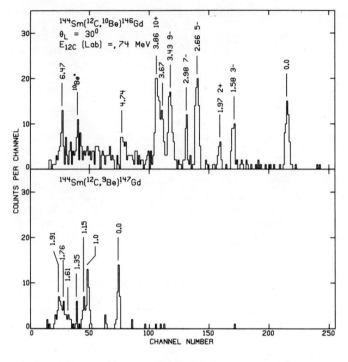

Fig. 4

        The odd-even mass differences for a number of closed-shell
nuclei are plotted in Fig. 5.   These differences are defined by
the following relations:

$$\delta_p(Z,A) = S_p(Z,A) {}^-S_p(Z+1,A+1) \equiv M.E.(Z-1,A-1)+M.E.(Z+1,A+1)-2M.E.(Z.A).$$

$$\delta_n(Z,A) = A_n(Z,A)-S_n(Z,A+1) \equiv M.E.(Z,A-1)+M.E.(Z,A+1)-2M.E.(Z,A)$$

where M.E.(Z,A), S(Z,A) are the mass excess and proton (neutron)
separation energies for nucleus (Z,A), respectively.

        For ^{146}Gd we find that:

$$\delta_p(^{146}Gd) = 3.75 \text{ MeV}$$

$$\delta_n(^{146}Gd) = 3.84 \text{ MeV}.$$

These values are quite similar to the observed proton and neutron
gaps in $^{208}_{82}Pb_{128}$ of 3.43 MeV and 4.21 MeV, respectively.

In Fig. 5, one can see that the odd-even mass differences for protons and neutrons are approximately equal at a specific shell closure point through the N or Z = 50 $1g_{9/2}$ shell closure. This is no longer true at the higher shell closures. Both the $1g_{7/2}$-$2d_{5/2}$ shell and the $1h_{11/2}$-$3s_{1/2}$-$2d_{5/2}$ shell have significantly larger energy gaps for protons than is observed for neutrons. This can be understood in terms of the widening of the potential well as A increases and the resulting effect on orbitals with different rms radii. The effect has been discussed by Bohr and Mottelson[16] for neutrons. For neutrons they show that at A = 114 the $1g_{7/2}$ orbital is predicted to be separated from the $3s_{1/2}$ orbital by approximately 1 MeV. At A = 146 the $1g_{7/2}$ orbital is predicted to have moved under the $2d_{5/2}$ orbital and the gap at N = 64 is estimated to be about 2 MeV.

The observed gaps as defined in Eq. (1) are a combination of single particle energy difference and a pairing term

$$\delta_i = \sqrt{\zeta_i{}^2 + 4\Delta_i{}^2} \qquad i = p \text{ or } n$$

where ζ_i is the single-particle energy differences and Δ_i is the pairing energy term. For $\Delta_i \sim 1.2$ MeV, the measured δ_n for ^{115}Sn implies $\zeta_n = 1.3$ MeV which is in good agreement with the above predictions.

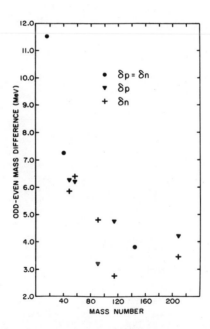

Fig. 5.   Experimental odd-even mass differences as a function of A.

We have performed calculations similar to those described by Bohr and Mottelson[16] using Becchetti and Greenlees[17] optical parameters, neglecting the energy dependence in the well depth. The computer code DWUCK-72[18] was used to calculated the proton and neutron binding. The results duplicated the qualitative effects for nuclei above ^{90}Zr but quantitatively were in error by up to 30%. It would appear the situation needs a detailed BCS calculation to accurately reproduce the experimental results.

With the results of this work and Ref. 16, of the five nuclei whose masses are needed to determine the size of the energy gap at Z = 64 and N = 82 only ^{147}Tb remains to be studied. The new information shows that a large gap (3.8 MeV) exists both for protons and neutrons in ^{146}Gd which is comparable to that observed in ^{208}Pb. The separation energies determined from these results are somewhat larger than had been previously believed and are somewhat surprising in terms of the structure arguments of Kleinheinz et al.[19] This enhances the need to know the masses of ^{147}Tb and ^{147m}Tb.

This work was performed in collaboration with Walt Benenson, Edwin Kashy, L. W. Robinson, S. Gales, L. H. Harwood, and R. M. Ronningen.

REFERENCES

1. E. P. Wigner, Proceedings of the Robert A. Welch Foundation Conference on Chemical Research, Edited by W. D. Millikan.
2. G. T. Garvey, in Nuclear Isospin, J. D. Anderson, S. D. Bloom, J. Cerny, and W. W. True, eds., Academic Press, New York (1969).
3. J. Janecke, in "Isospin in Nuclear Physics," D. H. Wilkinson, Ed., North-Holland Publishing Co., Amsterdam (1969).
4. W. Benenson and E. Kashy, Rev. Mod. Phys. 51:527 (1979).
5. E. Kashy, VI Conference on Atomic Masses and Fundamental Constants, E. Lansing, MI, 1979.
6. E. Kashy, W. Benenson, and D. Meuller, Measurement of Nuclear Masses Far From Stability in: "Atomic Masses and Fundamental Constants 5", J. H. Sanders and A. H. Wapstra, eds., Plenum Press, New York (1976).
7. P. M. Endt and C. Van der Leun, Nucl. Phys. A310:1 (1978).
8. P. Miller, H. Laumer, G. Stork, M. Mallory, H. Blosser, and J. A. Nolen, Jr., Michigan State University Cyclotron Laboratory Annual Report, (1977-78).
9. T. S. Bhatia, H. Hafner, J. A. Nolen, Jr., W. Saathoff, R. Sehuhmacher, R. E. Tribble, G. J. Wagner, C. A. Weidner, Phys. Lett. 76B:562 (1978).
10. G. C. Ball, W. G. Davies, J. S. Forster, and H. R. Andrews, Phys. Lett. 60B:265 (1976).

11. F. Naulin, C. Detraz, M. Bernas, E. Kashy, M. Langevin, F. Pougheon, and P. Roussel, Phys. Rev. C 17:830 (1978).

12. P. Kleinheinz, S. Lunardi, M. Ogawa, and M. R. Maier, Z. Physik A284:351 (1978).

13. W. P. Alford, R. E. Anderson, P. A. Batay-Csorba, R. A. Emigh, D. A. Lind, P. A. Smith, and D. C. Zafiatos, Bull. Am. Phys. Soc. 23:962 (1978).

14. I. Adam and K. S. Toth, Phys. Rev. 180:1207 (1969).

15. R. Firestone, R. C. Pardo, and W. C. McHarris, Bull. Am. Phys. Soc. 41:289 (1978).

16. A. Bohr and B. R. Mottelson, Nuclear Structure, Vol. I, Benjamin, Reading, Mass. (1969).

17. F. D. Becchetti and G. W. Greenlees, Phys. Rev. 182:1190 (1969).

18. P. D. Kunz, University of Colorado, unpublished.

19. P. Kleinheinz, R. Broda, P. J. Daley, S. Lunardi, M. Ogawa, and J. Blomqvist, Z. Physik A290:279 (1979).

MASS OF ^{9}C[†]

E. Kashy, W. Benenson, J.A. Nolen, Jr., and R.G.H. Robertson

Cyclotron Laboratory and Department of Physics
Michigan State University
East Lansing, MI 48824

ABSTRACT

A Q-value of −31.5762(30) MeV has been measured for the ^{12}C(^{3}He,^{6}He)^{9}C reaction. The mass of ^{9}C, when compared to its Coulomb analog levels in ^{9}B, ^{9}Be, and ^{9}Li, confirms a significant deviation from the quadratic form of the isobaric multiplet mass equation, with a cubic coefficient of d = 5.7 $\pm$ 1.5 keV.

I. INTRODUCTION

The properties of analog levels in nuclei have been of considerable interest in studies of microscopic and macroscopic nuclear properties: Examples are the mixing of $T_>$ levels with background $T_<$ levels in the first instance, and the determination of nuclear radii in the second. When three or more analog states are involved, it is expected[1] that the energies of the states in the various nuclei obey a quadratic relation, i.e., that the respective masses are described by the isobaric multiplet mass equation (IMME)

$$M(T_z) = a + bT_z + cT_z^2 .$$

To test that relation, four or more masses must be measured, which is possible when the isospin T is greater than or equal to 3/2. . A recent review of experimental results for quartets and quintets of states showed that only for the A = 9 quartet is a highly significant deviation from the quadratic dependence observed.[2] The present measurement then represents an effort to reduce the chance of an experimental error as the source of the effect. Previous measurements yielding the ^{9}C mass include

the ^{7}Be(^{3}He,n)^{9}C threshold measurements by Barnes et al.[3] and
Mosher et al.[4] with accuracies of ~5 keV and the Q-value of the
^{12}C(^{3}He,^{6}He)^{9}C reaction measured by Trentelman et al.[5] with an
accuracy of ~8 keV. The possibility of improving the latter
measurements using a magnetic spectrograph, electrostatic particle
separation, and photographic emulsion led to the present measurement.

II. EXPERIMENTAL PROCEDURE AND RESULTS

The method used has been described earlier.[6] In the present
work, the target consisted of a uniform mixture of 60 µg/cm^2
carbon isotope, with 40% of ^{13}C and 60% of ^{12}C. The incident
beam was 72 MeV ^{3}He from the Michigan State University Cyclotron.
The reaction products, i.e., ^{3}He, ^{4}He, and ^{6}He were detected
in the focal plane of the 90 cm Enge split-pole spectrograph,
using nuclear emulsions as track recorders. In order to separate
the rare ^{6}He from the abundant ^{4}He, an electrostatic deflector
was used in the region between the poles of the spectrograph.[7]
The ensuing separation was then proportional to m/q for a given
position in the focal plane, with the ^{6}He getting the largest
deflection. Since the abundant ^{3}H have the same m/q as ^{6}He,
they are subject to the same electric deflection, but they can
be differentiated from the ^{6}He due to the difference in their
specific ionization, i.e. their track brightness.

Fig. 1 shows the schematic of the positions at which the
various particle groups corresponding to accurately known transition were recorded on the photographic emulsion (L-42). In
order to make a Q-value measurement, the beam energy and reaction
angle must be measured. As can be seen from Fig. 1, the angle
is accurately determined from the position of the ^{3}He elastically
scattered from hydrogen in the target relative to those inelastic-
ally scattered from ^{12}C and ^{13}C. Similarly, the position of
the ^{6}He peak from the 3.350 MeV level of ^{10}C seen in the reaction
^{13}C(^{3}He,^{6}He)^{10}C* relative to the ^{4}He transitions represents a
sensitive measure of the beam energy. One can then calculate
the magnetic rigidity for all the well known transitions and
by a fitting procedure obtain an accurate description of magnetic
rigidity versus position on the plate. Since, in this measurement, the essential part is the comparison of the ^{13}C(^{3}He,^{6}He)^{10}C*
reaction leading to the 3.350 MeV level of ^{10}C with the ^{12}C(^{3}He,^{6}He)^{9}C
g.s. reaction, the two Q-values are closely tied; with a change
of 1.0 keV in the first reaction resulting in a 1.2 keV change
in the second. Thus a change in the mass excess of ^{10}C* would
result in an adjustment of the present results. We note finally
that the key advantage of the photographic emulsion in this work
lies in the simultaneous recording of the various particles of
interest. Table I shows the results of three plate exposures,

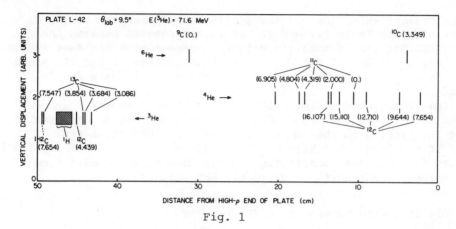

Fig. 1

with an average Q-value of -31.5762(30) MeV for the ^{12}C(^{3}He,^{6}He)^{9}C
reaction. The assumed Q-value was -18.5898 MeV for the ^{13}C(^{3}He,^{6}He)^{10}C*
reaction leading to the first excited state of ^{10}C at 3.350 MeV.

Table I. Q-values for the ^{12}C(^{3}He,^{6}He)^{9}C reaction.

Plate Number	Q value
L-41	-31574.0 (4.5)
L-42	-31576.8 (3.3)
L-43	-31578.3 (6.1)
Average	-31576.2 (2.4)

Other uncertainties:

^{10}C* Q value	1.2
^{12}C Target thickness corr.	0.8
^{11}C Levels	1.0
Fitting error estimate	1.4
Result: Q =	-31576.2 (3.2) keV

The results for all the precise mass measurements of ^{9}C
are shown in Table II, and rather good agreement between the
various results is seen. A weighted average yields a mass excess
of 28913.4(2.2) keV for ^{9}C in its ground state. Table III shows
the mass excesses of the quartet of states. The data for ^{9}B
and ^{9}Be lowest T = 3/2 states have been reviewed with a very
slight readjustment from originally published values[8,9] due to
small shifts in calibrant reactions. Table IV shows the results
of the fitting to the IMME. The quadratic IMME fit has a χ^2
of 14.4. If a cubic term of T_Z^3 is added, we have d = 5.7 ± 1.5 keV,
with the dominant contribution to the uncertainty coming from
the inner members of the quartet, ^{9}Be* and ^{9}B*.

Table II. Mass excess of ^{9}C (keV).

28918.2(5.1)	Barnes et al.[3]
28909.1(4.0)	Mosher et al.[4]
28915.6(8.0)	Trentelman et al.[5] †
28913.8(3.2)	Present work
28913.4(2.2)	Weighted mean

† Corrected for the new ^{6}He Mass, see ref. 12.

Table III. Mass excess of lowest T = 3/2 states with A = 9.

Nucleus	T_z	Mass excess (keV)
^{9}Li	3/2	24954.8(2.0)
^{9}Be	1/2	25740.0(1.7)
^{9}B	-1/2	27071.0(2.3)
^{9}C	-3/2	28913.4(2.2)

Table IV. Parameters of the IMME for the lowest A = 9 quartet
           (values and uncertainties in keV).

Coefficients	Quadratic fit	Cubic fit
a	26337.7(1.6)	26339.4(1.6)
b	-1320.7(0.9)	-1332.4(3.2)
c	265.1(1.0)	264.3(1.0)
d	----	5.7(1.5)
χ^2	14.4	----

Bertsch and Kahana[10] have calculated possible contributions
to a d-term and obtain d ~3.6 keV from 3-body Coulomb and Coulomb
plus charge dependent forces. Since the states in ^{9}Be and ^{9}B
are extremely narrow, i.e. 0.38 and 0.40 keV respectively,[11]
it is unlikely that they are shifted differentially in a significant way due to mixing with background levels, and this effect
would not then contribute to the observed d-term. We conclude
that at least in A = 9, we have a 3-body interaction the effect
of which is reflected in the experimental results.

This material is based upon work supported by the National
Science Foundation under Grant No. Phy 78-22696.

1. E.P. Wigner, in Proceedings of the Robert A. Welch Foundation
    Conferences on Chemical Research, Houston, Texas 1957, edited
    by A. Milligan, The Structure of the Nucleus, p. 67.
2. W. Benenson and E. Kashy, Reviews of Modern Physics, 51,
    527 (1979).
3. C.A. Barnes, E.G. Adelberger, D.C. Hensley, and A.B. McDonald,
    in: International Nuclear Physics Conference, edited
    by R.L. Becker, C.D. Goodman, P.H. Stelson, and A. Zucker
    (Acdemic Press Inc., New York, 1967) p. 261.
4. J.M. Mosher, R.W. Kavanagh, and T.A. Tombrello, Phys. Rev.
    C 3, 438 (1971).
5. G.F. Trentelman, B.M. Preedom, and E. Kashy, Phys. Rev.
    C 3, 2205 (1971).

6. J.A. Nolen, Jr., G. Hamilton, E. Kashy, and I.D. Proctor, Nucl. Inst. and Meth. 115, 189 (1974).

7. R.G.H. Robertson, T.L. Khoo, G.M. Crawley, A.B. McDonald, E.G. Adelberger, and S.J. Freedman, Phys. Rev. 17, 1535 (1978).

8. E. Kashy, W. Benenson, and J.A. Nolen, Jr., Phys. Rev. C 9, 2102 (1974).

9. E. Kashy, W. Benenson, D. Mueller, R.G.H. Robertson, and D.R. Goosman, Phys. Rev. C 11, 1959 (1975).

10. G. Bertsch and S. Kahana, Phys. Lett 33B, 193 (1970).

11. P.A. Dickey, P.L. Dyer, K.A. Snover, and E.G. Adelberger, Phys. Rev. C 18, 1973 (1978).

12. R.G.H. Robertson, E. Kashy, W. Benenson, and A. Ledebuhr, Phys. Rev. C 17, 4 (1978).

MASS MEASUREMENTS USING EXOTIC REACTION - $^{48}Ca(^{3}He, ^{11}C)^{40}S$

R. T. Kouzes and R. Sherr

Physics Department
Joseph Henry Laboratories
Princeton University
Princeton, New Jersey 08544

Accurate mass measurements have received a strong impetus in recent years to test model predictions based on local systematics, global fits and independent particle models among others. One widely used technique is precise reaction Q-value measurements giving an unknown mass relative to known masses and calibrant Q-values. We have previously applied this technique[1,2] to reach nuclei far from the valley of stability using reactions such as $^{70}Zn(^{3}He, ^{8}B)^{65}Co$ and $^{70}Zn(^{4}He, ^{7}Be)^{67}Ni$. Exotic reactions which involve the transfer of three or more nucleons can accurately measure masses which are otherwise unreachable except by heavy ion reactions which may suffer severe resolution limitations due to target thickness effects.

The mass of ^{40}S is the first of a series of mass measurements we are performing at Princeton to investigate models and reaction systematics. The $(^{3}He, ^{9-12}C)$ reactions on a ^{48}Ca target can reach the isotopes of sulfur from ^{39}S to ^{42}S. Exotic reactions with 80 MeV ^{3}He (see Table III) can potentially reach the T = 4 states in ^{40}Cl, ^{40}Ar, ^{40}K, and ^{40}Ca giving five of the nine members of the isotopic multiplet of which ^{40}S is the $T_Z = 4$ member.

The nucleus ^{40}S is known to have a lifetime of more than 10^{-7} seconds from a measurement performed by Artukh et al.[3] We have measured the Q-value of the $^{48}Ca(^{3}He, ^{11}C)^{40}S$ reaction to obtain the mass excess of ^{40}S using the Princeton University AVF cyclotron with the QDDD spectrometer (14.5 msr solid angle). A detector consisting of two resistive-wire gas-proportional counters and a plastic scintillator was used to obtain position, two energy loss measurements, and a scintillator signal for use in particle identification, with particle trajectory and time-of-flight as additional constraints. Data was taken and analyzed on a Data General Eclipse computer using the Princeton acquisition code ACQUIRE.[4]

A 97% enriched ^{48}Ca target of 90 μg/cm^2 thickness on a 20 μg/cm^2 carbon foil was employed. A 200 μg/cm^2 carbon foil and a 20 μg/cm^2 carbon on formvar foil were used in calibration reactions.

The technique of accurate measurement of the Q-value of the ^{48}Ca(^{3}He,^{11}C)^{40}S reaction is the same as used previously.[1,2] Our first runs at 8 degrees showed problems with background rejection, thus we made the final measurements at 12 degrees in the lab. Calibrations included elastic scattering on ^{12}C, ^{16}O, and ^{48}Ca, the ground states of the ^{12}C(^{3}He,^{4}He)^{11}C and ^{48}Ca(^{3}He,^{4}He)^{47}Ca reactions and the ground and first excited states of the ^{12}C(^{3}He,^{11}C)^{4}He reaction. The latter reaction was also useful in establishing particle identification. Consistency between all calibrants restricted the scattering angle to 12.0 ± 0.2 degrees and the beam energy to 80.32 ± 0.05 MeV.

Figure 1 shows the resulting position spectrum from the ^{48}Ca(^{3}He,^{11}C)^{40}S reaction and the spectrum of the carbon-on-formvar target for approximately equivalent thickness times charge. The particle identifier spectrum is also shown. We expect background from the ^{12}C target backing and from oxidation of the target. However, only ^{16}O on the target can produce a significant real ^{11}C background since all other Q-values are more negative. The correlation of the time-of-flight (TOF) parameter with scattering angle has been used to reduce the effective solid angle by about 20% to limit the ^{16}O induced background in the ^{40}S spectrum which was most intense near 9 degrees. Across the 7 degree horizontal acceptance of the QDDD solid angle, the TOF spectrum shows no structure for the peak we identify as the ground state.

The background spectrum does not show peaking and thus a flat background of one count per channel was subtracted from the ^{40}S spectrum. Since there is background in the spectrum and there is some structure to the left and right of the peak we identify as the ground state, it is possible we have misidentified the ground state peak. But the peak is of the expected width due to target thickness (about 120 keV or 2.3 channels) and the first excited state is expected to be around 1.5 MeV making misidentification of an excited state as the ground state unlikely. The structure at the left end of the ^{40}S spectrum we take to be the first excited state of ^{40}S plus the Doppler-broadened first excited state of ^{11}C (2.0 MeV).

Table I.   Error analysis where values are keV
           contribution to the ^{40}S mass uncertainty

Peak centroid uncertainty	21
Calibration uncertainty	17
Target thickness uncertainty (90 ± 20 μg)	16
Scattering angle uncertainty (± 0.2 degrees)	8
Beam energy uncertainty (± 50 keV)	12
Root mean square error	35

Table II.   Comparison of this measurement with model
            predictions for the ^{40}S mass excess in MeV.
            Abbreviations shown are used in Fig. 1.

Present		$-22.519 \pm .035$
Wapstra (systematics, Ref. 6)	W	-22.24
Meyers (Ref. 7)	M	-24.10
Groote and Takahashi (Ref. 7)	GR	-23.93
Liran and Zeldes (Ref. 7)	L	-22.41
Beiner, Lombard and Mas (Ref. 7)	B	-21.60
Janecke (Ref. 7)	J	-22.48
Comay and Kelson (Ref. 7)	C	$-22.49 \pm .41$
Monahan and Serduke (Ref. 8)		-22.82
Garvey (Ref. 9)	G	-22.43

It is of interest to note that Wilcox et al.[5] established only an upper limit of about 5 nb/sr at 10° for the reaction ^{48}Ca(α, ^{12}C)^{40}S at 110 MeV, whereas the cross section for the ^{40}S peak in our spectrum is about 60 nb/sr. (The Q-values are com-

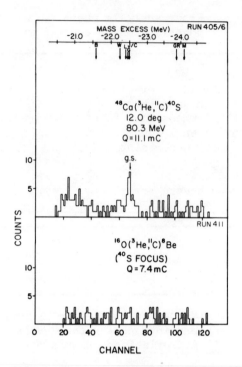

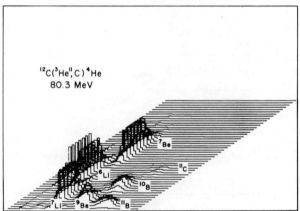

Figure 1.   The particle identifier and the spectrum of ^{40}S showing
             the mass excess scale and several mass predictions.
             (Table II defines the abbreviations.) One channel equals
             52 keV

Table III. Reaction List to Reach $T = 4$ States in Mass 40

Reaction	g.s. Q-value	Predicted ΔE_c	Predicted Excitation Energies
		$T = 4$ States	
$^{48}\mathrm{Ca}(^3\mathrm{He}, ^{11}\mathrm{C})^{40}\mathrm{S}$	-17416 (35)		0
$^{48}\mathrm{Ca}(^3\mathrm{He}, ^{11}\mathrm{B})^{40}\mathrm{Cl}$	-10413 (500)	6136	10375
$^{48}\mathrm{Ca}(^3\mathrm{He}, ^{11}\mathrm{Be})^{40}\mathrm{Ar}$	-14420 (6)	6470	23563
$^{49}\mathrm{Ti}(^3\mathrm{He}, ^{12}\mathrm{B})^{40}\mathrm{K}$	-13462 (2)	6805	28081
$^{48}\mathrm{Ti}(^3\mathrm{He}, ^{11}\mathrm{Be})^{40}\mathrm{Ca}$	-18886 (6)	7138	35749

parable and the angular momentum mismatch is very large in both cases.) If in both reactions the dominant mechanism was a ^{8}Be transfer or successive transfer of two α-particles one would expect the $(\alpha, ^{12}C)$ cross section to be very much larger than that for $(^3He, ^{11}C)$, rather than much smaller, as observed.

Table I gives the error analysis for this experiment showing that the peak centroid is the dominant error. This measurement gives a Q-value of -17416 keV for the $^{48}Ca(^3He, ^{11}C)^{40}S$ reaction. Using the value -44216 ± 4 keV for the ^{48}Ca mass excess we obtain -22519 ± 35 keV for the ^{40}S mass. Table II compares the current result to various mass predictions from models[7,8,9] and systematics.[6] Several of the model predictions can be seen to compare reasonably with the result of this measurement, which differs by 0.28 MeV from the value obtained from systematics reported in the 1977 mass evaluation compiled by Wapstra and Bos.[6]

The masses of the T = 4 multiplet for A = 40 can be computed using the Bansal-French-Zamick method for the Coulomb displacement energies of particle-hole states.[10] Assuming the configuration to be $[(d_{3/2}^{-4}, 0^+, 2) \times (f_{7/2}^4, 0^+, 2)](0^+, 4)$ we have calculated the displacement energies and the corresponding excitation energies of the T = 4 states. These are listed in Table III together with proposed reactions for exciting these states. All of these excited states are quite particle-unbound but the decays are T-forbidden and may therefore appear as isolated peaks on a continuous background, as has been observed in (p,t) spectra of such states. The choice of the ^{48}Ti target for exciting the $^{40}Ca(0^+, 4)$ state is suggested by the large amplitude $(\sqrt{36/70})$ of a component of this state which seems "natural" to reach by the $(^3He, ^{11}Be)$ reaction. The reaction $^{44}Ca(^4He, ^8He)$ would appear to be more promising but the component it would "naturally" excite has an amplitude of only $\sqrt{1/70}$.

We wish to acknowledge R. E. Tribble of Texas A & M University for providing the target used in this measurement.

REFERENCES

1. R. T. Kouzes and D. Mueller, Nuc. Phys. A307, 71 (1978).
2. R. T. Kouzes et al., Phys. Rev. C18, 1587 (1978).
3. A. G. Artukh et al., Nuc. Phys. A176, 284 (1971).
4. R. T. Kouzes, N.I.M. 155, 261 (1978).
5. K. H. Wilcox et al., LBL-4000 (Annual Report 1974) p. 92
6. A. H. Wapstra and K. Bos, Atomic and Nuc. Data Tables 19, 175 (1977).
7. S. Maripuu, ed., At. Data Nucl. Data Tables 17, (1976).
8. J. E. Monahan and F. J. D. Serduke, Phys. Rev. C17, 1196 (1978).
9. G. T. Garvey et al., Rev. Mod. Phys. 41 (Oct. 1969).
10. R. Sherr and G. Bertsch, Phys. Rev. C12, 1671 (1975).

MASS MEASUREMENTS WITH PION DOUBLE CHARGE EXCHANGE

H. Nann

Northwestern University, Evanston, Illinois 60201, USA
and
Indiana University, Bloomington, Indiana 47405, USA[*]

I.   INTRODUCTION

The first discussions[1] of pion double charge exchange (DCX) dating back to the early 1960's mention already the great potential of this reaction to study nuclei far from the line of stability. Early attempts by Gilly et al., at CERN failed because of the low pion intensities available at that time. The situation changed with the advent of high intensity pion beams at the 'meson factories.' DCX experiments by Burman et al.[3] at the Low Energy Pion (LEP) channel of the Clinton P. Anderson Meson Physics Facility (LAMPF) yielded the first experimental value for the mass of ^{16}Ne via the ^{16}O(π^+,π^-) reaction. However, these experiments suffered from poor energy resolution (FWHM $\sim$ 4 MeV) and the extracted mass excess for ^{16}Ne is not very accurate (M.E. = 24.4 ± 0.5 MeV). When the Energetic Pion Channel and Spectrometer (EPICS) facility at LAMPF became fully operational, the first pion DCX was studied there by the Northwestern University group.[4,5] In the course of these experiments, the mass excess of ^{18}C was measured via the ^{18}O$(\pi^-,\pi^+)^{18}$C reaction.[5] Since then not only our group but also a collaboration of groups[6] from the New Mexico State University, the University of Texas at Austin and the Los Alamos Scientific Laboratory measured masses with pion DCX.

II.   EXPERIMENTAL METHOD

The EPICS facility consists of a high intensity pion channel and a high resolution spectrometer. The pion beam incident onto the target is vertically momentum dispersed ($\sim$ 6 cm horizontal by $\sim$ 20 cm

---

*Present address

vertical). This feature has the disadvantage that large amounts of
target material have to be used. The spectrometer, as shown in
Fig. 1, consists of three quadrupole (Q_1-Q_3) and two dipole (D_5-D_6)
magnets. The quadrupole triplet images the dispersion plane of the
target one-to-one at its focus where a set of four position sensitive multiwire proportional chambers (MWPC) F_1-F_4 are positioned.
After bending through the dipoles, the particles are detected by
four drift-type MWPC's R_1-R_4 placed near the 'focal plane' of the
spectrometer. An energy loss spectrum is constructed by the on-line
computer from the x, y, θ, and ϕ information provided by the front
and rear MWPC's and the field settings of the channel and the spectrometer.

The EPICS facility was modified for the DCX runs to allow operation at very small angles. A circular bending magnet (labeled
"C" magnet in Fig. 1) was placed after the target on the rotating
frame of the spectrometer in order to separate the charge-exchanged
pions from the outgoing pion beam. This kept the singles rates in
the front detectors low.

Electron rejection was achieved by a propane-gas threshold
Cerenkov counter at the end of the entire detection system. In some
runs time of flight through the spectrometer was measured between
a scintillation detector, S_1, just in front of the first MWPC and
two scintillation detectors, S_2 and S_3, after the MWPC's at the
'focal plane.' An overall time resolution of $\sim$ 1.2 n sec was
achieved.

The pion spectra, see Figs. 2-4, are very clean with almost
absent background on the left of the ground state transitions. The
overall resolution obtained was about 350 keV.

Heavy-ion reactions can be used, of course, to reach the same
final nuclei as with pion DCX. The cross sections are of the same

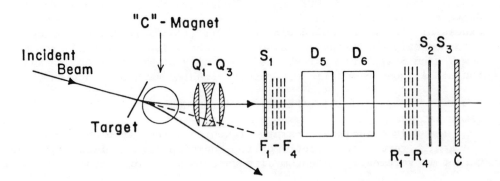

Fig. 1. Schematic of the experimental setup.

order of magnitude. However, because of the very clean pion identification which is possible, pion DCX is in many cases advantageous
over heavy-ion reactions.

The method of measuring masses used in the present experiments
compared the magnetic rigidities of pions from the same type of DCX
reaction where in one case a well-known mass is reached and in the
other the unknown is produced. By using a calibration reaction with
a Q-value similar to that of the reaction under study, both DCX

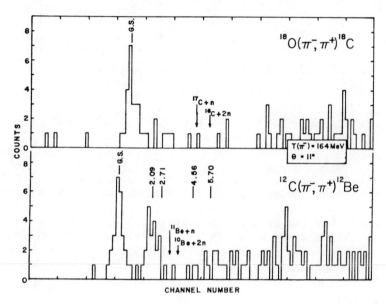

Fig. 2. Missing mass spectrum of the (π^-, π^+) reactions on ^{18}O and
^{12}C at $T(\pi^-)$ = 164 MeV and θ = 11°.

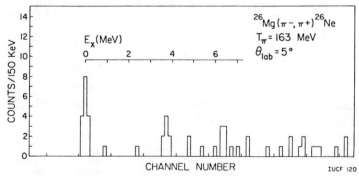

Fig. 3. Missing mass spectrum of the ^{26}Mg$(\pi^-, \pi^+)^{26}$Ne reaction at
$T(\pi^-)$ = 163 MeV and θ = 5°.

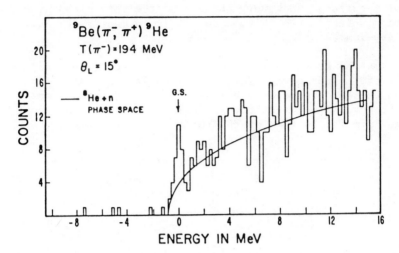

Fig. 4.   Missing mass spectrum of the ^{9}Be$(\pi^-,\pi^+)^9$He reaction at
          $T(\pi^-)$ = 194 MeV and θ = 15°.

reactions can be run under the same mechanical and magnetic settings
of the channel and the spectrometer. This method minimizes the error
dependence on the absolute calibration of the channel and spectrometer magnets, but requires an accurate calibration of the missing
mass scale. The missing mass scale was determined from elastic pion
scattering at the same spectrometer field setting as for the mass
measurements. Only the polarity and the energy of the channel were
changed.

     The error obtained is usually determined by a number of factors.
The most important ones are given in Table I. They are considered
to be lower limits for the present operational status of the EPICS
facility. The largest contributions are due to the statistical uncertainty in the determination of the peak centroides for the known
and unknown transitions and due to the differential nonlinearities
of the missing mass scale.

III.   DISCUSSION OF RESULTS

     As noted in the following discussion, both the (π^+,π^-) and the
(π^-,π^+) reactions have been employed to investigate the properties
of nuclei far from the line of stability.

Masses of T_Z = 3 Nuclei

     Very little is known about light T_Z = 3 nuclei. Before our
work,[5] only the mass excesses of ^{22}O and ^{28}Na have been reported
from experiments employing quite different techniques. The mass

Table I. Error analysis

Source of error	Effect on mass determination (in keV)
Centroid determination:	
calibration reaction	± 40
unknown reaction	± 40
Missing mass scale, differential nonlinearity	± 30
Relative energy loss in targets	± 20
Beam energy	± < 10
Reaction angle	± < 10
Overall	± 70

Table II. Summary of ^{18}C and ^{26}Ne mass predictions

Final nucleus reached by the (π^-,π^+) reaction	^{18}C	^{26}Ne	Ref.
Experimental mass excess (MeV)	24.91(15)	0.43(11)	present work
Predicted mass excess (MeV)			
Thibault and Klapisch	25.51	0.30	10
Jelly et al. (GK)	25.50	0.17	11
Jelly et al. (MSM)	24.57	−0.27	11
Groote et al.	25.09	0.13	13
Janecke	25.53	0.19	13
Garvey-Kelson	24.85	−0.06	present work

Table III. Summary of ^{9}He mass predictions

Author	Mass excess (MeV)	B(n) (MeV)	Ref.
This experiment	40.98(20)	−1.31(20)	
Janecke	42.03	−2.36	13
Jelly et al. (GK)	42.61	−2.94	11
Jelly et al. (MSM)	43.49	−3.82	11
Thibault and Klapisch	42.75	−3.08	10

of ^{22}O was determined by Hickey et al.[7] using the heavy-ion reaction
^{18}O(^{18}O,^{14}O)^{22}O, while the mass of ^{28}Na was measured by the alkali-
metal specific mass spectrometer techniques of Thibault et al.[8]

We measured the masses of ^{18}C and ^{26}Ne using the (π^-,π^+) DCX
reaction on ^{18}O and ^{26}Mg, respectively. Spectra are shown in Figs. 2
and 3. The ground state transitions are clearly seen with almost no
background. For both cases, the ^{12}C$(\pi^-,\pi^+)^{12}$Be reaction which has
a known Q-value of −26.100(15) MeV was used as a calibration reaction.

It should be mentioned that in the ^{18}C spectrum (see Fig. 2)
there might be an indication of an excited state at $\sim$ 2 MeV of excitation. In the ^{26}Ne spectrum (see Fig. 3) a peak is observed at
$\sim$ 3.75 MeV of excitation. Shell model calculations by Cole, Watt
and Whitehead[9] predict excited states in ^{26}Ne at 2.18, 3.95, 3.97
and 4.48 MeV. One can speculate that the peak observed at 3.75 MeV
excitation corresponds to the 2^+-4^+ doublet at 3.95-3.97 MeV. Better
statistics are needed to investigate these excited states.

The resulting mass excesses are given in Table II. For comparison, various predictions from the literature are included.

Study of ^{9}He

Because of the very clean discrimination of pions against other
particles, pion DCX is a very powerful tool for studying very light
nuclei (A < 12) near the proton or neutron drip line. Especially
for unbound systems, peaks from excited states of the detected particle and background from other types of particles reaching the
detector system are highly undesirable.

In the light of this remark we studied the nucleus ^{9}He with the
^{9}Be(π^-,π^+) He reaction. A missing mass spectrum is shown in Fig. 4.
The smooth curve corresponds to the phase space distribution for the
breakup into ^{8}He + n. The clear enhancement over the phase space
distribution can be attributed to states in ^{9}He. From this measurement we concluded that the ground state of ^{9}He is unstable to ^{8}He + n
decay by 1.31 ± 0.20 MeV. An indication of an excited state at $\sim$
1.6 MeV is also seen.

In Table III the mass excess and the neutron binding energy of
^{9}He are compared to various predictions from the literature. The
experimental mass excess is much less unbound than the predicted.

With this new mass excess value for ^{9}He, the mass excess of
^{10}He was calculated using the Garvey-Kelson transverse mass equation.
Its value of 49.40 MeV is stable against one-neutron decay (B(n) =
+0.93 MeV) but still unbound for two-neutron decay (B(2n) = −1.66
MeV).

Isobaric Mass Quintets

The isobaric multiplet mass equation (IMME) describes the dependence of the masses of an isobaric multiplet on its charge state by a quadratic equation:

$$M(T_Z) = a + bT_Z + cT_Z^2$$

The validity of this equation for mass quartets (T = 3/2) was recently reviewed by Benenson and Kashy.[15] With the exception of the ground state A = 9 mass quartet, the IMME holds extremely well. Tests of the IMME for T = 2 quintets[16-22] have only recently become possible when the masses of some of the fourth and fifth members were measured.

Burleson et al.[6] employed the (π^+, π^-) DCX reaction on ^{12}C, ^{16}O, ^{24}Mg and ^{32}S and measured the masses of the $T_Z = -2$ nuclei ^{12}O, ^{16}Ne, ^{24}Si and ^{32}Ar. They obtained the following mass excesses: for ^{12}O 32.059(100),* for ^{16}Ne 24.051(100),* for ^{24}Si 10.682(100),* and for ^{32}Ar -2.181(100)* MeV. These values can be compared to the results of KeKelis et al.[18] for ^{12}O and ^{16}Ne of 32.10(12) and 23.92(8) MeV, respectively, and of Tribble et al.[23] for ^{24}Si of 10.80(4) MeV.

In Table IV all the available information on the A = 12, 16, 24 and 32 mass quintets is collected. These masses were used to determine the coefficients of the IMME. The results are given in Table V. It can be seen that there is no essential evidence for higher order terms in T_Z than the quadratic.

IV.   CONCLUDING REMARKS

It is encouraging to note that new types of reactions are being employed to measure masses of nuclei far from stability. With the availability of high intensity pion beams, the pion DCX reaction has been demonstrated to be a powerful tool in the study of exotic nuclei. In the near future many new masses of nuclei on both sides of the line of stability will be measured via pion DCX using the EPICS facility at LAMPF.

ACKNOWLEDGEMENTS

The author is indebted to Professor G.R. Burleson for communication of and permission to quote his data prior to publication. This work was supported in part by the U.S. Department of Energy and the National Science Foundation.

---

*An error of 100 keV is assigned to these mass excess values by this author.

Table IV.   Properties of T=2 levels which are members of isospin
            quintets

A	J^π	T_Z	Nucleus	E_x(keV)	Mass Excess[a](keV)	Ref.
12	0^+	2	^{12}Be	g.s.	25078(15)	17
		1	^{12}B	12710(20)	26080(20)	24
		0	^{12}C	27595.0(24)	27595.0(24)	25
		-1	^{12}N	unknown	unknown	
		-2	^{12}O	g.s.	32076(77)	6,18
16	0^+	2	^{16}C	g.s.	13695(7)	12,26
		1	^{16}N	9928(7)	15610(7)	27
		0	^{16}O	22721(3)	17984(3)	27
		-1	^{16}F	unknown	unknown	
		-2	^{16}Ne	g.s.	23971(62)	6,18
24	0^+	2	^{24}Ne	g.s.	-5949(10)	12
		1	^{24}Na	5969.0(16)	-2448.5(18)	28
		0	^{24}Mg	15436.4(6)	1505.8(9)	28
		-1	^{24}Al	5955(10)	5903(9)	20
		-2	^{24}Si	g.s.	10784(37)	6,23
32	0^+	2	^{32}Si	g.s.	-24092(7)	28
		1	^{32}P	5073.1(9)	-19231.6(12)	38
		0	^{32}S	12050(4)	-13965(5)	28
		-1	^{32}Cl	5033(10)	-8295.6(52)	21
		-2	^{32}Ar	g.s.	-2181(100)	6

[a]The quoted value is the weighted mean of the results given in the
references.

Table V.   Coefficients of the IMME

A	a(MeV)	b(MeV)	c(MeV)	d(keV)	e(keV)	χ^2
12	27.595(2)	-1.748(16)	0.229(3)	–	4.0(74)	–
	27.595(2)	-1.764(28)	0.245(8)	4.0(74)	–	–
	27.595(2)	-1.751(14)	0.246(7)	–	–	0.29
16	17.984(3)	-2.575(13)	0.196(19)	–	4.8(34)	–
	17.984(3)	-2.594(11)	0.215(7)	4.8(34)	–	–
	17.983(3)	-2.587(10)	0.221(5)	–	–	1.97
24	1.506(1)	-4.182(10)	0.224(8)	6.0(85)	-2.1(45)	–
	1.506(1)	-4.175(5)	0.220(6)	–	0.9(12)	0.50
	1.506(1)	-4.177(4)	0.221(4)	2.2(23)	–	0.22
	1.506(1)	-4.177(4)	0.223(3)	–	–	1.07
32	-13.965(5)	-5.465(9)	0.199(8)	-3.3(84)	1.9(45)	–
	-13.965(5)	-5.468(3)	0.201(7)	–	0.3(15)	0.15
	-13.966(5)	-5.468(3)	0.202(5)	0.2(28)	–	0.18
	-13.966(3)	-5.468(3)	0.202(2)	–	–	0.19

REFERENCES

1. S. D. Drell, H. J. Lipkin, and A. de-Shalit, quoted in T.E.O. Ericson, Proc. CERN 1963 Conf. High Energy Physics and Nuclear Structure.
2. L. Gilly et al., Phys. Lett. 19, 335 (1965).
3. R. L. Burman et al., Phys. Rev. C 17, 1774 (1978).
4. K. K. Seth et al., Phys. Rev. Lett., to be published.
5. K. K. Seth et al., Phys. Rev. Lett. 41, 1589 (1978).
6. G. R. Burleson et al., submitted to Phys. Rev. Lett.
7. G. T. Hickey et al., Phys. Rev. Lett. 37, 130 (1976).
8. C. Thibault et al., Phys. Rev. C 12, 644 (1975).
9. B. J. Cole, A. Watt, and R. R. Whitehead, J. Phys. A7, 1399 (1974).
10. C. Thibault and R. Klapisch, Phys. Rev. C 9, 793 (1974).
11. N. A. Jelly, J. Cerny, D. P. Stahel, and K. H. Wilcox, Phys. Rev. C 11, 2049 (1975).
12. A. H. Wapstra and K. Bos, At. Data Nucl. Data Tables 19, 175 (1977).
13. A. H. Wapstra and K. Bos, At. Data Nucl. Data Tables 17, 474 (1976).
14. G. T. Garvey, W. J. Gerace, R. L. Jaffe, I. Talmi, and I. Kelson, Rev. Mod. Phys. 41, 51 (1969).
15. W. Benenson and E. Kashy, Rev. Mod., Phys., to be published.
16. R. G. H. Robertson, E. Kashy, W. Benenson, and A. Ledebur, Phys. Rev. C 17, 4 (1978); R. G. H. Robertson, W. Benenson, E. Kashy, and D. Mueller, Phys. Rev. C 13, 1018 (1976); R. G. H. Robertson, W. S. Chien, and D. R. Goosman, Phys. Rev. Lett. 34, 33 (1975).
17. D. E. Alburger, S. Mordechai, H. T. Fortune, and R. Middleton, Phys. Rev. C 18, 2727 (1978).
18. G. J. KeKelis et al., Phys. Rev. C 17, 1929 (1978).
19. D. M. Moltz et al., Phys. Rev. Lett. 42, 43 (1979).
20. J. Äystö et al., Phys. Lett. 82B, 43 (1979).
21. E. Hagberg et al., Phys. Rev. Lett. 39, 792 (1977).
22. R. E. Tribble, J. D. Cossairt, and R. A. Kenefick, Phys. Rev. C 15, 2028 (1977).
23. R. E. Tribble et al., private communication.
24. F. Ajzenberg-Selove, Nucl. Phys. A248, 1 (1975).
25. R. G. H. Robertson et al., Phys. Rev. C 17, 1535 (1978).
26. H. T. Fortune et al., Phys. Lett. 70B, 408 (1977).
27. F. Ajzenberg-Selove, Nucl. Phys. A281, 1 (1977).
28. P. M. Endt and C. van der Leun, Nucl. Phys. A310, 1 (1978).

PRODUCTION OF NEW ISOTOPES BY HEAVY ION FRAGMENTATION*

T. J. M. Symons

Nuclear Science Division, Lawrence Berkeley Laboratory
University of California, Berkeley, California   94720

In recent years, work at Dubna,[1] CERN,[2] Berkeley, Los Alamos[3] and elsewhere has increased our knowledge of light nuclei far from stability. In these experiments, heavy ion beams have steadily increased in importance with the use first of transfer reactions and more recently deep-inelastic reactions for isotope production. In this talk, we shall discuss the application of yet another class of reaction, namely heavy ion fragmentation, to the production of very neutron-rich light nuclei.

In heavy ion reactions at high energy ($E/A \gg 20$ MeV/nucleon) it has been found that an appreciable fraction of the total reaction cross section goes into events in which the projectile breaks up into one or more fragments which move forward in the laboratory at close to the beam velocity. These reactions are assumed to arise from peripheral collisions in which there is relatively small energy and momentum transfer between target and projectile.[4]

The fragments observed have a considerable dispersion in A/Z and the cross sections for production of very neutron rich nuclides are comparable to those obtaining for deep-inelastic scattering. However, these reactions also have some specific advantages over low energy experiments, as follows:

(i)   Since both projectile and fragment are moving at high velocity in the laboratory it is possible to use thick targets. For example, a target of 1 gm cm^{-2} Be is appropriate for the fragmentation of ^{48}Ca at 200 MeV/nucleon. This is between two and three orders of magnitude thicker than would be used in a typical deep-inelastic scattering experiment.

(ii)   Since the reaction products are produced close to zero
degrees in the laboratory, it is possible to collect almost the
full reaction cross section in a spectrometer of quite modest
acceptance. In the experiments described here, a zero degree
spectrometer of ∿1 msr acceptance was used and greater than 30%
of the reaction cross section was accepted even in unfavorable cases.

(iii)   Because all of the fragments are moving at close to the
same velocity in the laboratory, magnetic analysis separates
isotopes according to their A/Z values.   This means that exotic
species are separated from more abundant ones and the detectors
used are not required to detect the nuclides far from stability
in a very large background of more commonplace ones.

These three advantages combine to provide a technique of very
high efficiency that may be used even when the beam intensity is
small.

The experiments described here were carried out using the zero
degree spectrometer[5] at the Lawrence Berkeley Laboratory.   Beams of
^{40}Ar and ^{48}Ca were accelerated by the Bevalac and used to bombard
C and Be targets.   The arrangement of the apparatus is shown in
figure 1 and consists of a quadrupole doublet and two dipole magnets
followed by a (7 m) vacuum tank.   The fragments are double focussed

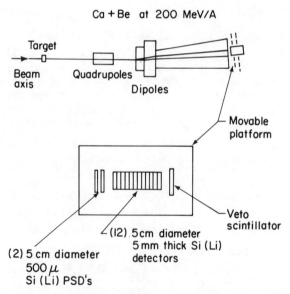

Fig. 1   Area Experimental Layout for Study of the Fragmentation
of 212 MeV/nucleon ^{48}Ca.

by the quadrupoles in the focal plane of the spectrometer and detected by semi conductor detector telescopes.

This telescope comprised two 500 μ thick, 6 cm diameter, position sensitive Si(Li) detectors for horizontal and vertical position measurement followed by 12,5 mm thick, 5 cm diameter, Si(Li) detectors for energy loss measurements. Finally, the telescope was backed by a plastic scintillator for rejection of light particles punching through the silicon.

The 5 mm thick detectors are of particular interest. These devices were developed by D. Greiner and J. T. Walton for use in a telescope from aboard the ISEE-II sattellite. The detectors were designed to resolve isotopes of Fe by energy loss measurements alone. This requires uniformity of thickness of the order of 0.2% (±10μ). Special techniques were developed by the Berkeley detector group under F. S. Goulding and J. T. Walton to fabricate these detectors to such fine tolerances. The detector pulse heights were digitised using a multi-adc system interfaced to CAMAC and the data were recorded event by event on magnetic tape using a PDP 11/40 computer.

The maximum beam intensity available was of the order of 4×10^7 particles/beam pulse ($\sim 10^7$ particles/second) which is very small in comparison to a typical low energy nuclear physics experiment. Accurate beam monitoring was achieved with a variety of scintillators and an ion chamber. Two scintillators were mounted directly in the beam, one of which counted individual beam particles. For the other, the photomultiplier tube leakage current was digitized using a current to frequency converter. This is valuable at intensities greater than measurable by direct counting. For the very highest intensities, scintillator telescopes measured the flux of secondary particles scattered from the target. Unfortunately, the beam intensity can show considerable variations from pulse to pulse. For this reason, all the monitor scalers were read out via CAMAC and written to magnetic tape after every beam pulse.

The beams used were 205 MeV/nucleon ^{40}Ar and 212 MeV/nucleon ^{48}Ca. The targets were natural C and Be (900 mg cm^{-2}) for the two experiments respectively.

The combination of the spectrometer and focal plane telescope provides a system capable of two independent measurements of the particle mass and change. First, the particles are identified by the energies deposited in the Si(Li) detectors. For each detector in the stack, a particle identification signal (I) is calculated using the formula

$$I_i = [(E_i + \Delta E_i)^n - E_i^n]/S_i \propto M^{n-1} Z^2$$

where ΔE_i is the energy that is lost in the ith detector, E_i is the total energy deposited in subsequent detectors up to the stopping detector, S_i is the thickness of the ith detector, n is a parameter which varies from element to element but is usually ~1.78, and M and Z are the particle mass and charge respectively. The I_i signals are then combined to form a weighted mean and χ^2 function defined by

$$\chi^2 = \sum_{i=1}^{s-1} \left(\frac{I_i - \overline{I}}{\varepsilon_i^2} \right)$$

where ε_i is the area on each I_i. This error is derived by assuming a certain detector resolution and differentiating the identification function appropriately. The mass resolution is improved considerably by rejecting particles with large values of χ^2. This eliminates not only events that misidentify due to fluctuations in the energy loss, but, most importantly, those that react in the detectors. At these energies ~ 30% of the incident particles will react in the silicon.

Secondly, the total energy, T, deposited in the telescope is combined with the particle deflection, D, in the spectrometer to form a second particle identification signal

$$I = {}^k/_{TD^2} - {}^T/_{2Z^2} \propto {}^M/_{Z^2}$$

where k is the spectrometer calibration constant. These two functions may be combined to calculate the charge and mass of the fragment unambiguously.

The results obtained from such an analysis are shown in fig. 2, which contains the mass spectra for 8 elements produced in the ^{48}Ca bombardment.[6] In each case a 30% χ^2 cut and a total energy cut have been applied as well as a cut on charge. Fourteen new isotopes have been identified from the data shown in this figure in addition to the two observed in our previous experiment using an ^{40}Ar beam.[7] The new isotopes observed in heavy ion fragmentation are

$$^{22}N, \quad ^{26}F, \quad ^{28}Ne, \quad ^{33,34}Mg, \quad ^{35,36,37}Al, \quad ^{38,39}Si, \quad ^{41,42}P, \quad ^{43,44}S$$

and ^{44,45}Cl, for each of which at least 10 counts have been observed. In addition we can confirm the recent observation of ^{37}Si, ^{40}P and ^{41,42}S by Auger et al.[6]

This represents a significant advance in this field although it must be emphasized that we have not yet reached the neutron drip line for any element heavier than beryllium. However, we are now almost at this point for several elements as may be seen in fig. 3,

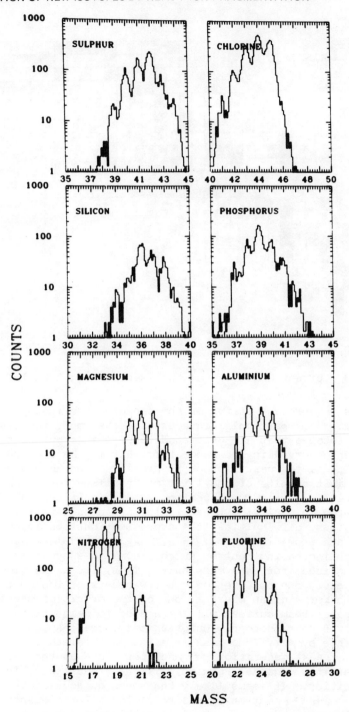

Fig. 2   Mass Spectra for Elements Produced in the
Fragmentation of 212 MeV/nucleon ^{48}Ca.

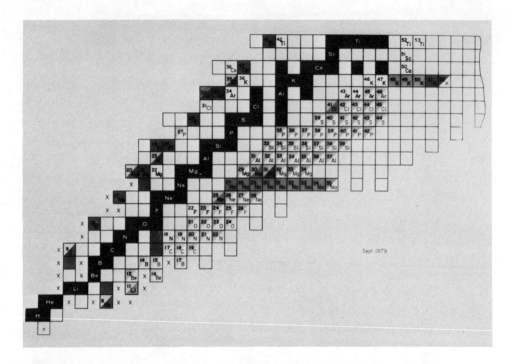

Fig. 3 Current Chart of the Nuclides for Light Nuclei.

which shows the present status of the table of the isotopes for
light nuclei.[8] Of particular interest are the empty **squares which**
represent isotopes that are predicted to be stable but which have
not yet been observed in the laboratory. It may be seen that for
oxygen and neon, the next isotope is predicted unstable and that
we are within two units of the limit of stability for boron, carbon,
nitrogen and fluorine. The experimental difficulties that will
be encountered in verifying these limits are considerable. To
illustrate these difficulties, the cross sections observed in this
experiment are shown in fig. 4. In all cases, the cross sections fall
steeply with increasing neutron number, sometimes by as much as an
order of magnitude for each mass unit. Thus, if, as is almost
always the case, one observes only a few counts of the most neutron-
rich isotope for a given element, one or two orders of magnitude
more integrated beam current will be needed to demonstrate the
<u>absence</u> of the next isotope. Unfortunately, the chances of
achieving this by very long runs at existing accelerators are slim.
However, the next generation of machines, such as the MSU coupled
cyclotron and the upgraded Bevalac, will have substantially higher
beam intensities. It seems likely that such an accelerator will
be able to reach the neutron drip line at least for elements up to
neon and also to produce any particle stable isotope with mass
less than 50.

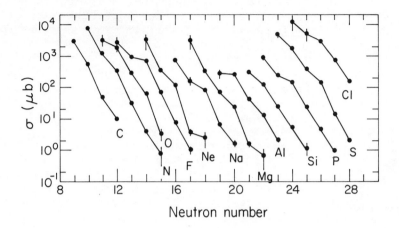

Fig. 4 Production cross sections for isotopes produced in the fragmentation of 212 MeV/nucleon ^{48}Ca.

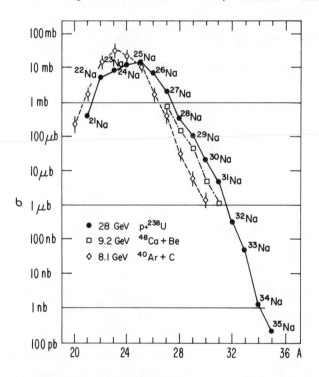

Fig. 5 Comparison of production cross sections for Na isotopes in the ^{48}Ca+Be, ^{40}Ar+C and p+^{238}U reactions.

Before leaving the technical aspects of these experiments, it is interesting to compare the cross sections obtained using different reactions. In fig. 5, we show the production cross sections for Na isotopes obtained using ^{40}Ar and ^{48}Ca projectiles as well as from proton spallation of ^{238}U.[2] Clearly, the more neutron rich ^{48}Ca is a much more effective projectile than ^{40}Ar, as prediced by the abrassion-ablation model. Also, the large yield obtained from the p + ^{238}U reaction indicates that further gains will be achieved by going to heavier projectiles.

In conclusion, these experiments have shown that the fragmentation of high energy heavy ions is a powerful new tool for the production of nuclei far from stability. This is not, of course, an end in itself. Rather, we should realise that nature has presented us with an outstanding new technique for the study of these nuclei. The application of these reactions to atomic mass measurements or secondary beam experiments will present an exciting challenge to our ingenuity. We can, I feel sure, look forward to learning the results of these studies at the next and subsequent conferences in this series.

Acknowledgements

It has been my pleasure to work with, and learn from my colleagues at the Lawrence Berkeley Laboratory whose skill and dedication ensured the success of these experiments.

*This work was supported by the Nuclear Science Division of the U.S. Department of Energy under contract No. W-7405-ENG-48.

References

1. A. G. Artukh, V. V. Avdeichikov, G. F. Gridnev, V. L. Mikheev, V. V. Volkov and J. Wilczynski, Nucl. Phys. A176, 284 (1971)
2. C. Thibault, R. Klapisch, C. Rigaud, A. M. Poskanzer, R. Prieels, L. Lessard and W. Reisdorf, Phys. Rev. C12, 644 (1975).
3. G. W. Butler, D. G. Perry, L. P. Remsberg, A. M. Poskanzer, J. B. Natowitz and F. Plasil, Phys. Rev. Lett. 38, 1380 (1977).
4. A. S. Goldhaber and H. H. Heckman, Ann. Rev. Nucl. Sci. 28 161 (1978).
5. D. E. Greiner, P. J. Lindstrom, H. H. Heckman and F. S. Bieser, Nucl. Instr. and Meths. 116, 21 (1974).
6. G. D. Westfall, T. J. M. Symons, H. J. Crawford, D. E. Greiner, H. H. Heckman, P. J. Lindstrom, J. Mahoney, D. K. Scott, A. C. Shotter, T. C. Awes, C. K. Gelbke and J. M. Kidd, to be published.
7. T. J. M. Symons, Y. P. Viyogi, G. D. Westfall, P. Doll, D. E. Greiner, H. Faraggi, P. J. Lindstrom, D. K. Scott, H. J. Crawford and C. McParland, Phys. Rev. Lett. 42, 40 (1979).
8. J. Cerny and J. Äystö, private communication, 1979.

SEARCH FOR LIGHT NEUTRON-DEFICIENT NUCLEI PRODUCED IN 800 MEV PROTON SPALLATION REACTIONS*

D. J. Vieira, G. W. Butler, and D. G. Perry
Los Alamos Scientific Laboratory, Los Alamos, NM 87545

A. M. Poskanzer
Lawrence Berkeley Laboratory, Berkeley, CA 94720

L. P. Remsberg
Brookhaven National Laboratory, Upton, N.Y. 11973

J. B. Natowitz
Texas A&M University, College Station, TX 77843

INTRODUCTION

Defining the limits of particle stability in the light mass region provides a challenge to both the experimentalist and the theorist. The surprising discoveries of the particle stability of the neutron-rich nuclei ^{11}Li (ref. 1), ^{14}Be (ref. 2), ^{19}C (ref. 3), and ^{32,34}Na (ref. 4), which were predicted to be unstable with respect to one or two neutron emission, pointed out significant deficiencies in the understanding of the nuclear mass surface for light nuclei far from the valley of β-stability. Today such measurements of very neutron-rich or very neutron-deficient nuclei remain one of the most critical tests of current nuclear mass theories.

Several recent experiments[5,6,7] have been successful in determining the particle stability of some 23 previously unknown neutron-rich isotopes of the elements nitrogen through chlorine by a positive identification in on-line particle spectra. However, no analogous

---
*This work supported by the U.S. Department of Energy.

search has yet been undertaken on the neutron-deficient side of
β-stability in this region. In this contribution we describe the
initial experiments performed at the Clinton P. Anderson Meson
Physics Facility (LAMPF) in which time-of-flight (TOF) techniques
have been used to search for new neutron-deficient nuclei at the
limits of proton stability.

EXPERIMENTAL METHOD

The current LAMPF experiments have concentrated on the neutron-
deficient isotopes of Mg to Ar, since the limits of proton stability
have previously been determined up through Na. Spallation products
resulting from 800 MeV proton bombardments of three medium mass
targets have been detected using an improved version of the dE/dx-
TOF techniques reviewed by Butler[8].

Fig. 1 shows the Thin Target Area at LAMPF where these measurements were performed. It consists of a scattering chamber with
several 5 m flight tubes. In the present experiment, spallation
residues produced by the intense (500 μA average current) 800 MeV

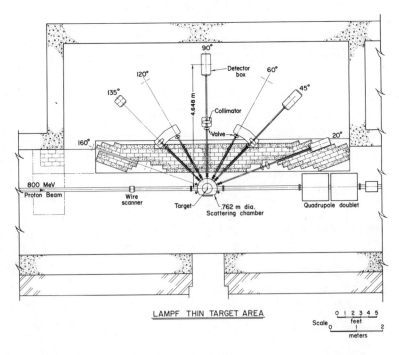

Fig. 1   A schematic diagram of the LAMPF Thin Target Area. The
scattering chamber is located 15 m upstream of the first
pion production target.

proton beam with targets of $^{nat}CaF_2$, ^{nat}Ni, and ^{92}Mo (thickness
~0.5 mg/cm²) were detected at 45°.² The observed energy distributions of the aluminum residues resulting from the three separate
targets are shown in Fig. 2. The Al yields are observed to decrease
rapidly with energy and for higher A targets a more gradual energy
dependence is found. These features are consistent with the interpretation of spallation being the dominant reaction mechanism.

   In order to acquire data where the spallation yields were
sufficiently large, a detection system capable of characterizing
reaction products to as low an energy as possible was necessary.
This system consisted of two fast-timing, secondary-emission channel
plate detectors[9] and a standard ΔE-E, gas-Si telescope[10]. A schematic layout of the experimental arrangement is shown in Fig. 3.
For each event nine parameters were recorded: the amplitudes of
the channel plate signals (CP1 and CP2), the ΔE-E energies of the
detector telescope, the pressure and temperature of the gas in the
detector telescope, and three time measurements - T_a which measures
the TOF between CP1 and CP2, T_b which measures the time difference
between CP1 and the next beam pickoff signal, and T_c which measures
the TOF between CP2 and the E detector From these measured para-

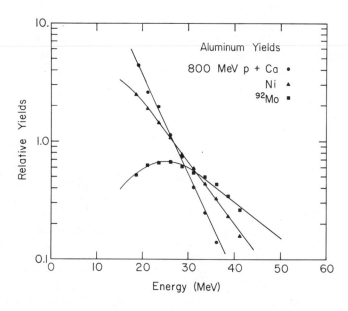

Fig. 2   The relative yields of aluminum (Z=13) observed at 45° (lab)
         in 800 MeV proton bombardments. The solid lines have been
         drawn to guide the eye.

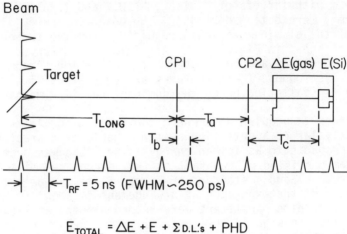

$$E_{TOTAL} = \Delta E + E + \Sigma D.L.'s + PHD$$

$$Mass = 2 E_{TOTAL}(T_{LONG}/D_{LONG})^2$$

Fig. 3   Schematic layout of the experimental arrangement, the LAMPF
beam microstructure, and the important experimental parameters. See text for a detailed description.

meters the following were determined: 1) the long flight path TOF
between the target and CP1 (T_{Long}), 2) the total kinetic energy
(E_{total}), 3) the mass (A), and 4) the atomic number (Z) of the
spallation residue.

The long flight time is given by

$$T_{Long} = nT_{RF} - T_b \tag{1}$$

where T_{RF} is the RF period between beam bursts, and n is the number
of RF periods in the interval between the creation of the residue
and the beam burst immediately following the CP1 signal. The integer
n is obtained by rounding off the quantity $(T' + T_b)/T_{RF}$ in which T'
is an estimate of the long flight path TOF calculated from the short
flight path measurement, T_a. Thus in this technique the long flight
path TOF was determined to high precision by combining a coarse time
measurement, T_a, which was used to determine the long flight time to
the nearest RF period, and a fine time measurement, T_b, which measured
the time when the event occurred to a fraction of a RF period. After
correcting T_a and T_b for time walk with respect to CP1 and CP2
amplitudes, a time resolution of 0.25 ns (FWHM) was obtained over

a short flight path (CP1-CP2) of 50 cm, while the long flight path
(target - CP1 = 4.3 m) time resolution was found to be 1.2 ns.

The total kinetic energy of the reaction product was obtained
by adding together the pressure and temperature-normalized ΔE-E
detector telescope energies. Dead layer corrections for energy
losses in the thin carbon foil ($\sim$20 $\mu g/cm^2$) of the channel plate
detectors and the polypropylene gas isolation window ($\sim$60 $\mu g/cm^2$)
on the detector telescope were made using the dE/dx table lookup
method. Due to the lack of an existing dE/dx table of sufficient
accuracy in this energy region, we generated our own table from the
data itself. This table consisted of ΔE entries for different E
and Z values that were determined by a peak finding routine. The
use of this ΔE-E table as a dE/dx table enabled good dead layer
energy loss corrections to be made over the entire energy and Z
region of interest. Finally, an estimate of the pulse height defect
(PHD) using the method of Kaufman, et. al.[11] was added to give the
total kinetic energy, E_{total}. Once the total kinetic energy and
the long flight time were obtained, the calculation of the final
mass was straightforward using the second equation given in Fig. 3.

The Z of each spallation residue was determined by two separate methods. In the first method the atomic number was obtained
using the ΔE-E table lookup approach which was then followed by a
mass correction to remove the mass dependence of the Z determination.
The second method used the ΔE-TOF table lookup method, where, in
an analogous fashion to the ΔE-E table, a ΔE-TOF table was generated
from the data. The latter method has the added attraction of being
mass independent in first order[12]. However, when we compared the
two methods we found the results to be nearly identical, so in this
experiment we arbitrarily chose to use the ΔE-E table lookup method
as the final determination of the Z of the reaction product.

Three major requirements were used to avoid any ambiguity in
the final mass determination and to reduce background events due to
random coincidences and other spurious effects. First, a window of
0.35 ns was placed on the time difference between T_c, the measured
TOF between CP2 and the E detector, and the estimated CP2-E flight
time that was calculated from T_a and the pertinent energies and
distances. Furthermore, the final flight time, T_{Long}, was required
to differ from the estimated long flight time, $T'_{Long} + T_b$, by less than
3.2 ns. Finally, a lower kinetic energy restriction was added. The
latter represented a compromise between statistics and adequate
charge and mass resolution. At low energies the mass resolution was
dominated by energy uncertainties which resulted primarily from
energy straggling in the dead layers[13]. The energy straggling
accounted for an estimated energy uncertainty of $\sim$200 keV out of a

total uncertainty of ~350 keV. The Z resolution was also found to be very energy dependent due to charge neutralization effects which cause adjacent Z values to converge at low energies. In this analysis a total kinetic energy threshold of 26 MeV was chosen, resulting in a mass and charge resolution (FWHM) of 0.4 amu and 0.4 charge units, respectively, for a typical spallation product like ^{26}Al.

RESULTS AND CONCLUSION

About ten million events of the elements N through Ca were collected from a Ni target in a one month period. This data was analyzed in the fashion previously described and Fig. 4 shows two of the observed mass spectra. All of the known neutron-deficient isotopes of Al and P are seen out to the currently known limits of particle stability, ^{23}Al and ^{27}P. Beyond these points we find some evidence that ^{22}Al and ^{26}P may be stable, however the statistics are too meager to be convincing. Furthermore, the Al spectrum is confused by five potentially spurious events observed below mass 22. Current predictions using Kelson-Garvey mass relationships predict ^{22}Al to be marginally bound by 250 keV, while ^{26}P is predicted to be proton unbound by 90 keV[14]. Future attempts to define the neutron-deficient limits of particle stability and, in particular, the possible stability of ^{22}Al and ^{26}P are expected.

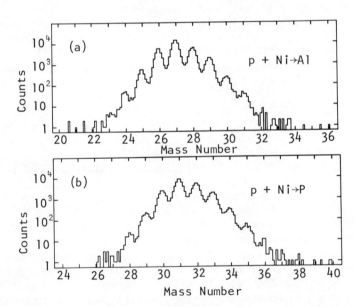

Fig. 4 Mass spectra observed in 800 MeV proton bombardment of ^{nat}Ni for the elements (a) Al and (b) P.

REFERENCES

1.   A. M. Poskanzer, S. W. Cosper, E. K. Hyde, and J. Cerny, Phys.
     Rev. Lett. 17, 1271 (1966).
2.   J. D. Bowman, A. M. Poskanzer, R. G. Korteling, and G. W.
     Butler, Phys. Rev. Lett. 31, 614 (1973).
3.   G. M. Raisbeck, P. Boerstling, P. W. Reisenfeldt, R. Klapisch,
     T. D. Thomas, and G. T. Garvey, in Proceedings of the 3rd
     International Conference on High Energy Physics and Nuclear
     Structure, New York, 1969, edited by S. Devons (Plenum, New
     York, 1970) p. 341.
4.   R. Klapisch, C. Thibault, A. M. Poskanzer, R. Prieels, C.
     Rigaud, and E. Roeckl, Phys. Rev. Lett. 29, 1254 (1972); C.
     Thibault, these proceedings.
5.   G. W. Butler, D. G. Perry, L. P. Remsberg, A. M. Poskanzer,
     J. B. Natowitz, and F. Plasil, Phys. Rev. Lett. 38, 1380 (1977).
6.   P. Auger, T. H. Chiang, J. Galin, B. Gatty, D. Guerreau, E.
     Nolte, J. Pouthas, X. Tarrago, and J. Girard, Z. Physik, A 289,
     225 (1979).
7.   T. J. M. Symons, Y. P. Viyogi, G. D. Westfall, P. Doll, D. E.
     Greiner, H. Faraggi, P. J. Lindstrom, D. K. Scott, H. J.
     Crawford, and C. McParland, Phys. Rev. Lett. 42, 40 (1979);
     T. J. M. Symons, these proceedings.
8.   G. W. Butler, in Proceedings of the 3rd International Conference
     on Nuclei Far From Stability, Corsica, 1976, CERN Report 76-13,
     p. 15.
9.   A. M. Zebelman, W. G. Meyer, K. Halbach, A. M. Poskanzer, R. G.
     Sextro, G. Gabor, and D. A. Landis, Nucl. Instrum. Methods 141,
     439 (1977).
10. M. M. Fowler and R. C. Jared, Nucl. Instrum. Methods 124, 341,
     (1975).
11. S. B. Kaufman, E. P. Steinberg, B. D. Wilkins, J. Unik, A. J.
     Gorski, and M. J. Fluss, Nucl. Instrum. Methods 115, 47 (1974).
12. L. H. Harwood, R. I. Cutler, K. W. Kemper, and R. V. Leclaire,
     Nucl. Instrum. Methods 154, 317 (1978); S. C. Luckstead, Ph.D.
     thesis, Washington State University, 1978 (unpublished).
13. H. Schmidt-Böcking, in Lecture Notes in Physics, edited by
     K. Bethge (Springer-Verlag, Berlin, 1978) Vol. 83, p. 81.
14. I. Kelson and G. T. Garvey, Phys. Lett. 23, 689 (1966);
     J. Jänecke, At. Data Nucl. Data Table 17, 455 (1976); A. H.
     Wapstra and K. Bos, At. Data Nucl. Data Table 19, 177 (1977).

ATOMIC MASSES - COMMENTS ON MACRO-MICRO MASS-FORMULAE[*]

W. Freudenreich, E.R. Hilf[+], K. Takahashi

Institut für Kernphysik
Technische Hochschule Darmstadt
D-6100 Darmstadt, Germany

INTRODUCTION

The research on atomic masses started from early accidental measurements and simple empiric rules. The next step was to aim at a systematic collection of the masses of all nuclei, stable against particle emission, and at semi-empiric models to guide the experiments with predictions.

The prediction power of the many semi-empiric mass formulae is somewhat saturating to the point that neither a refit of their parameters to new experimental data nor an increase of the already pretty high numbers of parameters may yield significant improvements.

The more microscopic models, such as the deformed Hartree-Fock or its simplification to the energy density formalism may soon gain a prediction-power as of the semi-empiric approaches.

The present note deals with a simple single-particle model to extract both the average "macroscopic" trends and a microscopic shell-correction, with the aim to keep the number of parameters very small. So that one can calculate and study additional effects such as pairing, neutron-proton correlations as corrections, without the fear that part of them are already submerged into a mass-formula with too many parameters.

*Work supported by the German Bundesministerium für Forschung und Technologie, Grant 06 DA 702.
+While at the Dept. of Physics, Univ. of Washington, Seattle, USA.

BAG MODEL

The concept of the nucleon-bag model to calculate the nuclear mass one may borrow from the quark-bag model[1,2] proposed to describe the nucleon, now that the quantum-chromodynamics to treat the three-body problem has not been developped.

The three quarks are assumed to move freely in a bag, confined by an external pressure B (to represent the confining interactions). The nucleon mass m is then expressed by the sum of the kinetic energy and the volume energy $(B\,V_b)$ of the bag with the radius r_b being determined by the equilibrium condition $\partial_V\, m = 0$. Even if we simply assume the quarks to move non-relativistically and approximate the kinetic energy T_q by the uncertainty principle,

$$(1) \qquad T_q \simeq \frac{\hbar^2}{2\cdot m_q \cdot r_b^2} \qquad ,$$

we have the quark mass m_q of o.1 to 5 GeV for the estimates of r_b of 1 to .15 fm, respectively .

Analogously, a nucleon-bag model would assume two independent non-interacting Fermi-gases in a bag of radius R with the constant potential-well depth V_o inside the bag. Then the single-particle average level-density for each component has the asymptotic form[3]

$$(2) \qquad \bar{\rho}(\varepsilon) = a \cdot V \cdot \sqrt{\varepsilon} \; - b \cdot S + \ldots \qquad ,$$

where V and S are the volume and the surface area. For a perfect Fermi-gas the coefficients are

$$(3) \qquad a = \frac{\nu}{4\pi^2}\left(\frac{\hbar^2}{2m^*}\right)^{-3/2} , \qquad b = \frac{\nu}{16\pi}\left(\frac{\hbar^2}{2m^*}\right)^{-1} ,$$

where ν is the spin-isospin degeneracy.

The level-density can be scaled[4] by introducing a scaling energy

$$(4) \qquad \alpha := \frac{\hbar^2}{2m^*}\cdot\left(\frac{6\pi^2}{\nu\cdot V}\right)^{2/3}$$

so that, with $e := \varepsilon/\alpha$, (2) reads in dimensionless variables

$$(5) \qquad \bar{\varrho}(e) = \frac{3}{2}\cdot\sqrt{e} \; - c\cdot s + \ldots \qquad ,$$

where

$$(6) \qquad c(\nu) := \frac{1}{8}\left(\frac{9\pi\nu}{2}\right)^{1/3} , \qquad s := S/V^{2/3} ,$$

with $s = (36\pi)^{1/3}$ for a sphere.

The scaled Fermi-energy e_F for n nucleons is gained by integrating (5) and inverting to second order,

$$(7) \qquad e_F = n^{2/3} \left(1 + \tfrac{2}{3} c \cdot s \cdot n^{-1/3} \right) ,$$

and the total scaled kinetic energy for one component results in

$$(8) \qquad T/\alpha =: T = \tfrac{3}{5} n^{5/3} \left(1 + \tfrac{5}{6} c \cdot s \cdot n^{-1/3} \right) .$$

Assuming[5] the Fermi-energy for the saturating $N = Z$ nuclei to be the same as for the nuclear matter,

$$(9) \qquad \mathcal{E}_F = \alpha \, e_F = \frac{\hbar^2}{2 m^* r_0^2} \left(\frac{9\pi}{2\nu} \right)^{2/3} ,$$

one has the nuclear scaling-energy of the following form

$$(10) \qquad \alpha = \frac{\hbar^2}{2 m^* r_0^2} \left(\frac{9\pi}{8} \right)^{2/3} A^{-2/3} \left(1 - \tfrac{2}{3} c \cdot s \cdot A^{-1/3} \right) ,$$

and the total kinetic energy thus reads

$$(11) \qquad T = \frac{\hbar^2}{2 m^* r_0^2} \left(\frac{9\pi}{8} \right)^{2/3} \left(\tfrac{3}{5} A + \tfrac{1}{10} c \cdot s \cdot A^{2/3} \right).$$

The second term yields the surface tension a_s which, for $m^* = m$ is half of the experimental value. If one would have more roughly equaled[3] the potential-volume V_0 with the matter-volume $r_0^3 A$, then $a_s \sim 5/2$ of the experimental value would have resulted.

For $N \neq Z$ nuclei, after expanding to second order in $A^{-1/3}$ and $I = (N - Z)/A$, and using (10) one gets

$$(12) \qquad T(Z,N) = a_t A + a_s A^{2/3} + a_I I^2 A + a_{SI} I^2 A^{2/3} + \ldots$$

with

$$(13) \qquad a_t = \tfrac{3}{5} \left(\frac{9\pi}{8} \right)^{2/3} \frac{\hbar^2}{2 m^* r_0^2} $$

and for spherical nuclei

$$(14) \qquad \frac{a_s}{a_t} = (3\pi^2)^{1/3} / 8 , \qquad \frac{a_I}{a_t} = \tfrac{5}{9} , \qquad \frac{a_{SI}}{a_s} = \tfrac{2}{9} $$

so that the ratio of the symmetry and surface coefficients is

$$(15) \qquad \frac{a_I}{a_s} = 40 / (9 \cdot (3\pi^2)^{1/3}) \simeq 1.44 ,$$

which is close to the experimental value of about 1.32, in contrast to 0.29 as gained by the identification[3] of the potential- and density-volumes.

The nucleon-bag model for the nucleus thus reads

(16) $$M(Z,N) = M_r - BA + T + E_c + E_p ,$$

where M_r is the rest mass of all nucleons. For the Coulomb-energy E_c and the pairing-term E_p the simple ansatzes

(17) $$E_c \simeq \frac{3}{5} Z^2 e^2 /(r_c A^{1/3}), \qquad E_p \sim \begin{cases} \delta_p/\sqrt{A} & \text{odd-odd} \\ 0 & \text{odd-A} \\ -\delta_p/\sqrt{A} & \text{even-even} \end{cases}$$

may be sufficient. A fit of (16) to known 1240 atomic masses ($20 \leq A \leq 250$) with regard to the 5 parameters, B, $(m^*/m)r_0^2$, s, r_c and δ_p , yields with $s/(36\pi)^{1/3} \simeq 1.16$ an rms of 2.84 MeV and thus is equivalent to the macroscopic Weizsäcker-Bethe formula which, with the same pairing term, reproduces the same. This is no surprise since if one retains only the leading terms of T, see (12), the mass formula (16) is equivalent to the Weizsäcker-Bethe formula.

There all the shell- and deformation-effects are smeared out. By the way, if one would try to improve (16) by using the droplet-model[6] mass-formula with its 14 parameters but also no shell-effects, the result of rms=2.65 MeV indicates that with regard to predicting masses the droplet-formula alone may contain correlated terms. Only the addition of a shell-correction term does improve the result if it is accompanied by a refit of the smooth part. However, for explaining the isotope-shifts of the nuclear radii the droplet-model[7] was a mayor step forward.

GROSS THEORIES

Superseding the simple bag-model, there are two traditional ways.
1) to simplify the wave-function drastically but elaborate with various assumptions on a two-body force and with various techniques of solving the many-body problem in order to calculate at least average saturation properties of nuclear matter (equilibrium-binding per particle, -density, -incompressibility). The known attempts somewhat fail to give the "experimental" results. They end up with too high a density or a lack of binding (of some MeV/n) or both.

This general shortcoming may be due to neglecting too much the internal structure of the nucleons. For curiosity we give a simple estimate, evaluating qualitatively the quark-bag model to give the saturation properties of nuclear matter. In the MIT-model[1] a radius of 1 fm is claimed for the 3-quark bag. Thus these bags as constituents of the nucleus would touch. So Baym[8] speculated that most-energetic quarks, say$(3-n_b)$ quarks, may percolate (hop) from one bag through the touching surface of the bags to the neighbour quark-bags. The two conditions that quarks have to be always in any bag (confinement) and that the bags should be colour-singlets can be fulfilled, analogous to the problem of those conducting electrons in a metal which percolate from one atom to the next ones, with the Wigner-Seitz-cells staying neutral and the electrons being bound.

Let the kinetic energy for the n_b quarks staying in a quark-bag be given by (1). Those $n_n = (3 - n_b)$ quarks of each bag, which are assumed to percolate, may be found in any of the bags, but not in the intermediate space (thus they are always confined), and let them form a perfect Fermi-gas in the space of volume $A \cdot V_b$ and the surface area $A \, S_b$ so that $s_b = S_b / V_b^{2/3} = (36\pi)^{1/3} A^{1/3} d$. Here d is a factor measuring to what extent the surface area as a boundary is reduced due to the contacts of the bags. Then applying (1) and (8) one gets for the total kinetic energy

$$(18) \quad T_{tot}/A = \frac{5}{3} n_b \frac{\hbar^2}{2 m_q r_b^2} + \frac{\hbar^2}{2 m_q r_b^2} \left(\frac{3}{5} c_1 n_n^{5/3} + \frac{3}{4} c_2 n_n^{4/3} d \right),$$

where

$$(19) \quad c_1 := \left(\frac{9\pi}{2} \right)^{2/3} \quad \text{and} \quad c_2 = c_1 \left(6\pi^2 \right)^{1/3} / 4 .$$

A minimization of T_{tot}/A with respect to n_n for given r_b yields the number of percolating quarks, which come out to be rather small,

$$(20) \quad \left(n_n \right)_{equ}^{1/3} = \frac{c_2}{2 c_1} \left(\sqrt{d^2 + \frac{20}{3} \frac{c_1}{c_2^2}} - d \right) .$$

The radius of a quark-bag r_b in nuclear-matter density $\rho = \left(\frac{4\pi}{3} r_0^3 \right)^{-1}$ should be smaller than or equal to r_0, (dense packing), whereas for low densities, $r_0 \gg r_{bo}$, the quark-bag radius should attain its vacuum value r_{bo}. A smooth interpolation with the desired limits is

$$(21) \quad r_b =: r_{bo} \cdot D^\nu , \qquad D := \left(r_0 / r_{bo} \right)^{1/\nu} / \left(1 + r_0 / r_{bo} \right)^{1/\nu},$$

where ν is a free parameter. With increasing nuclear-matter density both the bag-contact window-area and the inverse of the bag-radius r_b^{-1} rises. We assume quantitatively a relation

$$(22) \quad d = D^{2\nu/\tau}$$

with another parameter $\tau > 0$.

Trying different simple (positive) values of ν and τ one realizes that nuclear-matter saturates as a consequence of the balance between the reduction of the total kinetic energy due to the percolating quarks (feeling less surface-boundaries than if bound in a quark-bag) and the increase of kinetic energy of the bound quarks inside the bags (the radii of which shrink with increasing matter-density). The matter is by far too compressible since we do not introduce any hard-core-type behaviour of nucleons. As an example, for $\nu = 3/2$ and $\bar{\pi} = 1/2$ one gets $\Delta T/A = -15$ MeV for the nuclear-matter saturation-density. ($\varrho = \varrho_0/8$: -11.7 MeV, $\varrho = 125\varrho_0$: -16.8 MeV).

2) The second way to supersede the simple bag-model is to do a full-scale shell-model calculation. Although the results are rather promising, it has been performed up to the sd-shell nuclei only. For heavier nuclei, one may try to set up analytic structures by arguing from a simpler lowest-seniority scheme and fit parameters. In this way, Zeldes[9] could, with 178 parameters, name and demonstrate several pairing and j-coupling features of the nuclear mass-surface. This method proved to be very successful, although the nuclear shape effects are mocked up into the shell-parameters and it gives no radii since it is not self-consistent. Here even the smooth part is hidden in the "shell"-effects.

The gross theories try to combine the advantages of the two ways by keeping smooth average-trend analytic expressions but develop simple microscopic models to set up analytic shell-corrections. For example, an analytic shell-correction term derived from bunching an averaged smooth single-particle spectrum at the magic numbers, with the magic gap widths and their positions as parameters, (in total 39), was given elsewhere[10]. Although the rms was then as low as o.65 MeV, there were too many parameters to study now additional effects due to residual-interactions such as pairing, "mutual support of magicities" and so on.

In the following, we thus propose a simple shell-correction, which makes use of Woods – Saxon single-particle levels and of the scaling-relations with only a few parameters.

AN APPROACH TO A NEW MICROSCOPIC SHELL CORRECTION

For simplicity the neutron and proton shell-corrections are assumed to be additive,

(23)
$$M(Z,N) = \overline{M}(Z,N) + \widetilde{M}_Z(Z,N) + \widetilde{M}_N(Z,N).$$

A simple way to gain a microscopic shell-correction is to calculate the single-particle energies ε_i of a Woods-Saxon potential (assumed here to be spherical) and set as usual

(24)
$$\widetilde{M} = \Sigma t_i - \overline{\Sigma t_i},$$

where the single-particle kinetic energy is approximated by

(25) $\qquad t_i \simeq |V_o| - |\varepsilon_i| - \langle V_c \rangle$,

with V_c being the depth of the Woods-Saxon potential and $\langle V_c \rangle = 2 E_c$ as a measure of the average Coulomb-energy per proton inside the nucleus. The values of the potential parameters are taken from the droplet-model prediction.

The average trend $\overline{\Sigma t_i}$ we estimate by assuming that the asymptotic average single-particle level-density (2) and the scaling relation (5) holds, so that with (8) one gets

(26) $\qquad \widetilde{M} = (\alpha_{ws} - \bar{\alpha}) \cdot T(n)$, $\qquad \alpha_{ws} := \Sigma t_i / \tau$

However, the Woods-Saxon potential depends on three variables R, a, V_o , whereas the Schrödinger-equation can only be scaled with respect to the radius R , resulting in a strong A-dependent potential-depth and -diffuseness in contrast to the experimentally known approximate constancy of both. Thus, as a compromise we stick to the ansatz (26) but allow for an adjustment of $\bar{\alpha} \sim (mR^2)^{-1}$ with regard to a slight dependence on $A^{-1/3}$ and I ; the first check of the ansatz

(27) $\qquad \dfrac{m^*}{m} \cdot R_{scal}^2 = a \cdot \left(1 + \dfrac{b}{A^{1/3}}\right) \cdot \left(1 + c \cdot I\right) \cdot A^{2/3}$

yields that $a \simeq r_{oo}$ and thus $R_{scal} \approx R_{DM}$ (droplet model) and $m^* = m$ is already a fairly good choice.

In order to inspect visually whether $\bar{\alpha}$ with (27) represents the smooth average trend of the Woods-Saxon single-particle levels, fig. 1 shows an example where the Woods-Saxon levels are compared with the Fermi-gas level-density expression. Each section of the parabola contains the same number of particles as the degeneracy of the corresponding Woods-Saxon level. For the "official magic numbers" (--due to N. Zeldes) the discrete levels are connected to the centres of gravity of the referential sections. The inclusion of $\langle V_c \rangle$, and of the surface term in the level-density formula are very effective in matching the general trends. Similar agreements by eye have been observed for nuclei in a wide range of the Z-N -plane.

The discussion preceeding (27) still does not guarantee a success of the ansatz (26). Here we encounter an unanswerable question: what are the experimental neutron and proton shell effects? As a first step of the iteration, we have tried two methods: 1) take the theoretical values[10] with 39 parameters, 2) define the total "experimental" shell energy as $M_{exp} - \widetilde{M}$ and then assume for the moment that the separation into two components could simply be done by using the ratio $\tau(N) / \tau(Z)$. Then we have determined $m^* \cdot R_{scal}^2$ for each nucleus and for each nucleon system

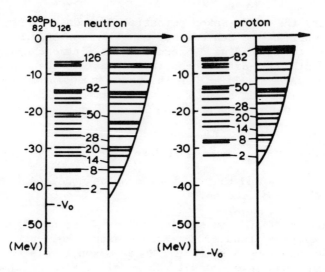

Fig. 1.   Comparison of the single-particle levels of ^{208}Pb for
          the Woods-Saxon potential (left) and the smooth average
          single-particle level-density

from the above-defined $\widetilde{M}$ and α_{ws}. The fact that the resulting values are a smooth function ("without shell effects") implies that the ansatz (26) should work. These steps are inevitable since the ansatz (26) contains a very small difference times a big number so that just fitting around the parameters could bring a spurious solution of their values: For example, it is very easy to get the rms as low as 1.5 MeV with extremely big neutron- and proton-shell energies with opposite signs.

The extension of the above method to deformed nuclei is planned. As we have reported in this note the present status of our study of tiny shell effects, let us conclude it by citing the motto of the State Michigan to compliment the organizers: "If you seek a delightful peninsula, behold it here"

ACKNOWLEDGEMENTS:

Two of the authors (E.R. Hilf and K. Takahashi) want to thank the organizers of the Topical Research Summer Institute on Nuclear Collisions at the Department of Physics, University of Washington, Seattle, USA for the opportunity to collaborate there on this subject.

REFERENCES

1. A.Chodos, R.L.Jaffe, K.Johnson, C.B.Thorn, and V.A.Weisskopf, New extended model of hardrons, Phys.Rev. D9:3471(1974)
2. G.E.Brown, and M.Rho, The little bag, State Univ. of N.Y. at Stony Brook, preprint(1979)
3. D.L.Hill, and J.A.Wheeler, Nuclear Constitution and Interpretation of fission phenomena, Phys.Rev. 89:11o2(1953)
4. E.R.Hilf, Oberflächenspannung und Thermodynamik des perfekten Gases, Zeitschr.f.Naturforsch. 25a:1191(197o)
5. E.R.Hilf, and G.Süssmann, Surface tension of nuclei according to the Fermi-gas model, Phys.Lett. 21:654(1966)
6. W.D.Myers, and W.J.Swiatecki, Average nuclear properties, Ann.Phys(N.Y.)55:395(1969)
7. W.D.Myers, Droplet model isotope shifts and the neutron skin, Phys.Lett. 3oB:451(1969)
8. G.Baym, Confinement of quarks in nuclear matter, preprint ILL-78-MS-1 (1979)
9. N.Zeldes, Recent shell and subshell effects in masses, in Proc. Int.Workshop on Gross properties of nuclei and nuclear excitations VII vol.B, 1979, ed. H.v.Groote, Technische Hochschule Darmstadt, Germany
1o.H.v.Groote, E.R.Hilf, and K.Takahashi, a new semi-empirical shell-correction to the droplet-model, Atomic Data and Nucl. Data Tables, 17:418(1976)

THE SHELL CORRECTION METHOD AND ITS APPLICATION TO NUCLEAR MASSES

Ingemar Ragnarsson*

CERN
Geneva
Switzerland

ABSTRACT

The calculation of nuclear masses within the Nilsson-Strutinsky method is briefly reviewed. In numerical calculations, the standard liquid drop or droplet model is used for the macroscopic energy and the microscopic energy is calculated from the single-particle orbitals of the modified oscillator potential. Results for nuclei with A$\geq$90 are presented. Except for some closed shell nuclei, the discrepancies between experimental and theoretical shell effects seldom exceed 1 MeV. For medium-heavy nuclei and outside the regions where the parameters have been fitted, the general trends appear to be much better described within the liquid drop than within the droplet model.

1. INTRODUCTION

In Nilsson-Strutinsky type calculations, the nuclear mass is being made up of two parts, a smoothly varying macroscopic part and a microscopic part caused by quantum mechanical effects. The macroscopic part is described for example by the liquid drop or droplet models. For these models, the parameters have earlier been fitted with empirical shell corrections. Here, we will instead use shell energies extracted from the single-particle levels of the modified oscillator (M.O.) potential with the potential parameters fitted to reproduce different nuclear properties. In principal, it would then be desirable to make a fit also of the macroscopic part

---

*On leave of absence from the Department of Mathematical Physics, Lund, Sweden.

but here we will confine ourselves to take the parameters of the liquid drop model from ref.[1] and those of the droplet model from ref.[2]. The final comparison between theoretical and experimental masses is then also in some way a test of these parameter fits.

2. SUMMARY OF MASS FORMULA

In the liquid drop (L.D.) model[3], the nuclear binding energy is described as a sum of a volume, a surface and a Coulomb energy:

$$E_{L.D.} = -a_1(1-\kappa_v I^2)A + a_2(1-\kappa_s I^2)A^{2/3}B_s(\alpha_k) +$$
$$+ \frac{3}{5}\frac{e^2 Z^2}{R_c}\left[B_c(\alpha_k) - \frac{5\pi^2}{6}(\frac{d}{R_c})^2\right] ; \qquad (1)$$
$$I = (N-Z)/A$$

In this formula, $B_s(\alpha_k)$ and $B_c(\alpha_k)$ are the surface and Coulomb energies of a sharp surface nucleus in units of their corresponding values for spherical shape. The different deformation degrees of freedom are described by α_k and the second term in the Coulomb energy is a diffuseness correction (shape independent) with d being the diffuseness depth. As mentioned above, the parameters of ref.[1] will be used:

$$a_1 = 15.4941 \text{ MeV}$$
$$a_2 = 17.9439 \text{ MeV}$$
$$\kappa_v = \kappa_s = 1.7826$$
$$R_c = 1.2249 \text{ A}^{1/3} \text{ fm}$$
$$d = 0.546 \text{ fm}$$

The expression (1) is valied for an odd-even nuclei while for an even-even nucleus $11/\sqrt{A}$ should be subtracted.

The droplet model (D.M.) is a refinement of the liquid drop model in the sense that the expansion in $A^{-1/3}$ and I^2 is carried out to higher order. It has recently been reviewed by Myers[2] and as we will use exactly the same parameters, we refer to this reference for the different constants employed.

From the experimental masses, it is evident that there is an extra binding for nuclei with N = Z. In the droplet model there is a term depending on $|N-Z|/A$ (a Wigner term) which accounts for this while no similar term is present in liquid drop model used here. As we will only discuss nuclei with N considerably larger than Z, this should be of no importance.

The single-particle energies are calculated from a modified oscillator potential which for symmetric shapes takes the form[4] (in our calculation, no deep minima which would really influence the masses are found for $\gamma\neq0$ and the γ-degree of freedom is therefore neglected):

$$V_{M.O.} = \frac{1}{2}\hbar\omega_o\,\rho^2\left[1-\frac{2}{3}\varepsilon\,P_2 + 2\varepsilon_4 P_4 + 2\varepsilon_6 P_6 + \ldots\right] -$$
$$\kappa\hbar\omega_o\left[2\vec{\ell}_t\cdot\vec{s} + \mu(\vec{\ell}_t^2 - <\vec{\ell}_t^2>)\right] \qquad (2)$$

Subsequently, the shell energy

$$E_{shell} = {\sum}'e_\nu(\alpha_k) - <{\sum}'e_\nu(\alpha_k)> \qquad (3)$$

is calculated by the Strutinsky prescription[5]. The pairing
correlations are treated in the BCS approximation and the
corresponding correction energy δE_{pair} is given by

$$\delta E_{pair} = E_{pair} - <E_{pair}>$$

where $<E_{pair}>$ is the average pairing energy[4] which must be subtracted not to be counted twice as it is already included in the
macroscopic energy.

The different terms are now added and the nuclear binding
energy is given as the minimum with respect to deformation of this
sum. It is then convenient to make a distinction between deformation
dependent and deformation independent terms:

$$E = E_{macr}^{even-even}(\alpha_k=0)$$

$$+ \min_{\alpha_k}\{E_{macr}(\alpha_k)-E_{macr}(\alpha_k=0)+E_{shell}(\alpha_k)+\delta E_{pair}(\alpha_k)+E^{odd}(\alpha_k)\}$$

The effect of the odd particle is for odd-Z and odd-N nuclei
approximated by $E^{odd}=\Delta_p$ or Δ_n as calculated from the BCS-equations.
The expression within brackets above is referred to as the
(collective)potential energy, the total energy or when the minimum
is taken as the shell energy contribution to the nuclear mass. This
shell energy contribution is the quantity we will use when comparing
experimental and theoretical masses. Note that it includes the
deformation dependent part of the macroscopic energy and in our
definition also the effect of the odd particle(s).

3. THE VARIATION OF THE NUCLEAR POTENTIAL ENERGY

As an illustration, we show in fig. 1 the nuclear potential
energy (the total energy) as a function of ellongation ε for
neutron-rich nuclei ranging from ^{240}O to the superheavy region. For
each ε, the energy has been minimized with respect to a necking
coordinate which is a combination of ε_4 and ε_6[6]. In this plot, for
each nucleus, the minimal energy with respect to ε corresponds to
the shell energy contribution to the nuclear mass. We note that
for nuclei in the vicinity of closed shells, this contribution is

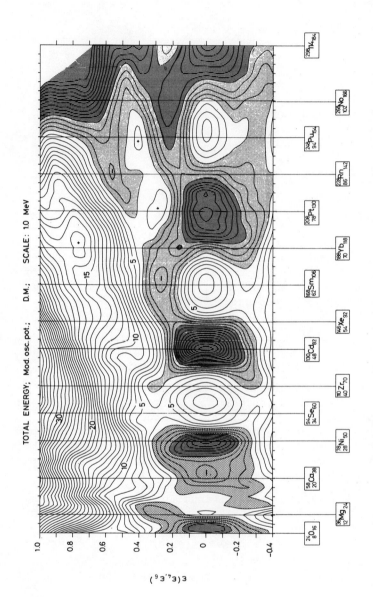

Fig. 1. The total energy, including droplet model energy, shell and pairing corrections, for neutronrich nuclei in the form of a contour map as a function of elongation $\varepsilon(\varepsilon_4, \varepsilon_6)$ and proton number. For each ε, a minimisation with respect to a combination of ε_4 and ε_6 has been performed. The modified oscillator potential has been used with different κ- and μ-values for the different shells. In this figure, for a specific nucleus, the minimum with respect to deformation corresponds to the theoretical shell energy contribution to the nuclear mass.

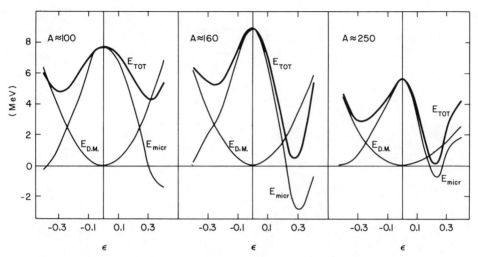

Fig. 2.   The decomposition of the total energy into one macroscopic and one microscopic term is shown for a typical nucleus from each of the three deformed regions: neutronrich $A \cong 100$, rare-earths and actinides.

strongly negative. On the other hand, in the middle of shells there is a strong positive shell effect for spherical shape and the nucleus will find it advantageous to go deformed. If the lightest nuclei are excluded, three such deformed regions are seen, namely the $A \cong 100$ region, the rare earth and the actinide regions. It is interesting to note that in the latter regions, the shell energy contribution to the mass (i.e. the minimum of the total energy of fig. 1) is near zero while it is 3-5 MeV in the $A \cong 100$ region.

This difference is illustrated in some more detail in fig. 2. There the microscopic ($E_{shell} + \delta E_{pair}$) and macroscopic contributions are shown separately for one typical nucleus from each of the three mentioned regions. In all three cases, the microscopic energy has a maximum for spherical shape and a minimum for positive ϵ-values. However, the typical frequency of the variation with respect to deformation is different. Thus, in the heavy regions, the minimum is situated at $\epsilon \cong 0.25$ and $\epsilon \cong 0.30$ respectively where the droplet model energy is still not very high. In the lighter region on the other hand, the minimum of the microscopic energy is found around $\epsilon \cong 0.4$. Thus, to really exploit this minimum, the nucleus must go to quite large deformations where however the droplet energy is quite high (6-8 MeV). The final choice for the nucleus will be a compromise with the total energy minimum at a smaller deformation where however the macroscopic energy is still rather large and the microscopic energy is not really minimized (this is certainly true also in the heavier regions but to a much smaller extent).

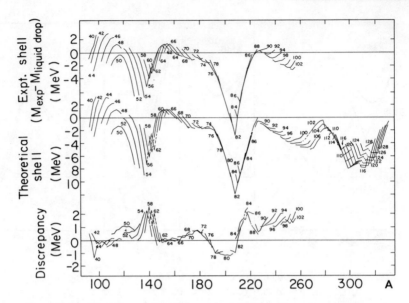

Fig. 3.   Experimental and theoretical masses for nuclei ranging
          from Z=40 to the heaviest ones. The results are based on
          the M.O. potential and the L.D. model[1]. For N>82, the
          parameters of the M.O. as given in ref.[4] have been used
          and for smaller neutron numbers the A=140 extrapolated
          (Z≥50) and the A=110 modified parameters respectively[8].
          The pairing calculations have been performed with
          $G(\frac{P}{N}) \cdot A$ = 19.2±7.4(N-Z)/A and $2\sqrt{15Z(N)}$ orbitals included
          in the heavier region (N>82) and with $G_P \cdot A$=22.0, $G_N \cdot A$=
          22.0-18.0(N-Z)/A and $2\sqrt{10Z(N)}$ orbitals included in the
          lighter region.

4. NUCLEAR MASSES IN THE A≥90 REGION

        A comparison between experimental and theoretical masses in
a wide region of nuclei is shown in fig. 3. The experimental
masses includes the even-even nuclei with Z≥40 and N≥52 of the
1971 mass evaluation[7]. The theoretical masses are given for these
same nuclei and are additionally extended up to the superheavy
region. They are based on results from the Lund-Berkeley-Warsaw
collaboration[4,8,9]. For the macroscopic energy, the liquid drop
model as described above[1] has been used. The general features of
the experimental as well as the theoretical shell effects are
analogous to those of fig. 1 with strongly peaked minima around
spherical closed shells and broad maxima in deformed regions.
There is some general disagreement in the translead region which
could either be due to the shell energy or to the liquid drop
energy. Furthermore, there are some problems with the shell effects
around [208]Pb and around neutron number N=82. However, in view of

the fact that the energy levels have been fitted for deformed
nuclei, the discrepancies around closed shells are surprisingly
small. Indeed, if energy levels are used which are more nearly
considered to coincide with the experimental ones around ^{208}Pb,
much larger discrepancies are obtained in this region[10].

5. THE NEUTRON-RICH A≈100 REGION OF NUCLEI

After the 1971 mass evaluation, a lot of new data have become
available. Of special interest are the measurements by Epherre et
al.[11] of Rb- and Cs-isotopes ranging from the very neutron-
deficient to the very neutron-rich side. The two-neutron separation
energies for the Rb-isotopes clearly shows that between neutron
numbers 58 and 60 a transition from spherical to deformed shapes
take place[11].

However, before comparing the masses in this region to theory,
we will discuss some other ground state properties to get a
feeling of how well these nuclei are understood. The transition to
deformed shapes for the Rb-isotopes agrees well with the very
sudden drop in 2^+ energies between N=58 and N=60 for the Zr and Sr
isotopes[12,13,14] and the measured B(E2)-values which for ^{102}Zr and
^{100}Sr indicates deformations around ε=0.3 [14]. In addition, the
large E_4+/E_2+ -ratios (3.23 for ^{100}Sr) only allows for axially
symmetric shapes. In fig. 4, the potential energy surfaces for the
Zr-isotopes taken from a recent calculation[15] are shown. The
transition from spherical to deformed shapes is well accounted for
and also the deformation of ^{102}Zr is in good agreement with
experiment. Very similar results, both experimentally and
theoretically are obtained for the Sr-isotopes[14]. A problem is that
the theoretical calculations give an oblate minimum which is almost
as deep as the prolate one (or even somewhat deeper for N=60).
However, the agreement between experimental and theoretical
quadrupole moments is much worse on the oblate side and it seems
safe to conclude that the neutronrich Sr:s and Zr:s are strongly
deformed prolate nuclei. Let us also mention that the recently
measured spectroscopic properties of ^{97}Rb [16] suggests that this
nucleus has a similar large prolate deformation[6].

It has been suggested that the excited 0^+ state of the
spherical Zr-isotopes might correspond to a shape isomeric state
with a similar deformation as the ground state of the heavier Zr-
isotopes (see e.g. ref.[17]). However, in fig. 4 one observes that
for ε≅0.3, the energy stays almost constant independent of neutron
number. On the other hand, in the region N=50-58, the energy of the
ground state (ε=0) varies by several MeV, a variation which is
confirmed by the measured masses (see below). The potential energy
surfaces of fig. 4 then inevitably leads to the conclusion that
the excitation energy of a deformed 0^+ state should increase in a

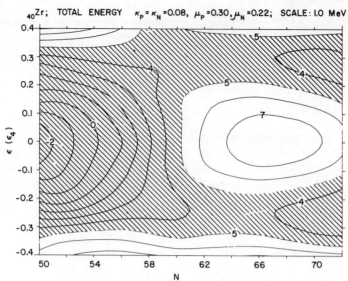

Fig. 4.   The total energy of the neutronrich Zr-isotopes plotted
          in the form of a contour map as a function of elongation
          $\varepsilon(\varepsilon_4)$ and neutron number. The κ- and μ-values of the M.O.
          are given in the figure. The pairing constants were chosen
          $G(\frac{P}{N}) \cdot A = 19.2 \pm 7.4(N-Z)/A$ with $2\sqrt{10Z(N)}$ orbitals included in
          the calculation. For each neutron number, the minimum with
          respect to deformation gives the shell energy contribution
          to the mass. Note how this quantity raises steadily in the
          spherical region and the stays almost constant when the
          nucleus goes deformed.

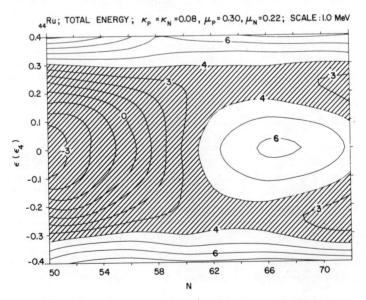

Fig. 5.   Same as fig. 4 but for the Ru-isotopes.

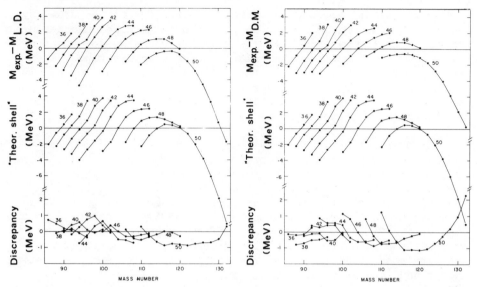

Fig. 6a.   Theoretical and   Fig. 6b.   Same as fig. 6a but with
           experimental masses   the liquid drop energy
           for nuclei with Z≤50   replaced by the droplet
           and N≥50. The M.O.   model energy[2].
           parameters are those
           of fig. 4 and for the
           macroscopic energy, the
           liguid drop[1] has been
           used.

monotonic way when the neutron number decreases. This is however
not the case and it means that at least for nuclei close to N=50,
the excited 0^+ state cannot correspond to a very large deformation.
However, in the transitional region around N=58-60, the different
0^+ states might very well be formed because of shape isomeric
minima (c.f. ref.[18]).

    In fig. 5, the potential energy surfaces for the Ru-isotopes
are shown. In agreement with experiment[12] the transition from
spherical to deformed shapes is less abrupt and the deformations
smaller than for the Zr-isotopes.

    We now continue to discuss the nuclear masses in this region
of nuclei. In figs. 6a and b, a comparison between the experimental
masses of the 1977 mass evaluation[19] and calculated shell energy
contributions are shown. For the macroscopic energy, either the
liquid drop (fig. 6a) or the droplet model (fig. 6b) has been used.
One immediately observes that the general agreement between theory
and experiment is much better within the liquid drop than within

the droplet model approach. However, it is difficult to draw any
general conclusions from this fact as the large discrepancies are
generally in the vicinity of closed shells where we do not have
too strong confidence in the calculated shell corrections. Even
so, it is interesting to note that within the liquid drop model,
the discrepancy is never larger than 1 MeV, not even for the $_{50}$Sn
isotopes and that for the doubly-closed ^{132}Sn, there is almost
exact agreement between theory and experiment.

A complication in the mass comparison of figs. 6 is the N=56
subshell closure which is observed in the experimental masses for
Kr, Sr and Zr but not for the heavier isotopes. The parameters
used in the theoretical calculations of figs. 6 exhibits this
subshell closure but this has as a consequence that the heavier
isotopes are not very well described around N=56. Compare here the
mass fit in fig. 3 where somewhat different parameters with a
smaller N=56 gap were used[8]. The agreement for the Zr:s is then
not very good while for heavier isotopes, the discrepancies between
theory and experiment are quite small.

6. NUCLEAR MASSES FAR FROM STABILITY

The mass measurements for the very neutron-rich Rb isotopes
were mentioned above and in fig. 7, these masses are for even N
compared to theory. As was also found in ref.[11] large discrepancies
are observed in the droplet model case for N=60 and 62. Indeed, as
was discussed in connection with fig. 2, one would not expect
larger shell energy contributions for even nuclei in this region
than around 5 MeV. This is also consistent with the results of
figs. 6 where for the nuclei shown neither $M_{exp}-M_{L.D.}$ nor
$M_{exp}-M_{D.M.}$ is in any case above 4 MeV. In the shell effects of fig.
7, the odd-particle effect is also included but we still arrive at
the conclusion that for ^{99}Rb, $M_{exp}-M_{D.M.}$ is at least 2 MeV larger
than what is reasonable. Thus, the discrepancies for ^{97}Rb and ^{99}Rb
appear to be at least to a large extent due to the droplet model
parameters.

The theoretical and experimental[11] masses of the even-N Cs
isotopes are shown in fig. 8. In this region, it has earlier been
found that the nuclear ground state properties are reasonably well
described within the modified oscillator approach[9,20]. What concerns
the masses, one observes in fig. 8 that except for some difficulties
with the shell effects around N=82, the general experimental trends
are rather well described when the liquid drop model is used.
However, for the droplet model there are large discrepancies away
from stability in a similar way as for the Rb:s. The general
conclusion which can be drawn is that for mediumheavy nuclei far
from stability and outside the regions where the parameters have
been fitted, the droplet model predicts too strong binding while
the trends of the liquid drop model appears to be in approximative

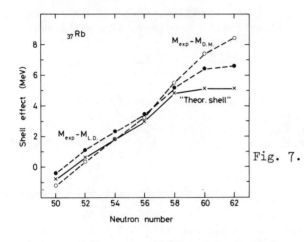

Fig. 7. For even-N neutron-rich Rb-isotopes, the theoretical shell energy contributions to the nuclear mass are compared to experimental ones. The latter are given both within the droplet model and the liquid drop model. As in other figures, the theoretical shell energy contribution includes the deformation dependent part of the macroscopic energy, the shell + pairing corrections and for these odd nuclei also the effect of the odd particle (approximated by Δ_N). Within the accuracy of the figure, these theoretical points are the same in the droplet as in the liquid drop model.

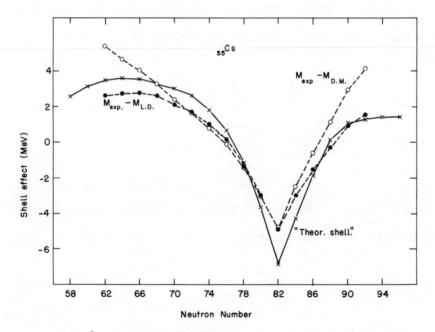

Fig. 8. Same as fig. 7 but for even-N Cs isotopes.

agreement with experiment. Going back to figs. 6 one observes that this conclusion is also consistent with the discrepancies observed there.

ACKNOWLEDGEMENT

The cooperation with the late professor Sven Gösta Nilsson has strongly influenced this paper. His inspiration, support and friendship is warmly appreciated. Valuable discussions with M. Epherre, P.G. Hansen, R. Klapisch, W.D. Myers and Z. Szymański are acknowledged.

REFERENCES

1. W.D. Myers and W.J. Swiatecki, Ark. Fys. 36 (1967) 343.
2. W.D. Myers, Droplet Model of Atomic Nuclei (1977 IFI/Plenum-New York) and At. Data and Nucl. Data Tables 17, (1976) 411.
3. W.D. Myers and W.J. Swiatecki, Nucl. Phys. 81 (1966) 1.
4. S.G. Nilsson, C.F. Tsang, A. Sobiczewski, Z. Szymański, S. Wycech, C. Gustafsson, I.L. Lamm, P. Möller and B. Nilsson, Nucl. Phys. A131 (1969) 1.
5. V.M. Strutinsky, Nucl. Phys. A95 (1967) 420.
6. I. Ragnarsson, Proc. Int. Symp. on Future Directions in Studies of Nuclei far from Stability, Vanderbilt University, 9-12 Sept. 1979.
7. A.H. Wapstra and N.B. Gove, At. Data and Nucl. Data Tables 9 (1971) 265.
8. I. Ragnarsson, Proc. Conf. on Properties of Nuclei far from the Region of Beta Stability, Leysin, 1970 (CERN 70-30, 1970) 847.
9. I. Ragnarsson, A. Sobiczewski, R.K. Sheline, S.E. Larsson and B. Nerlo-Pomorska, Nucl. Phys. A233 (1974) 329.
10. M. Brack, J. Damgaard, A.S. Jensen, H.C. Pauli, V.M. Strutinsky and C.Y. Wong, Rev. Mod. Phys. 44 (1972) 320.
11. M. Epherre, G. Audi, C. Thibault, R. Klapisch, G. Huber, F. Touchard, H. Wollnik, Phys. Rev. C19 (1979) 1504.
12. R.C. Jared, H. Nifenecker and S.G. Thompson, Proc. Symp. on the Physics and Chemistry of Fission, Rochester 1973 (IAEA, Vienna) Vol. 2, p. 211.
13. H. Wollnik, F.K. Wohn, K.D. Wünsch and G. Jung, Nucl. Phys. A291 (1977) 355.
14. R.E. Azuma, G.L. Borchert, L.C. Carraz, P.G. Hansen, B. Jonson, S. Mattsson, O.B. Nielsen, G. Nyman, I. Ragnarsson and H.L. Ravn, Phys. Lett. B, in press.
15. P. Arve and I. Ragnarsson, current work.
16. R. Klapisch, priv. comm., 1979.
17. R.K. Sheline, I. Ragnarsson and S.G. Nilsson, Phys. Lett. 41B (1972) 115.
18. T.A. Khan, W.D. Lauppe, K. Sistemich, H. Lawin, G. Sadler and

H.A. Selic, Z. Physik <u>A283</u> (1977) 105.
19. A.H. Wapstra and K.H. Bos, At. Data and Nucl. Data Tables <u>19</u> (1977) 177.
20. C. Ekström, G. Wannberg and J. Heinemeier, Phys. Lett. <u>76B</u> (1978) 565;
C. Ekström, L. Robertsson, G. Wannberg and J. Heinemeier, Phys. Scripta <u>19</u> (1979) 516.

MASSES FROM INHOMOGENEOUS PARTIAL DIFFERENCE EQUATIONS;

A SHELL-MODEL APPROACH FOR LIGHT AND MEDIUM-HEAVY NUCLEI

Joachim Jänecke

The University of Michigan

Ann Arbor, Michigan 48109

GENERAL CONSIDERATIONS

Any mass equation $M(N,Z)$ may be expressed as a special solution of an inhomogeneous partial difference equation. One only
has to introduce an appropriate difference operator D and calculate
$\kappa(N,Z) \equiv D\,M(N,Z)$. The inhomogeneous partial difference equation

$$D\,M(N,Z) = \kappa(N,Z) \qquad (1)$$

contains the original mass equation as a special solution. The
most general solution of eq. (1) will include solutions of the
homogeneous equation which can, for example, be used to generate
shell correction terms.

Third-order partial difference operators related to $\partial^3/\partial^2 N\partial Z$
and $\partial^3/\partial N\partial Z^2$ were found to be particularly useful.

We define[2] difference operator $^{m,n}\Delta$ by

$$^{m,n}\Delta f(N,Z)=f(N,Z)-f(N-m,Z-n). \qquad (2)$$

We then define

$$I = -^{1,0}\Delta\,^{0,1}\Delta \qquad (3)$$

$$D_T = {}^{1,-1}\Delta\ I = -^{1,-1}\Delta\,^{1,0}\Delta\,^{0,1}\Delta \qquad (4)$$

$$D_L = {}^{1,1}\Delta\ I = -^{1,1}\Delta\,^{1,0}\Delta\,^{0,1}\Delta. \qquad (5)$$

Here, $I_{np}(N,Z)\equiv I\,M(N,Z)$ is the operational definition of the well
known effective neutron-proton interaction,[1,2] and D_T and D_L are
the transverse and longitudinal difference operators. The quantities $D_T M(N,Z)$ and $D_L M(N,Z)$ describe the variation of I_{np} with
neutron excess and nucleon number, respectively. The operators

can approximately be written as

$$I \approx -\partial^2/\partial N \partial Z \tag{6}$$

$$D_T \approx -(\partial/\partial N - \partial/\partial Z)(\partial^2/\partial N \partial Z) \tag{7}$$

$$D_L \approx -(\partial/\partial N + \partial/\partial Z)(\partial^2/\partial N \partial Z). \tag{8}$$

If it is assumed that $I_{np}(N,Z)$ is independent of T_z or A, one obtains the homogeneous partial difference equations $D_T M(N,Z)=0$ or $D_L M(N,Z)=0$ which represent the well known Garvey-Kelson mass relations.[3] They have the solutions

$$M(N,Z) = G_1(N)+G_2(Z)+G_3(N+Z) \tag{9}$$

and $$M(N,Z) = F_1(N)+F_2(Z)+F_3(N-Z). \tag{10}$$

Here, $G_i(k)$ and $F_i(k)$ are arbitrary functions which can be obtained from a χ^2-adjustment to the known masses.[3,4] Eqs. (9) and (10) describe the symmetry and Coulomb energies of nuclei poorly.

The most general solutions of the inhomogeneous equations

$$D_T M(N,Z) = \tau(N,Z) \tag{11}$$

or $$D_L M(N,Z) = \lambda(N,Z) \tag{12}$$

consist of a special solution of the inhomogeneous equations plus the most general solutions of the homogeneous equation. The functions $\tau(N,Z)$ and $\lambda(N,Z)$ are exactly defined as they represent

$$\tau(N,Z) = D_T M_{exact}(N,Z) \tag{13}$$

$$\lambda(N,Z) = D_L M_{exact}(N,Z). \tag{14}$$

They are, of course, not generally known and have to be replaced by expressions based on theoretical considerations. They are finite[5] but on the average less than 100 keV. They display a rather complicated dependence on N and Z which becomes apparent when one differentiates even simple liquid-drop-model Coulomb and symmetry energy expressions. Generally valid homogeneous difference equation can therefore not be derived by further differentiating eqs. (11) or (12).

SHELL-MODEL MASS EQUATION

A general solution of the transverse difference equation (11) has been derived with $\tau(N,Z)$ based on theoretical shell-model expressions for the symmetry and Coulomb energies. However, instead of explicitly integrating eq. (11), a much simpler procedure was adopted by assuming an analytical form of an inhomogeneous solution and properly adjusting a few shell-model parameter. The function $\tau(N,Z)$ can then be obtained, if so desired, by simple differentiation. The most general solution of eq. (11) was given the form

$$\Delta M(N,Z) = \Delta M_n N + \Delta M_H Z + E_c(Z,N) + S(N,Z) \qquad (15)$$
$$+G_1(N) + G_2(Z) + G_3(Z+N).$$

For each of the regions of nuclei displayed in fig. 1, the Coulomb energy $E_c(Z,N)$ was represented by a shell-model expression[6] with parameters derived from an _independent_ adjustment to experimental Coulomb displacement energies.[7] Subsequently, shell-model symmetry energy expressions[8,9] $S(N,Z)$ were introduced for each of the regions of fig. 1 and the parameters deduced from the experimental mass values. The functions $G_i(k)$, finally, were obtained from a χ-square adjustment to all mass values[10] with $2 \leq Z,N \leq 50$. As the Coulomb energy is explicitly given by $E_c(Z,N)$, the remaining terms must be charge-symmetric. Therefore, $S(Z,N) = S(N,Z)$ and $G_1(k) = G_2(k)$. The latter equation provides a useful test as solutions may be obtained with or without the constraint $G_1 = G_2$, and its validity can be tested. Unlike eq. (9), eq. (15) is valid for N-Z positive, zero and negative. The reason that eq. (9) is limited to N>Z is a large value of $\tau(N,Z)$ at N=Z due to a cusp in the symmetry energy.

The Coulomb energy expression

$$E_c(Z,N) = (\frac{A_0}{A})^{\sigma/3} (\lambda + \mu p + \nu p(p-1) + 2\pi[\frac{p}{2}]) \qquad (16)$$

was used for the diagonal regions of fig. 1. Here, $p=Z-Z_0$. For the off-diagonal regions

$$E_c(Z,N) = E_c(Z_0,N) + E_c(Z,N_0) - E_c(Z_0,N_0) \qquad (17)$$

was used. Eq. (17) implies that the isotope shift of the nuclear charge radius is equal to that of the nearest magic proton number Z_0. The coefficients obtained for eq. (16) from the Coulomb energy data[7] are given in table 1.

The symmetry energy $S(N,Z)$ for the diagonal regions was expressed as

$$S(N,Z) = \alpha(A)T(T+1) + \tilde{\alpha}(A)(T(T+1))^2. \qquad (18)$$

When neutrons and protons occupy the same shell-model orbits, the seniority coupling scheme in the isospin formalism[8] demands a dependence on T(T+1). The second term in eq. (18) represents higher-order corrections (like rotational bands). The coefficients $\alpha(A)$ and $\tilde{\alpha}(A)$ (see table 2) were represented by polynomials divided by A. Only for the 1p shell was $\tilde{\alpha}(A)$ found significantly different from zero. Eq. (18) is related to the systematics of excitation energies of isobaric analog states (see figs. 8.21 and 8.22 of ref. 11). Fig. 2 displays the polynomials $A \times \alpha(A)$ together with values deduced from the energy corrected masses of neighboring isobars.

The particle-hole interaction[9] $H_{ij} = -a + b \vec{t}_i \vec{t}_j$ was used to express the symmetry energy for the off-digagonal regions as

Table 1. Coefficients for eq. (16)

Region	Z_O	A_O	σ	λ	μ	ν	π
(1,1)	0	2	0.0	0.0	0.0	336.9	90.0
(2,2)	2	10	0.0	853.8	919.5	246.1	97.2
(3,3)	8	28	0.9	12122.2	3145.4	222.9	23.2
(4,4)	20	48	0.3	70197.6	7177.0	164.5	44.0
(5,5)	28	78	0.9	122295.2	8707.1	156.2	30.3

Table 2. Coefficients for eq. (18)

Region	A_O	α_O	α_1	α_2	$\tilde{\alpha}_O$	a_p
2,2	10	5663.2	494.3	-48.4	-163.0	8578.7
3,3	28	2299.8	17.8	0.0	0.0	19482.7
4,4	48	1316.9	4.8	10.3	0.0	29767.5
5,5	78	935.1	0.0	0.0	0.0	30071.4

$$\alpha(A)=(A_O/A) \ \Sigma \ \alpha_i (A-A_O)^i \qquad\qquad \tilde{\alpha}(A)=(A_O/A) \ \Sigma \ \tilde{\alpha}_i (A-A_O)^i$$

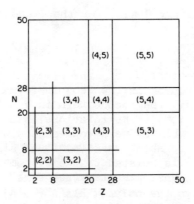

Fig. 1.   Shell-model regions

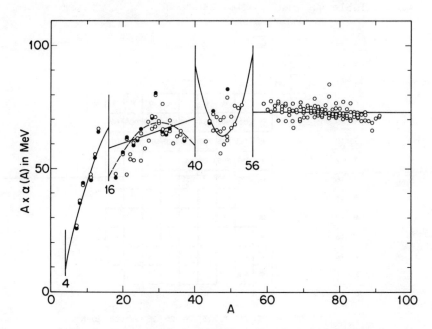

Fig. 2. Symmetry parameters a(A) in S(N,Z)=[a(A)/A]T(T+1) from the Coulomb energy corrected masses of neighboring isobars and comparison with eq. (18).

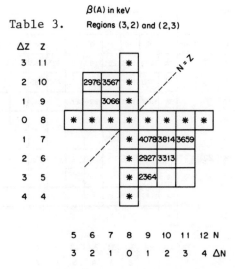

Table 3. β(A) in keV
Regions (3,2) and (2,3)

β (A) in keV

Table 4. Regions (4,3) and (3,4)

ΔZ	Z	17	18	19	20	21	22	23	24	25	26	27	28	29	N
----	----	----	----	----	----	----	----	----	----	----	----	----	----	----	
3	23				*										
2	22		1738	1777	*										
1	21			1992	*										
0	20	*	*	*	*	*	*	*	*	*	*	*	*	*	
1	19				*	2004	1942	1854	1918	1761	1829	1806	1869		
2	18				*	1765	1775	1735	1773	1739	1819	1783	1817		
3	17				*	1797	1885	1826	1830						
4	16				*	2025	1920		1859						
5	15				*										
6	14				*										
7	13				*										
8	12				*										
9	11				*	1451									
10	10				*										
		3	2	1	0	1	2	3	4	5	6	7	8	9	ΔN

N = Z

β (A) in keV

Table 5. Region (5,4)

ΔZ	Z	27	28	29	30	31	32	33	34	35	36	37	38	N
----	----	----	----	----	----	----	----	----	----	----	----	----	----	
1	29		*											
0	28	*	*	*	*	*	*	*	*	*	*	*	*	
1	27		*	1262	1003	1251	1183	1256	1323	1288	1342	1311		
2	26		*	970	1027	1108	1188	1182	1292	1265				
3	25		*	1184	1100	1216	1294	1232						
4	24		*	1164	1222	1260	1350							
5	23		*	1361	1332	1329								
6	22		*	1299	1395									
7	21		*	1331	1379									
8	20		*	1119	1300									
9	19		*											
		1	0	1	2	3	4	5	6	7	8	9	10	ΔN

N = Z

Table 6

$$\beta(N,Z) = \frac{-2}{\Delta N \Delta Z} \left[B(N,Z) - B(N_0,Z) - B(N,Z_0) + B(N_0,Z_0) \right.$$
$$\left. - a_{p1}(\delta_{ee} - \delta_{oo}) - a_{p2}(\delta_{ee} + \delta_{oo}) \right]$$

Regions	a_{p1} keV	a_{p2} keV	$\langle \beta(N,Z) \rangle$ keV
(3,2) and (2,3)	-239	146	3307 ± 530
(4,3) and (3,4)	-131	310	1824 ± 113
(5,4)	-181	265	1239 ± 107

$$\beta(N,Z) \approx \beta(Z,N) \approx \text{CONST}$$

Table 7. Coefficients for eq. (19)

Regions	A_0	β_0	β_1	α_{p1}	α_{p2}
(3,2) and (2,3)	19	3365.2	210.8	-239	146
(4,3) and (3,4)	38	1860.2	39.8	-131	310
(5,4)	63	1259.9	17.1	-181	265

$$S(N,Z) = S(N_O,Z) + S(N,Z_O) - S(N_O,Z_O) + \tfrac{1}{2}\beta(A)\ \Delta N \Delta Z \quad (19)$$

with $\Delta N = |(N-N_O)|$ and $\Delta Z = |(Z-Z_O)|$. Here $\tfrac{1}{2}\beta(A)=a-\tfrac{1}{4}b$ is the average of the T=0 and T=1 energies without 2T+1 weighting encountered[9] in an n-p formalism when neutrons and protons are in different orbits. The individual values for $\beta(A)$ obtained from the experimental semi-magic and magic mass values are displayed in tables 3, 4 and 5. They are approximately constant and charge-symmetric (see table 6). Using two- or three-parameter fits for $\beta(A)$, similar to $\alpha(A)$, gives the parameters shown in table 7. The use of a higher-order term (not given here) similar to that in eq. (18) was not justified due to the limited data.

The functions $G_1(N)$, $G_2(Z)$ and $G_3(Z+N)$, finally, were obtained for fixed $E_C(Z,N)$ and $S(N,Z)$ with and without the constraint $G_1(k) = G_2(k)$. The results are very similar for the two cases. Fig. 3 displays the three functions. The standard deviation for the masses of all nuclei with $2\leq Z,N\leq50$ is 250 keV, for the Coulomb

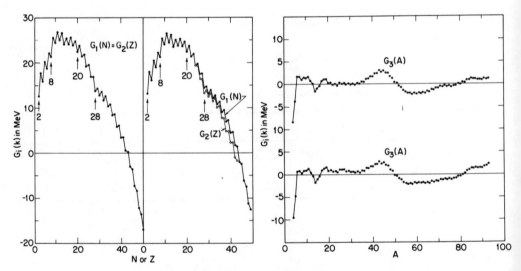

Fig. 3.   Functions $G_1(N)$, $G_2(Z)$ and $G_3(A)$ obtained with and without the constraint $G_1(k)=G_2(k)$ for the mass equation $\Delta M(N,Z)=\Delta M_n N + \Delta M_H Z - K\,A + E_C(Z,N) + S(N,Z) + \eta(2N^2+2Z^2 - A^2) + G_1(N) + G_2(Z) + G_3(A)$ [K=11500 keV; η=26.3728 keV]. Pairing energies with coefficients from tables 2 and 7 may be explicitly included.

energies it is about 150 keV. Masses of unknown neutron-rich and proton-rich nuclei and the limits of stability have been obtained (see fig. 4). Better inhomogeneous terms $\tau(N,Z)$ in eq. (11) based on more detailed theoretical considerations and new mass measurements of very neutron- and proton-rich nuclei will permit improved mass predictions in the future. An estimate of the accuracy of the mass predictions can be obtained from the dependence of the predictions on the few parameters entering into the inhomogeneous term.

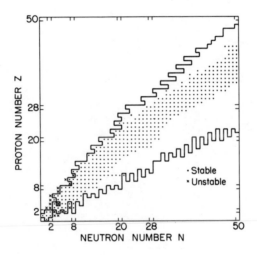

Fig. 4.   Limits of stability.

Work supported in part by the National Science Foundation.

References

1. J. Jänecke and B. P. Eynon, Nucl. Phys. A243 (1975) 326
2. J. Jänecke and B. P. Eynon, At. Data and Nucl. Data Tables 17 (1976) 467
3. G. T. Garvey, W. J. Gerace, R. L. Jaffe, I. Talmi and I. Kelson, Rev. Mod. Phys. 41 (1969) S1
4. J. Jänecke, At. Data and Nucl. Data Tables 17 (1976) 455
5. J.Jänecke and H. Behrens, Phys. Rev. C9 (1974) 1276
6. K. T. Hecht, Nucl. Phys. A104 (1968) 280
7. W.J. Courtney and J. D. Fox, At. Data and Nucl. Data Tables 15 (1975) 141

8. A. de-Shalit and I. Talmi, Nuclear Shell Theory (Academic
    Press, N.Y., 1963)
9. R. K. Bansal and J. B. French, Phys. Lett. 11 (1964) 145;
    L. Zamick, Phys. Lett. 19 (1965) 580
10. A. H. Wapstra and K. Bos, At. Data and Nucl. Data Tables 19
    (1977) 177
11. J. Jänecke, in Isospin in Nuclear Physics, ed. D. H. Wilkinson,
    North-Holland (1969) p. 298

ATOMIC MASS DATA AND n-p RESIDUAL INTERACTION IN DEFORMED NUCLEI

P.C. Sood and R.N. Singh

Physics Department
Banaras Hindu University
Varanasi 221005 (India)

Extensive efforts have been continously made for the precise determination of the atomic masses and for seeking reliable descriptions - semiempirical as well as those based on semi-microscopic and microscopic theories - of the nuclear binding energies. However the interrelationship of these quantities with low energy excitation spectra of the involved nuclei and/or with the effective interactions giving rise to these spectra has seldom been investigated. In the following we present results of a study exploiting such an interrelationship and show how the mass data may be advantageously used for obtaining a characterization of the residual neutron-proton interaction and hence for prediction of the low energy spectra of doubly odd deformed nuclei.

We consider the odd-odd deformed nucleus to be composed of the deformed even-even core with the unpaired valence nucleons (a proton and a neutron) coupled to it. Thus the band-head energy in an odd-odd nucleus may be written as

$$E_{pn}^{K}(Z,A) = E_{p}^{\Omega_{p}}(Z,A-1) + E_{n}^{\Omega_{n}}(Z-1,A-1) + \Delta E_{rot} + \Delta E_{int} \tag{1}$$

where the last term represents the contribution from the n-p interaction V_{np} and

$$\Delta E_{rot} = \frac{\hbar^2}{2I} [K^{\pm} - \Omega_{p} - \Omega_{n}] \tag{2}$$

Ω 's being the projections of the particle angular momenta j_p and j_n on the symmetry axis and K being the corresponding projection of the total angular momentum I. The Gallagher-Moszkowski[1] rule uniquely (with the solitary exception of ^{166}Ho) defines that, out of the two $K^{\pm}$ states, the one with the parallel projections Σ_p and Σ_n of the particle spin momenta lies lower in energy. E_p and E_n are taken as the single (quasi)-particle energies from the observed spectra of the neighboring odd-A nuclei. With this prescription one expects that the rotation-particle coupling (RPC) effects are already taken into account. Practically all the investigations so far have dealt with the splitting energy ΔE_{GM} of the K^+ and the K^- states (the so-called Gallagher-Moszkowski GM pair). A recent summary of such investigations may be found in references 2 and 3. However, no serious effort has so far been made to evaluate $<V_{np}>$ and hence to predict the band-head energies of the two quasiparticle states in these nuclei. For instance a recent study[4] simply adopts

$$\Delta E_{int} = (80 \pm 50) \text{ keV} \tag{3}$$

for the actinide region. We propose[5] that the mass data combined with the GM splitting energy can lead to a characterization of the V_{np} and thence to the band-head energies.

In common with other studies[2] we adopt a zero range interaction potential[6]

$$V_{np} = -4\pi g \delta(\vec{r}_p - \vec{r}_n)[(1-\alpha) + \alpha \vec{\sigma}_p \cdot \vec{\sigma}_n] \tag{4}$$

where g is the interaction strength parameter, and α provides a measure of the spin-spin contribution. One usually replaces the parameter g by a new parameter W with the inclusion of the oscillator frequency (from the wavefunctions) such that W has the dimensions of the energy. The interaction energy is then given as

$$\Delta E_{int} = <\psi|V_{np}|\psi>$$

$$= W(1-\alpha)A_W(K) + \alpha W A_\sigma(K) \tag{5}$$

where A_W and A_σ represent the matrix elements for the Wigner (spin-independent) and the spin dependent terms respectively in the potential (4). The splitting energy of the GM pair given by

$$\Delta E_{GM} = 2\alpha W A_\sigma (K^{\pm}) \qquad (6)$$

is by itself incapable of determining α and W separately and hence of defining the interaction potential. However we may rewrite Eqs. (1-2) in terms of the masses as follows:

$$M(Z,A) = M(Z,A-1) + M(Z-1,A-1) - M(Z-1,A-2)$$

$$+ \frac{\hbar^2}{2I} [K_G^{\pm} - \Omega_p - \Omega_n] + \langle\psi_G|V_{np}|\psi_G\rangle \qquad (7)$$

where the subscript G refers to the ground state of the odd-odd nucleus. Thus we obtain an 'experimental' measure of the residual interaction energy ΔE_M in the ground state from the atomic mass data, the corresponding theoretical value being given by Eq. (5)

$$\Delta E_M \equiv \langle\psi_G|V_{np}|\psi_G\rangle = W(1-\alpha)A_W(K_G) + \alpha W A_\sigma(K_G) . \qquad (8)$$

Combining Eqs. (6) and (8) we obtain the following expressions for the interaction parameters

$$\alpha = \left[1 + \left(\frac{2\Delta E_M}{\Delta E_{GM}} - 1\right) \frac{|A_\sigma(K_G)|}{|A_W(K_G)|} \right]^{-1} \qquad (9)$$

$$W = \frac{\Delta E_{GM}}{2\alpha|A_\sigma(K_G)|} . \qquad (10)$$

The matrix elements $A(K_G)$ can be evaluated using the appropriate symmetrized wavefunctions involving $D_{MK\chi}^I(\Omega_p)\chi(\Omega_n)$ with the Nilsson orbitals assigned for the valence nucleons from the corresponding odd-mass data. These are then combined with ΔE_M obtained from the recent mass data[7] and ΔE_{GM} for the ground state GM pair from a recent compilation[2] leading to the evaluation of α and W from Eqs. (9-10) and further to the energies of the various two-quasi-particle states through Eqs. (1) and (5).

We find that a correlation exists between the experimental ΔE_M and ΔE_{GM} (the statistical correlation coefficient being 0.8 for the ten rare earth nuclei given in Fig. 1) of the form

$$\Delta E_{GM} = k \, \Delta E_M \tag{11}$$

with $K = 0.305$ obtained through a least squares fit. The predicted values of the GM splitting energy as obtained from the mass data with Eq. (11) are shown in Fig. 1 in comparison with the experiment.

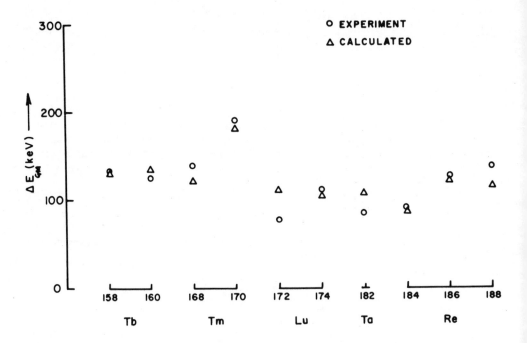

Figure 1   The calculated (vide Eq. (11)) splitting energies of the ground state GM pairs for some rare earth nuclei are compared with the experimental values.

This empirical correlation may be physically understood as follows. In the asymptotic limit of large deformation, where Σ is a good quantum number along with Ω, it can be mathematically shown[5] that

$$A_W(K_G) = A_\sigma(K_G) \qquad (12)$$

and this result combined with Eq. (9) gives in this limit

$$\Delta E_{GM} = 2\alpha \ \Delta E_M \qquad (13)$$

which is seen to be the same as Eq. (11) with $2\alpha = k$. Thus we deduce a common value of $\alpha \approx 0.15$ in the asymptotic limit for these rare-earth nuclei. Combining Eqs. (10) and (13) we further obtain

$$W = \Delta E_M \Big/ \big| A_\sigma(K_G) \big| \qquad (14)$$

and thus are in a position to predict the excitation energies of the various two quasiparticle states starting with the mass data.

For quantitative predictions of the excitation energies of non-rotational states in deformed nuclei we however use the mass data along with the splitting energy of the ground state GM pair as the input data to evaluate α and W through Eqs. (9-10) and to calculate the interaction energy using Eq. (5) and the excitation energies of the various two particle states using Eqs. (1-2). The results following this procedure for the nucleus ^{238}Np are shown in Fig. 2 in comparison with the recent experimental data.[4] Our predictions as shown do not include the states involving the configuration $5/2^+$[622] due to some computational difficulty; this results in no prediction corresponding to 278 keV 5^+ and 343 keV 5^- states. The agreement on the whole is found to be very good.

It may be added that these studies can help in proper configuration assignment to the observed levels. For example the 301 keV 6^- observed state was earlier stated[8] to correspond to $5/2^-$[523↑]$_p$ + $7/2^+$[624↓]$_n$ whereas Ionescu et al[4] consider this unlikely on the ground that the excitation energies of these configurations in the corresponding odd-mass nuclei add up to around 500 keV. The latter investigators[4] suggested the assignment $5/2^+$[642↑]$_p$ + $7/2^-$[743↑]$_n$ for which the odd particle band head energies add to a value close to the observed 300 keV. Our calculations place the 6^- state of the earlier[8] configuration at 322 keV and the 6^- state of the latter[4] configuration to lie far outside the energy scale. Thus it is evident that the prediction of band head energies with the exclusion of the properly calculated

residual interaction contribution can lead to quite erroneous
configuration assignments to two quasiparticle states.

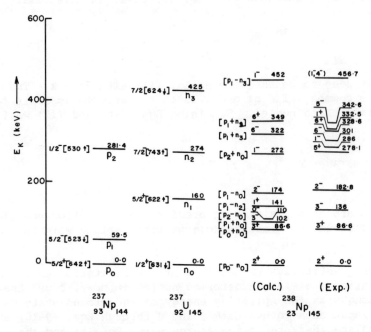

Figure 2   The calculated energies of the two quasiparticle states
           in ^{238}Np arising from the indicated single particle
           states in the neighboring odd-A nuclei are shown in
           comparison with the available experimental information
           (ref. 4) for this nucleus.

        Thus far we have assumed that the interaction parameters
α and W evaluated from the masses and the ground state data using
Eqs. (9-10) are characteristic of a given nucleus and may be
used as such for all the configurations appearing in the excited
states. We have further investigated this point and find that
this is not always the case. For illustration we discuss the
case of ^{174}Lu. For this nucleus the ground state configuration
is 7/2[404↓]$_p$ + 5/2[512↑]$_n$ giving rise to $K^{\pi} = 1$ as the ground
state and $K^{\pi} = 6$ at 170 keV in accordance with the GM rule.
Combining this with the mass data we get

        α = 0.24 : W = 4.65 MeV.   (15)

Assuming these interaction parameters to be the same for all
configurations for this nucleus, the next excited pair of levels
arising from the $7/2[404\downarrow]_p \pm 1/2[521\downarrow]_n$ configuration is calculated to lie at 566 KeV (4^-) and 630 keV (3^-) in comparison
with the experimental values of 365 keV and 433 keV respectively.
It is seen that although the calculated GM splitting (64 keV)
agrees well with the observed value (68 keV), the predicted band
head energies are too large. One finds that the GM pair of
4^- -3^- states for the latter configuration appears as the ground
pair in ^{172}Lu. Using the mass data and the observed splitting
therein we obtain

$$\alpha = 0.16 \quad ; \quad W = 8 \text{ MeV}. \tag{16}$$

Accepting that these interaction parameters should be used wherever
this configuration appears, we calculate the 4^- and the 3^- states
of this configuration to lie at 390 keV and 465 keV in ^{174}Lu
to be compared with the observed energies of 365 keV and 433 keV.
The agreement obtained with the experiment supports the view
that the residual interaction parameters are configuration dependent
and cannot be treated as constant within a given nucleus. However
this does not pose a serious difficulty, since the knowledge
of mass data and ground state GM splitting energies of a whole
series of nuclei is sufficient to determine the interaction parameters of practically all configurations of interest in well-
deformed rare-earth and actinide nuclei, and this enables us to
predict the positions of such configurations appearing as the
excited states in other nuclei.

In summary we conclude that the mass data combined with
the ground state properties provides valuable guidance for the
energies of, and the assignment of configurations to, the two
quasiparticle states in doubly odd deformed nuclei.

References

1. C.J. Gallagher and S.A. Moszkowski, Phys. Rev. 111 (1958)
   1282.
2. J.P. Boisson, R. Piepenbring and W. Ogle, Physics Reports
   26C (1976) 99.
3. D. Elmore and W.P. Alford, Phys. Rev. C14 (1976) 583.
4. V.A. Ionescu, Jean Kern, R.F. Casten, W.R. Kane, I. Ahmad,
   J. Erskine, A.M. Friedman and K. Katori, Nucl. Phys. A313
   (1979) 283.
5. R.N. Singh and P.C. Sood, Nucl. Phys. and Solid State Phys.
   (India) 21B (1978), and to be published.
6. N.I. Pyatov, Izv. Akad. Nauk. SSSR 27 (1963) 1436 (translation:
   Bull. Acad. Sc. USSR, Phys. Ser. 27 (1963) 1409).

7.   A.N. Wapstra and K. Bos, Atomic and Nuclear Data Tables
     19 (1977) 175.
8.   F. Asaro, M.C. Michel, S.G. Thompson and I. Perlman, Proc.
     Rutherford Jubliee Int. Conf., Manchester 1961, ed. J.B.
     Birks, (Heywood, London) p. 311.

MUTUAL SUPPORT OF MAGICITIES

K.-H. Schmidt

Gesellschaft für Schwerionenforschung, GSI
Darmstadt, Federal Republic of Germany

D. Vermeulen

Institut für Kernphysik, Techn. Hochschule
Darmstadt, Federal Republic of Germany

ABSTRACT

A strong correlation of neutron- and proton shell strengths
is demonstrated by a systematic comparison of nuclear binding
energies. Both, neutron- and proton shells show a maximum strength
at doubly magic nuclei and generally fall down drastically with
increasing number of additional particles or holes. This mutual
support of shell strengths is not reproduced by usually applied
theoretical shell corrections. Therefore, the results of shell
model calculations, especially in the vicinity of doubly magic
nuclei, concerning ground state shell corrections, fission barriers
and decay energies, are expected to show appreciable shortcomings.
If the variation of shell strengths is taken into account, the
interpretation of the separation energies in terms of single
particle energies is modified. This modification could solve the
'lead anomaly', the difficulty in describing the ground state
shell corrections and the separation energies around ^{208}Pb simultaneously.

INTRODUCTION

Neutron-proton interactions have been discussed for a long
time (see e.g. refs. [1,2,3]), mainly in connection with spectroscopic data. However, the role of neutron-proton interactions in
the description of ground state masses is not yet as clear. Usually
applied theoretical shell corrections do not include this effect.

though in several cases it has been pointed out that the single
particle gap of a closed proton shell may be changed if the number
of neutrons varies and vice versa (see e.g. refs. [4],[5]). Only in a
few cases, e.g. in the semiempirical mass table of Liran and
Zeldes[6] and in a shell correction approach of Hilf and v. Groote[7],
the variation of shell strengths with occupation numbers is included.

In the last years, the knowledge about the binding energies
of nuclei has enlarged considerably. Closed shells can now be
followed over a long range of nuclei. E.g. for proton rich N = 126
isotones far from the valley of beta stability, remarkable discrepancies have been observed between empirical and theoretical alpha
decay energies[8], which were explained by shortcomings of usually
applied shell corrections. In the present work, a survey of the
strengths of several closed shells all over the nuclide chart is
given.

Recently, an explanation for the lead anomaly, the discrepancy in describing the separation energies and the ground state
shell corrections in the vicinity of ^{208}Pb consistently, was given
by Werner et al.[9] in terms of a quasi particle interaction. The
variation of shell strengths, discussed in the present work,
enables us to look at the lead anomaly from a quite different
point of view.

EXPERIMENTAL EVIDENCE FOR THE MUTUAL SUPPORT OF MAGICITIES

As a closed shell is characterized by a kink in the binding-
energy surface as a function of the neutron and the proton number,
respectively (see fig. 1), the shell strength can be deduced from
the sharpness of this kink in the binding-energy surface which is
given by double mass differences. In the shell model, the single
particle neutron shell gap G_n for a nucleus with Z protons and N
neutrons is given be the double mass difference

$$
\begin{aligned}
- G_n(Z,N) &= \Delta S_n(Z,N) \\
&= S_n(Z,N+1) - S_n(Z,N) \\
&= M(Z,N+1) - 2\,M(Z,N) + M(Z,N-1)
\end{aligned} \tag{a}
$$

or, in order to avoid the even-odd structure, (see fig. 5):

$$
\begin{aligned}
- G_n(Z,N) &= 1/2\ \Delta S_{2n}(Z,N) \\
&= 1/2\ (\ S_{2n}(Z,N+2) - S_{2n}(Z,N)\ ) \\
&= 1/2\ (\ M(Z,N+2) - 2\,M(Z,N) + M(Z,N-2)\ ),
\end{aligned} \tag{b}
$$

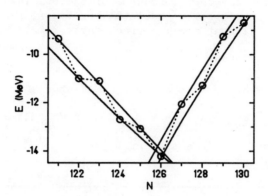

Fig. 1: Experimental masses of Pb-isotopes relative to the values of the droplet model without shell corrections[10]. The full lines are fitted polynomials to the masses of even and odd isotopes below and above N = 126, respectively.

if rearrangement and pairing corrections are neglected. This is demonstrated in fig. 2. A similar relation holds for the single particle proton shell gap G_p. A survey of this quantity in the vicinity of magic nuclei is given in figs. 3 and 4. The empirical masses are taken from ref.[11]. In order to restrict on the single particle properties, the reduced values $\Delta S_{2n}^{red}/2$ are shown which

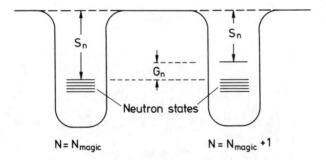

Fig. 2: Schematic diagram, demonstrating that in a shell model descriptions of nuclei without pairing corrections and rearrangement corrections the neutron shell gap G_n is given approximately by the difference of neutron separation energies S_n:

$$-G_n = S_n(Z, N_{magic}+1) - S_n(Z, N_{magic})$$

are corrected for the curvature ΔS_{2n}^{DM} of the binding energy surface as predicted by the droplet model (DM) (ref. [10]) including the Wigner term but without shell corrections:

$$\Delta S_{2n}^{red} = \Delta S_{2n} - \Delta S_{2n}^{DM}$$

$$= (\ M(Z,N+2) - 2\ M(Z,N) + (M(Z,N-2)\ ) \qquad\qquad (c)$$

$$- (\ M_{DM}(Z,N+2) - 2\ M_{DM}(Z,N) + M_{DM}(Z,N-2)\ ).$$

ΔS_{2p}^{red} is defined in a similar way.

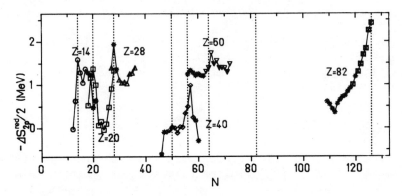

Fig. 3: Half of the difference ΔS_{2n}^{red} between two-neutron separation energies, treated as a function of the proton number Z. The contribution of the droplet mass surface[10] including the Wigner term has been subtracted. The neutron shells are indicated by different symbols. Stars correspond to the cases when the mass (ref. [11]) is taken from systematics and not from experiment.

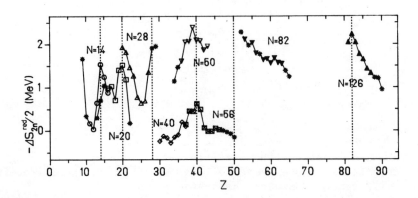

Fig. 4: Same as fig. 3 for half the difference ΔS_{2p}^{red} between two-proton separation energies, treated as a function of the neutron number N.

Figs. 3 and 4 demonstrate that the strengths of nearly all shells change drastically over the nuclide chart. The 82 proton shell, e.g., nearly has disappeared completely in this representation for the most neutron deficient known lead isotopes. Maxima of the double mass differences occur near the proton numbers 14, 20, 28, 40, 50, and 82 as well as near the neutron numbers 14, 20, 28, 40, 50, 56, 64, and 126. Most of them coincide with the well known magic numbers. In all cases where empirical binding energies are available, fig. 3 and 4 suggest that the single particle shell gap of one kind of nucleons tends to grow if the other kind of nucleons approaches a shell closure. Thus, the conclusion may be drawn that the magicities of neutrons and protons generally support each other.

PAIRING CORRELATIONS

The characteristics of pairing correlations are of special interest in this context because they are generally involved in mass systematics and may mask other trends to be investigated. Pairing is especially important in considering shell strengths because the pairing gap tends to counteract the shell correction.

As an example, the values ΔS_n and $\Delta S_{2n}/2$ are compared for isotones with the magic number N = 126 and for isotones with the non magic number N = 130 in fig. 5. For non magic nuclei, the experimental $\Delta S_{2n}/2$ values nearly coincide with the pure droplet

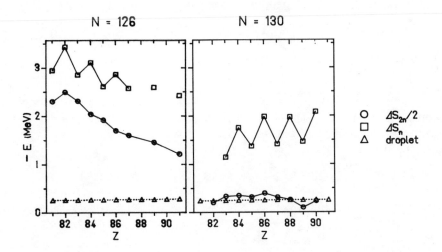

Fig. 5:   Double mass differences ΔS_n and $\Delta S_{2n}/2$ for the N = 126 and N = 130 isotones, demonstrating the influence of shell structure and of the pairing interaction on these quantities. The experimental masses are taken from refs. [77],[72],[8]).

estimation of this quantity. The ΔS_n-values are expected to deviate from the droplet contribution by about twice the neutron pairing gap. Additionally, the ΔS_n-values show an odd-even structure which is usually explained by an attractive interaction between the unpaired proton and the unpaired neutron in odd-odd nuclei (see e.g. discussion of fig. 2.5 in ref. [13]).

Also for N = 126 isotones there is a difference between the values of ΔS_n and $\Delta S_{2n}/2$. According to BCS-calculations of Mosel[14], the proton pairing correction of Pb-isotopes and the neutron pairing correction of 126-neutron isotones in the vicinity of ^{208}Pb are predicted to vanish. Consequently, rather the values of ΔS_n should represent the single particle shell gap than the values of ΔS_{2n}. The empirical masses shown in fig. 1, however, indicate that the neutron pairing correlation of ^{208}Pb does not disappear completely and that in this case the pairing corrected kink in the binding energy surface at N = 126 is represented by a value which lies between the ΔS_n and the $\Delta S_{2n}/2$ values.

DISCUSSION

Modified Interpretation of Separation Energies

So far we identified the shell gap with the difference of separation energies (relations a and b). This is justified, if the neutron shell correction δU_n does not depend on the proton number and vice versa. Then we have the simple relations:

$$\delta U_n(Z,N+1) - 2 \, \delta U_n(Z,N) + \delta U_n(Z,N-1) = - G_n^{red}$$

$$\delta U_p(Z,N+1) - 2 \, \delta U_p(Z,N) + \delta U_p(Z,N-1) = 0$$

(d)

The mutual support of magicities, as demonstrated by figs. 3 and 4, leads to an additional reduction of the binding energies of the neighbours of doubly magic nuclei. Consequently, relations (d) no longer hold, and the absolute values of ΔS_n and ΔS_p are enlarged by a correction term C:

$$\Delta S_n = \Delta S_n^{DM} + \delta U_n(Z,N+1) - 2 \, \delta U_n(Z,N) + \delta U_n(Z,N-1)$$

$$+ \, \delta U_p(Z,N+1) - 2 \, \delta U_p(Z,N) + \delta U_p(Z,N-1)$$

(e)

$$= \Delta S_n^{DM} - G_n^{red} - C_n$$

The magnitude of C will be discussed later, but generally we expect

$$G_n < |\Delta S_n| \quad \text{and} \quad G_p < |\Delta S_p|.$$

The discrepancy is expected to grow with the amount of the variation of the shell strength.

    This correction must also be applied to the data in figs. 3 and 4 if shell gaps are to be deduced, but it is obvious that it may only change the magnitude of the slopes but not the sign. Thus, the general statement of a mutual support of magicities still holds.

Connections to the Lead Anomaly

    In fig. 6, the ground state shell corrections δU of some nuclei around N = 126 are shown. The experimental values are compared with theoretical shell corrections, which were obtained from Nilsson model calculations with Strutinski renormalisation and BCS-pairing correlations (ref. [15]). In order to illustrate qualitatively the influence of the variation of shell strengths, we show the result of a simple calculation. In this calculation, the shell correction was obtained by a simplified Strutinski procedure, using a uniformly distributed level sequence which was bunched at a shell closure. For simplicity, the shell gaps around ^{208}Pb were assumed to follow an exponential law:

$$G_n^{red} = G_p^{red} = 2.1 \text{ MeV} * \exp(-(|Z-82| + |N-126|)/20) \qquad (f)$$

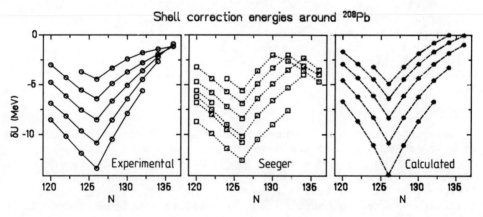

Fig. 6: Comparison of shell correction energies around N = 126. The lines connect the values of Pb, Po, Rn, Ra, and Th-isotopes. $\delta U_{experimental}$ is determined as the difference of the empirical mass[11] and the droplet mass[10] without shell corrections. δU_{Seeger} is the result of a Nilsson-model calculation with Strutinski renormalisation and BCS pairing-correlations[15]. For $\delta U_{calculated}$ the mutual support of magicities was taken into account (see text).

With these parameters, the experimental shell corrections
as well as the separation energies in the vicinity of ^{208}Pb can
approximately be reproduced. The main feature of these calculated
shell corrections is the flattening of the valley at N = 126
with increasing distance from ^{208}Pb, which is not reproduced by
Seeger's calculation. This flattening already sets in for the
spherical nuclei, directly neighboured to ^{208}Pb (compare figs. 3
and 4) and cannot be explained by the onset of a static deformation.

Usually, the single particle shell gaps are determined by
relation a). In ref. [13] the following values are given for ^{208}Pb:

$$G_n = 3.44 \text{ MeV} \quad \text{and} \quad G_p = 4.23 \text{ MeV}.$$

In relation (f), however, appreciable smaller values for the shell
gaps were introduced:

$$G_n = G_n^{red} + G_n^{DM} = 2.1 \text{ MeV} + 0.26 \text{ MeV} = 2.36 \text{ MeV} \quad \text{and}$$

$$G_p = G_p^{red} + G_p^{DM} = 2.1 \text{ MeV} + 0.88 \text{ MeV} = 2.98 \text{ MeV}.$$

Consequently, $C_n = 1.08$ MeV and $C_p = 1.25$ MeV. Brack et al.[16]
introduced an artificial reduction of the shell gaps for ^{208}Pb
of about the same amount (1.5 MeV) in order to reproduce the
experimental shell effects.

We think that we can explain this reduction by a modified
interpretation of the relation between separation energies and
single particle levels. The consideration of the mutual support
of magicities requires a correction term which reduces the ground
state shell correction of ^{208}Pb by about 1/3, compared to a shell
model calculation which is adapted to the separation energies[16].

Consequences for Mass Predictions

The shell correction approaches of Myers[10] and v. Groote
et al.[17] as well as Nilsson model calculations with Strutinski
renormalisation[15] yield a nearly constant value for the strength
of each shell. The same holds for the result of a spherical
Hartree-Fock calculation[18]. Therefore, the predictions of these
models for several quantities are expected to show some shortcomings. The extension of the peak of additional nuclear binding
energy due to the ground state shell correction around doubly
magic nuclei is smaller than calculated. By this effect the
calculated fission barriers are influenced directly[19]. Especially
the suspected island of superheavy nuclei may be considerably
smaller than expected. In addition, calculated decay energies,
e.g. the Q_α and Q_β values of superheavy nuclei, are also
affected.

CONCLUSION

It has been shown that the correlation between proton and neutron magicities shows clearly up in nuclear ground state masses. The shell correction approaches usually applied do not account for this effect. It has been pointed out that the problems in describing the ground state shell correction of ^{208}Pb and the separation energies around ^{208}Pb consistently may be solved, if the mutual support of neutron and proton magicities is taken into account.

ACKNOWLEDGEMENT

We wish to thank Mr. W.D. Myers for providing us with his computer code for calculating droplet properties of nuclei. Fruitful discussions with P. Armbruster and W. Reisdorf are greatfully acknowledged.

REFERENCES

1. A. de-Shalit, M. Goldhaber, Phys. Rev. 92 (1953) 92
2. B.L. Cohen, Phys. Rev. 127 (1962) 597
3. G. Scharff-Goldhaber, J. Phys. A: Math., Nucl. Gen., 7 (1974) L121
4. N. Zeldes, Ark. f. Fysik 36 (1966) 361
5. V.A. Kravtsow, N.N. Skachkov, Nucl. Data Tables A1 (1966) 491
6. S. Liran, N. Zeldes, Atomic Data and Nucl. Data Tables 17 (1976) 411
7. E.R. Hilf, H. v. Groote, Proc. of the Int. Worksh. on Gross Prop. of Nuclei and Nuclear Excitations III, Hirschegg, 1975 p. 230
8. K.-H. Schmidt, W. Faust, G. Münzenberg, H.-G. Clerc, W. Lang, K. Pielenz, D. Vermeulen, H. Wohlfahrt, H. Ewald, K. Güttner, Nucl. Phys., A318 (1979) 253
9. E. Werner, K. Dietrich, P. Möller, R. Nix, Contribution to the International Symposion on Physics and Chemistry of Fission, Jülich 1979, IAEA-SM-241-C26
10. W.D. Myers, Droplet model of atomic nuclei, (IFI/Plenum, New York, 1977)
11. A.H. Wapstra, K. Bos, Atomic Data and Nucl. Data Tables 19 (1977) 185
12. M. Epherre, private communication
13. A. Bohr, B.R. Mottelson, Nuclear Structure, Vol. I, (W.A. Benjamin, Inc., New York, 1969)
14. U. Mosel, Phys. Rev. C6 (1972) 971
15. P.A. Seeger, Int. Conf. on the Prop. of Nuclei far from Stability, Leysin 1970, Vol. 1, p. 217
16. M. Brack, J. Damgard, A.S. Jensen, H.C. Pauli, V.M. Strutinski, C.Y. Wong, Rev. Mod. Phys. 44 (1972) 320
17. H. v. Groote, E.R. Hilf, K. Takahashi, Atomic Data and Nucl. Data Tables 17 (1976) 418

18. M. Arnould, private communication
19. K.-H. Schmidt, Proc. of the Int. Worksh. on Gross Prop. of
     Nuclei and Nuclear Excitations VII, Hirschegg 1979, p. 102

TRENDS IN NUCLEAR MASS SYSTEMATICS

Nissan Zeldes

Racah Institute of Physics
Hebrew University of Jerusalem
Jerusalem, Israel

INTRODUCTION

We describe empirical trends in nuclear masses and relate
them to successively higher approximations of the shell model,
emphasizing regularities reaching beyond the lowest-seniority
approximation.

EMPIRICAL TRENDS

1.   Overall View

The nuclear mass surface as function of nucleon numbers roughly
resembles a valley above the N-Z plane, as described in fig. 1 of
ref. 1.   Closer scrutiny reveals splitting of the surface into four
sheets, one through each type of nuclei with given parities of N
and Z.   This is discussed in section 2.   The four surfaces generally vary smoothly with N and Z.   This is discussed in sec. 3 and
4.   However, there are sudden slope discontinuities along lines
of constant N and Z at the respective neutron and proton magic
numbers and along the line Z=N.   The latter discontinuity is referred to as Wigner Term in nuclear mass equations.

Fig. 1 of ref. 2 shows isotopic sections of the mass surface
illustrating splitting according to parity and slope discontinuities
at neutron magic numbers N=20,126.   Slope discontinuities at Z=N
are illustrated in fig. 4 below.

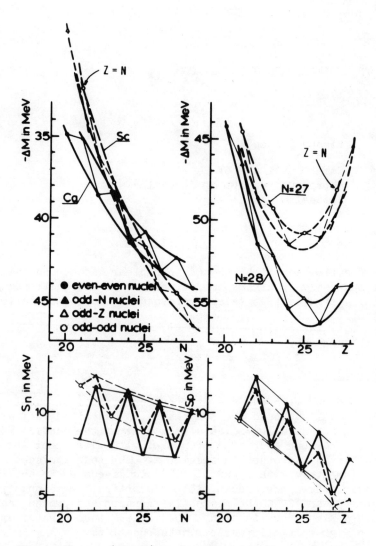

Fig. 1. Mass defects (above) and corresponding nucleon separation
         energies (below) for the Ca and Sc isotopes (left) and
         for the N=27,28 isotones (right) in the $1f_{7/2}$ shell.

2.   Relative Position of the Four Mass Surfaces

The relative position of the four sheets is clearly illustrated by sections cutting through nuclei situated on two of
them at a time. Thus, isotopic sections, like in the upper left
part of fig. 1 (and fig. 7) most directly demonstrate that the
even-even surface lies below the odd-N one, and the odd-odd surface lies above the odd-Z one. Similarly, isotonic sections, like
in the upper right part of fig. 1, directly show that the even-
even surface lies below the odd-Z one, and the odd-odd surface lies
above the odd-N one. Combining these results one infers that both
odd-A surfaces lie between the even-even and the odd-odd surfaces.

Fig. 1 likewise shows, that the odd-odd surface is nearer to
the mean odd-A surface than the even-even surface is. This is
directly demonstrated by comparing the magnitudes of oscillation of
the corresponding nucleon separation energies in the lower part of
the figure. This magnitude of oscillation is equal to twice the
vertical distance between the corresponding mass parabolas in the
upper part of the figure.

The above conclusions concerning the relative position of
odd-A and even-A surfaces hold in both diagonal and non-diagonal
shell regions.[†]

Direct information about the relative position of the two odd-A
and likewise of the two even-A surfaces is furnished by isobaric
(A=Const) and isodiapheric (I=Const) sections of the mass surfaces.
We consider separately diagonal and non-diagonal regions.

The upper part of fig. 2 shows isobaric sections in the non-
diagonal shell region $50 \leq N \leq 82$, $28 \leq Z \leq 50$. One observes the large
splitting between the two even-A parabolas, and the considerably
smaller one between the two odd-A ones, of which the odd-Z parabola
lies higher. The magnitude of oscillation of a Q_β line in the
lower part of the figure is equal to twice the vertical distance
between the corresponding parabolas in the upper part of the figure.

Isodiapheric sections of the mass surface[3] are considerably
longer. The upper part of fig. 3 shows partial isodiapheric
sections for I=12,13. The left and right sharp slope discontinuities are associated with the respective magic numbers N=5 and
Z=50. The section between them belongs to the same non-diagonal
shell region as fig. 2. Throughout this region the odd-Z surface

---

[†]One conveniently distinguishes diagonal shell regions in the N-Z
plane, where the valence neutrons and protons fill the same major
shell, and non-diagonal regions, where they fill different shells.

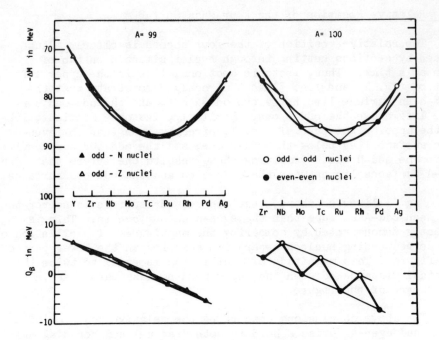

Fig. 2. Mass defects (above) and Q_β-values (below) for A=99
(left) and A=100 (right) isobars.

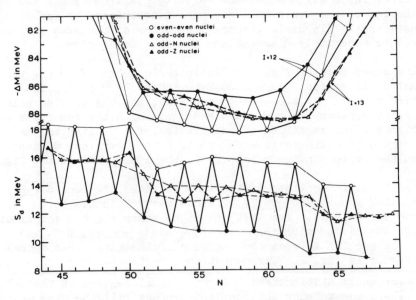

Fig. 3. Mass defects (above) and deuteron separation energies
(below) for I=12,13.

is above the odd-N one, like in fig. 2. On the other hand, the
odd-N surface is higher immediately outside these magic-number
boundaries. The situation is summarized by the Suess-Jensen rule[4]
stating that when a magic number is crossed towards heavier
nuclei, the corresponding odd-A surface becomes lower.

Deuteron separation energies are shown in the lower part of
fig. 3. The magnitude of oscillation of a S_d line is equal to
twice the vertical distance between the corresponding isodiapheric
ΔM lines in the upper part of the figure.

We turn to diagonal regions. The upper part of fig. 4 shows
isobaric sections in the $1f_{7/2}$ shell, after subtracting from the
experimental masses the charge-non-symmetric Coulomb energy and sum
of neutron-proton mass differences. One observes the slope discontinuity at Z=N mentioned above. Additionally, one observes the
large splitting between two even-A isobaric parabolas, and the
considerably smaller splitting for odd-A, like in fig. 2. The
slope discontinuity at Z=N and the splitting according to parity
are reflected in corresponding discontinuity and oscillation of
the Q_β lines in the lower part of the figure.

The figure illustrates the highly charge-symmetric nature of
the purely-nuclear interaction. In particular, the relative position of the odd-N and odd-Z parabolas changes in a charge-symmetric
way on crossing the cusp at Z=N: to the left of the cusp, where
Z<N, odd-Z nuclei are higher, whereas odd-N nuclei are higher to
its right, where N<Z.

Fig. 5 shows parts of isodiapheric sections with I=-3,3 and
corresponding deuteron separation energies. The larger circles
correspond to nuclei in the diagonal 1d2s (left) and $1f_{7/2}$ (right)
shell regions. As a rule, odd-Z nuclei are higher for I=3 where
Z<N, and charge-symmetrically odd-N nuclei lie higher for I=-3
where N<Z, similarly to the A=47 splitting in fig. 4. This is also
directly demonstrable by the zig-zag oscillation of the S_d lines.
Similar splitting occurs as a rule for other isodiapheres in the
1p, 1d2s and $1f_{7/2}$ diagonal shell regions.

We finally mention, that odd-odd Z=N nuclei are situated on
a line above the rest of the odd-odd surface. This is directly
demonstrated by the fact that the amplitude of oscillation of
isodiapheric S_d lines is larger for I=0 than for other I values[6].

3.   Quadratic and T Dependence

In a given non-diagonal shell region, and also on each side of
the Z=N line in a diagonal region, each of the four mass surfaces
varies smoothly, usually locally quadratically with N and Z.

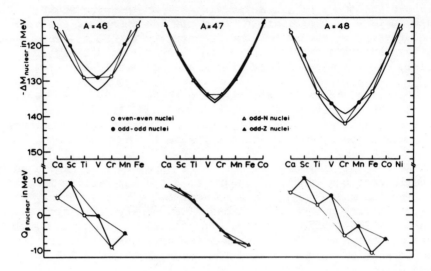

Fig. 4.   Mass defects (above) and Q_β-values (below) for A=46,47,48
          isobars in the $1f_{7/2}$ shell.   Coulomb energy[5] and neutron-
          proton mass differences, 0.391I MeV, were subtracted
          from the experimental masses.

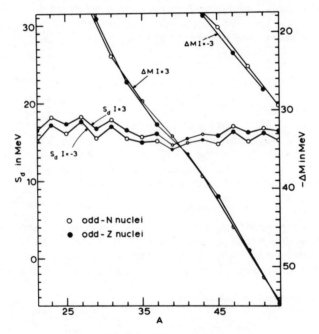

Fig. 5. ΔM and S_d values for I=-3,3.

This is most directly demonstrated by the linear systematics of mass differences between nuclei of the same parity type like S_{2n}, S_{2p}, $Q_{\beta\beta}$ and Q_α[7].

The T-dependence of the Wigner Term at Z=N is for A≤90 well described by a $T(T+1)$ expression[6,8].

Interestingly, the quadratic dependence on N and Z in diagonal regions changes on crossing the Z=N line, similarly to the splitting of odd-A isobars mentioned above. This is illustrated by considering systematics of two-nucleon separation energies as function of nucleon numbers, where the slopes $\partial S_{2n}/\partial N$ and $\partial S_{2p}/\partial Z$ are proportional to the respective coefficients of N^2 and Z^2 in the expression of the nuclear energy as function of N and Z.

Fig. 6 shows isotopic S_{2n} and isotonic S_{2p} lines of $1f_{7/2}$ nuclei. The lines display discontinuity associated with the Wigner Term, similarly to the Q_β lines in fig. 4, and the slope of each line is different on the two sides of the discontinuity: the magnitude of $\partial S_{2n}/\partial N$ is larger in the N<Z part of the shell region than in the N>Z part, and charge-symmetrically the magnitude of $\partial S_{2p}/\partial Z$ is larger in the Z<N part. This demonstrates that the quadratic part of the nuclear energy is charge-symmetrically different on the two sides of the Z=N line. The same regularity is found as well in the lighter 1p and 1d2s shell regions.

4.   Distortion of the Quadratic Dependence

Globally, the above quadratic dependence on N and Z is sometimes distorted by superposition of oscillating functions of N and Z^9, symmetric or antisymmetric with respect to particle-hole conjugation in the major shell, and vanishing at the magic numbers.

Since the oscillating terms differ somewhat for the four mass surfaces, the mutual distances of the latter likewise oscillate somewhat, mainly in a particle-hole symmetric way, between magic numbers. This is shown in fig. 2-5 of ref. 10.

Variation of distance between mass surfaces results in odd-even staggering of mass differences between nuclei lying on the same surface. Fig. 7 illustrates the absence of staggering in S_{2n} of the Rh nuclei which lie on two parallel isotopic parabolas, and the occurrence of staggering in S_{2n} of Ge, where the distance between isotopic mass parabolas decreases towards N=50. Odd-even staggering in S_{2p}, $Q_{\beta\beta}$ and Q_α systematics similarly results from inter-surface distance variations.

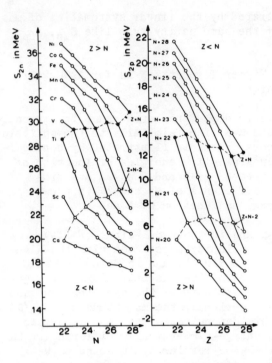

Fig. 6. Isotopic S_{2n} lines
(left) and isotonic
S_{2p} lines (right)
in the $1f_{7/2}$ shell.

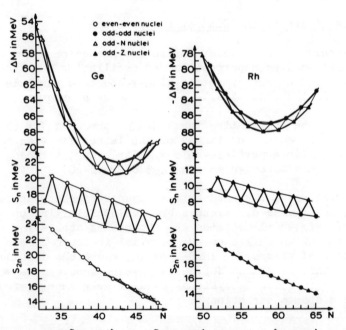

Fig. 7. Isotopic ΔM (above), S_n (middle) and S_{2n} (below) values
for Ge (left) and Rh (right) isotopes.

THEORETICAL CONSIDERATIONS

5. Independent-Particle Motion

The slope discontinuities of the mass surface at the magic numbers are associated with sudden decrease of nucleon binding energies in the heavier region. They are most naturally explainable by assuming independent-nucleon motion in a central potential.

6. Pairing and Isopairing

On the other hand, the odd-even splitting of nuclear masses cannot be obtained by such independent-particle motion. Additionally, independent-particle calculations with realistic potentials yield energy gaps at magic numbers which are too large[11] as compared to the experimental slope discontinuities. Likewise, such calculations are presumably[12] unable to account for the observed variations[13] of magic number discontinuities with N and Z.

In the next approximation of the shell model one considers average residual interactions within each configuration. For effective interactions with strong T=1 pairing and T=0 aligning and quasipairing components[14] the ground state has lowest possible isospin $T=\frac{1}{2}|I|$ and a large lowest-seniority (maximum-pairing) component.

The lowest-seniority mass equation, part E_{pair} of ref. 9, is in a given non-diagonal shell region a quadratic function of N and Z with four-fold pairing term, and it reproduces the splitting according to parity and the quadratic variation described in sec. 2 and 3.

7. Beyond the Lowest-Seniority Approximation

On the other hand, in a diagonal region the lowest-seniority equation has $a(a-1)$ and $T(T+1)$ quadratic terms, and a pairing term not distinguishing between odd-N and odd-Z nuclei. Thus it reproduces the rough three-fold splitting into even-even, odd-A and odd-odd surfaces (with odd-odd Z=N nuclei above the rest of the odd-odd surface), the T-dependence of the Wigner Term and a quadratic variation with N and Z on each side of the Z=N line. On the other hand, it cannot account for the finer two-fold splitting of odd-A nuclei and for the different coefficients of N^2 and Z^2 on each side of the Z=N line.

As a matter of fact, the above two-fold splitting of the odd-A surface in diagonal regions can be described by particle-hole-symmetric oscillating function, like those mentioned in sec. 4. Likewise, different coefficients of N^2 and Z^2 result by adding to

the lowest-seniority equation a sum of particle-hole symmetric and
anti-symmetric terms. Such oscillating trends cannot be reproduced
by the quadratic lowest-seniority equation, neither in diagonal nor
in non-diagonal regions.

The above lowest-seniority approximation can be improved upon
both by considering higher seniority states within a given configuration, and by including configuration mixing, which is the next
approximation in the shell model scheme. Both these extensions
seem to be required in nuclei with both valence neutrons and
protons, whereas mixed configurations in semi-magic nuclei are mainly
in states of lowest seniority.

Large shell-model calculations with mixed seniorities and mixed
configurations generally result in oscillating trends superposed
on the lowest seniority mass equation. In particular one expects
to reproduce the above odd-A splitting and different coefficients
of N^2 and Z^2 in diagonal regions.

I am grateful to Drs. B.A. Brown, H. von Groote and H. Wildenthal for discussions concerning the inadequacy of the lowest-seniority approximation.

1.   Cappeller,U., Nuclear Masses and their Determination,
              1957, Pergamon Press, 27.

2.   Zeldes, N. and Liran, S., 1976,Atomic Masses and Fundamental
              Constants 5, Plenum Press, 264.

3.   Kravtsov, V.A. and Skachkov, N.N., 1966, Nuclear Data A1, 491.

4.   Suess, H.E. and Jensen, J.H.D., 1951, Ark. Fys 3, 577.

5.   Myers, W.D. and Swiatecki, W.J., 1966, Nuclear Physics 81, 1.

6.   Zeldes, N. and Taraboulos, A., 1978, notas de fisica Vol. 1
              No. 9, 271.

7.   Wapstra, A.H., and Bos, K., 1977, ADNDT 17, 431.

8.   Janecke, J., 1965, Nuclear Physics 73, 97.

9.   Liran, S. and Zeldes, N. 1976, ADNDT 17, 431.

10.   Bohr, A., and Mottelson, B.R., 1969, Nuclear Structure, Vol. I,
              A. Benjamin,Inc.

11. Beiner, M., Flocard, H., Giai, N.V. and Quentin, P., 1975,
         Nuclear Physics A238, 29.

12. Zeldes, N., 1979, Gross Properties of Nuclei and Nuclear Excit-
         ations VII, B. Nuclei far off β-Stability, INKA-Conf.-
         79-001,TH Darmstadt, 56.

13. Schmidt, K.H., These Proceedings.

14. Schiffer, J.P. 1971, Annals of Physics 66, 798.

ATOMIC MASS FORMULAS WITH EMPIRICAL SHELL TERMS

Masahiro Uno and Masami Yamada

Science and Engineering Research Laboratory
Waseda University
3-4-1 Okubo, Shinjuku-ku, Tokyo, Japan 160

INTRODUCTION

An advance was made in the study of empirical mass formulas [1,2] by including new experimental data and by applying a new statistical treatment. Our mass formula consists of a gross part and proton and neutron shell parts. The shell parameter values were determined by a new statistical method, whose essential point was to take into account the intrinsic error of the mass formula. Main characteristics of the formula thus obtained are presented.

MASS FORMULA

We write the mass excess of the nucleus with Z protons and N neutrons as

$$M_E(Z,N) = M_{Eg}(Z,N) + P_Z(N) + Q_N(Z) - \Delta M_{odd-odd}(A). \qquad (1)$$

Here, $M_{Eg}(Z,N)$ is a smooth function of Z and N, while $P_Z(N)$ and $Q_N(Z)$ are not necessarily smooth with respect to the subscripts Z and N. The last term is a small correction for odd-odd nuclei.

The gross part $M_{Eg}(Z,N)$ is expressed in MeV as

$$M_{Eg}(Z,N) = 7.68023A + 0.39120I + a(A)\cdot A + b(A)\cdot|I|$$
$$+ c(A)\cdot\frac{I^2}{A} + E_c(Z,N) - 14.33\times10^{-6}Z^{2.39}, \qquad (2)$$

with $I = N - Z$, $A = N + Z$,

$$a(A) = a_1 + a_2 A^{-1/3} + a_3 A^{-2/3} + a_4 A^{-1},$$

$$b(A) = bA^{-2/3},$$

$$c(A) = c_1 + c_2 A^{-1/3} + c_3 A^{-2/3}/(1 + c_4 A^{-1/3}).$$

The Coulomb energy $E_c(Z,N)$ is taken as that of the trapezoidal charge distribution[1] with the radius parameter $r_0 = 1.13$ fm, and the last term of Eq.(2) is the binding energy of electrons in an atom. The correction term for odd-odd nuclei is

$$\Delta M_{odd-odd}(A) = \frac{11719.21}{(A + 31.4113)^2} - \frac{1321495}{(A + 48.1170)^3}.$$

We assume two functional forms for the shell terms $P_Z(N)$ and $Q_N(Z)$:
(1) Constant form

$$P_Z(N) = P_Z, \qquad Q_N(Z) = Q_N, \tag{3}$$

(2) Linear form

$$P_Z(N) = P_Z^0 + (N-N_Z^0)P_Z^1 \qquad (4a), \quad Q_N(Z) = Q_N^0 + (Z-Z_N^0)Q_N^1. \tag{4b}$$

Here, P_Z, Q_N, P_Z^0, P_Z^1, Q_N^0, Q_N^1 are adjustable parameters, of which the subscript Z runs from 2 to 102 and N from 2 to 156, while N_Z^0 and Z_N^0 are the constants representing the β-stability line. In determining the values of these parameters we only use odd-A and even-even nuclei.

Values of the gross-part parameters, a_i, b, c_i were obtained through the procedure described in Ref. 1. In the case of linear shell form, we have

$$a(A) = -17.080 + 30.138A^{-1/3} - 31.322A^{-2/3} + 24.192A^{-1},$$

$$b(A) = 15.0A^{-2/3},$$

$$c(A) = 35.493 - 90.356A^{-1/3} + 87.726A^{-2/3}/(1 + 0.45255A^{-1/3}).$$

In the case of constant shell form, the values, which will be published elswhere, are slightly different. Shell parameter values were determined by a new statistical treatment explained in the next section.

A NEW STATISTICAL TREATMENT

The shell parameter values should be adjusted so as to reproduce the whole set of "experimental shell energies", $M_{Eexp}(Z,N) - M_{Eg}(Z,N)$, as accurately as possible. In practice, however, there has been a problem associated with large variety of the experimental errors ranging from keV to MeV. In this section, we propose a new statistical treatment which solves this problem by dealing justly with the experimental masses with various errors. We explain it for the case of linear shell form.

We now divide the experimental shell energies $M_{Eexp}(Z,N) - M_{Eg}(Z,N)$ into the proton and neutron shell terms as $P_Z(N) + Q_N(Z)$, and to obtain the values of these two terms we use an iteration method. We first assume some initial values of neutron shell parameters Q_N^0 and Q_N^1, and fit the "experimental" proton shell energies,

$$M_{Eexp}(Z,N) - M_{Eg}(Z,N) - Q_N^0 - (Z - Z_N^0)Q_N^1, \qquad (5a)$$

to $P_Z^0 + (N - N_Z^0)P_Z^1$, and then we fit the "experimental" neutron shell energies,

$$M_{Eexp}(Z,N) - M_{Eg}(Z,N) - P_Z^0 - (N - N_Z^0)P_Z^1, \qquad (5b)$$

to $Q_N^0 + (Z - Z_N^0)Q_N^1$ in which the values of Q_N^0 and Q_N^1 may be different from those in (5a), and we repeat this procedure until the parameter values converge. The essential point of our statistical treatment lies in this fitting procedure. Let us explain it for the case of fitting the quantity (5a) to $P_Z^0 + (N - N_Z^0)P_Z^1$. In this case Z is fixed and N takes various values. For simplicity we write $N - N_Z^0$ as x_i and the quantity (5a) as y_i, whose subscript i distinguishes isotopes.

The ordinate y_i has an error, which we denote by η_i in the sense of standard deviation, while the abscissa x_i has no error. (How to determine η_i will be shown later.) With these variables we can write the error function Φ_{exp} of y_i as

$$\Phi_{exp}(y_i; \eta_i, Y_i) = \frac{1}{\sqrt{2\pi}\,\eta_i} \exp[-(y_i - Y_i)^2/2\eta_i^2], \qquad (6)$$

where Y_i represents the true value of y_i. If our theory was exact, these true values could be fitted exactly to the theoretical line, $y = P_Z^0 + P_Z^1 \cdot x$. From experiences, however, we know that this is not the case; whatever values we may choose for the parameters P_Z^0 and P_Z^1, there will remain certain errors. We take this error into account in our statistical treatment as the intrinsic error of our mass formula. We denote this intrinsic error by α_Z, with which we may write the error function Φ_{th} as

$$\Phi_{th}(Y_i; P_Z^0, P_Z^1, \alpha_Z) = \frac{1}{\sqrt{2\pi}\,\alpha_Z} \exp[-(Y_i - P_Z^0 - P_Z^1 \cdot x_i)^2/2\alpha_Z^2]. \quad (7)$$

It shows how the true values Y_i are distributed around the theoretical values $P_Z^0 + P_Z^1 \cdot x_i$. By the multiplication theorem, we can write the probability with which a series of experimental data y_i occur:

$$\Psi(y_i\text{'s}; \eta_i\text{'s}, P_Z^0, P_Z^1, \alpha_Z) =$$

$$\prod_i [\int_{-\infty}^{\infty} \Phi_{exp}(y_i; \eta_i, Y_i)\Phi_{th}(Y_i; P_Z^0, P_Z^1, \alpha_Z)dY_i], \quad (8)$$

where, we have taken account of the fact that the values of Y_i's are known only probabilistically. In Eq.(8), the product is taken over all isotopes with fixed Z. In addition to the intrinsic error α_Z, we take account of the errors coming from the uncertainties of the shell parameter values, which we refer to as extrinsic errors $\Delta P_Z^{ext}(N)$. Here, we note that the limit $\alpha_Z \to 0$ corresponds to the ordinary weighted least-squares method and the limit $\alpha_Z \to \infty$ corresponds to the equally-weighted least-squares method.

Next, we explain the procedure of determining the most probable values of shell parameters (denoted by $P_{Z_0}^0$ and $P_{Z_0}^1$), the extrinsic error $\Delta P_Z^{ext}(N)$, and the intrinsic error α_Z. The values $P_{Z_0}^0$ and $P_{Z_0}^1$ are given as those of P_Z^0 and P_Z^1 maximizing the probability Ψ in accordance with the principle of the probability theory. The extrinsic errors are obtained from Ψ by regarding it as the distribution function of P_Z^0 and P_Z^1. Actually, they are calculated by diagonalizing the variance matrix because Ψ is not normal with respect to P_Z^0 and P_Z^1. As for α_Z, one might try to determine its value by maximizing Ψ, but we adopt the following procedure starting from the original meaning of α_Z, that is, it is the intrinsic error of the theoretical line. First, we calculate the expectation value of the quantity $\sum_j (Y_j - P_Z^0 - P_Z^1 \cdot x_j)^2$, i.e.

$$<\sum_j \Delta Y_j^2> = \frac{\int_{-\infty}^{\infty} dP_Z^0 \int_{-\infty}^{\infty} dP_Z^1 \int_{-\infty}^{\infty} \cdots \int_{-\infty}^{\infty} \sum_j (Y_j - P_Z^0 - P_Z^1 \cdot x_j)^2 H \prod_i dY_i}{\int_{-\infty}^{\infty} dP_Z^0 \int_{-\infty}^{\infty} dP_Z^1 \int_{-\infty}^{\infty} \cdots \int_{-\infty}^{\infty} H \prod_i dY_i},$$

with

$$H = \prod_i [\Phi_{exp}(y_i; \eta_i, Y_i)\Phi_{th}(Y_i; P_Z^0, P_Z^1, \alpha_Z)]. \quad (9)$$

Since the quantity $<\sum_j \Delta Y_j^2>$ means the sum of the squares of the residuals, we tentatively equate it to the number of the data n_Z times α_Z^2 :

$$<\sum_j \Delta Y_j^2> = n_Z \alpha^2 \quad (10)$$

This may be regarded as an equation in α_Z, and we denote its solution by α_Z'. Now, we give consideration to the fact that we do not have data on the whole set of the nuclei to which our mass formula is to be applied. Thus, α_Z' may be different from α_Z in a systematic way; possibly, the frequency distribution of α_Z' over Z may be somewhat broader than that of α_Z. Actually, α_Z' takes zero for many Z's, but absence of the intrinsic error is unlikely even for a single Z value. Thus, we determine α_Z from α_Z' by

$$\alpha_Z^2 = 0.9\alpha_Z'^2 + 0.1\overline{\alpha}_Z^2 , \tag{11}$$

where $\overline{\alpha}_Z^2$ is the root mean square of $\alpha_Z'^2$ with respect to Z.

This procedure to determine the values of P_{Z0}^0, P_{Z0}^1, $\Delta P_Z^{ext}(N)$, α_Z, is followed in obtaining the values of Q_{N0}^0, Q_{N0}^1, $\Delta Q_N^{ext}(Z)$, α_N for neutron shell terms. At this stage we can specify the error of the quantity (5a) or (5b), i.e. η_i. The error of (5a) is $\{\epsilon_i^2 + \alpha_N^2 + [\Delta Q_N^{ext}(Z)]^2\}^{1/2}$, where ϵ_i is the experimental error. The error of (5b) is given by a similar expression.

After the iteration converges, we can calculate the theoretical error $\delta M(Z,N)$ as

$$\delta M(Z,N) = \{\alpha_Z^2 + \alpha_N^2 + [\Delta P_Z^{ext}(N)]^2 + [\Delta Q_N^{ext}(Z)]^2$$
$$+ [\tfrac{1}{3}\Delta M_{odd-odd}(A)]^2\}^{1/2}, \tag{12}$$

where the last term in curly brackets is added for odd-odd nuclei only.

RESULTS AND DISCUSSION

We took the experimental masses from the table of Wapstra and Bos[3]. We also used their systematics values assigning appropriate errors. Since convergence of our iteration method was rather slow, we imposed a condition that $(P_{Z0}^0)^2$ and $(Q_{N0}^0)^2$ should not be too large; this condition is sufficiently weak to avoid any appreciable change of the results.

We give the most probable values of shell parameters P_{Z0}^0, P_{Z0}^1, Q_{N0}^0, Q_{N0}^1 in Figs.1~4, and the intrinsic errors α_Z and α_N in Figs. 5 and 6. Only those for linear shell form are shown here.

The root mean square of intrinsic errors α_Z or α_N in keV is 153 for even Z, 211 for odd Z, 141 for even N, and 187 for odd N in the case of linear shell form, while it is 319 for even Z, 398 for odd Z, 326 for even N, and 341 for odd N in the case of constant shell form. There is a correlation between α_Z(or α_N) and the number of the data n_Z(or n_N) although fluctuation is large; α's increase with n's , and this tendency is clearer in the case of constant shell form.

The presence of this correlation must be kept in mind when we extrapolate the mass formula. The error of an extrapolated mass is expected to be larger than that given by Eq.(12). However, how much we should increase the errors is hard to answer from theoretical considerations only. Then, we resort to comparison with experiments.

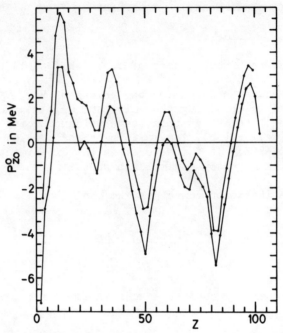

Fig.1. The most probable value of the proton shell parameter P_Z^0
        for linear shell form

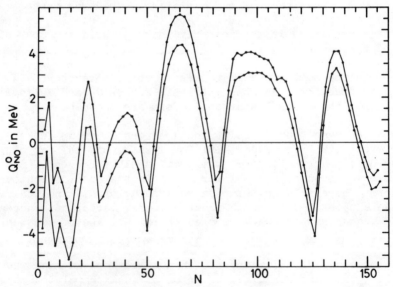

Fig.2. The most probable value of the neutron shell parameter Q_N^0
        for linear shell form

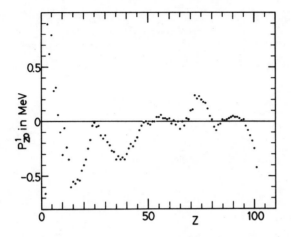

Fig.3. The most probable value of the proton shell parameter P_Z^1
      for linear shell form

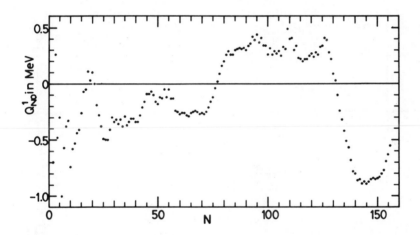

Fig.4. The most probable value of the neutron shell parameter Q_N^1
      for linear shell form

The mass data of Rb and Cs isotopes published recently[4] are suitable
for this purpose, because they were not used as our input data and
also because they require large extrapolations. The comparisons are
shown in Figs.7 and 8 with two kinds of theoretical errors: the
smaller ones are given by Eq.(12), and the larger ones include
additional errors due to the extrapolation. In the case of linear
shell form, this additional errors is taken to be one third of the

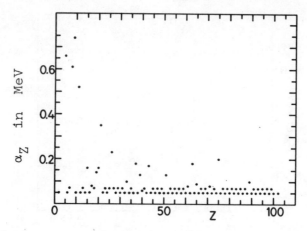

Fig.5. Intrinsic error of the proton shell term for linear shell
       form

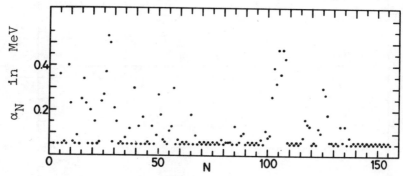

Fig. 6.   Intrinsic error of the neutron shell term for linear shell
           form

"maximum" increment, which is the increase of the error expected
when the intrinsic error is wholly due to the neglect of second
order shell form. In the case of constant shell form, the additional
error is somewhat more complicated. The errors given by Eq. (12)
are apparently too small in the case of extrapolation, but the
errors including the above mentioned additional errors seem to be
reasonable.

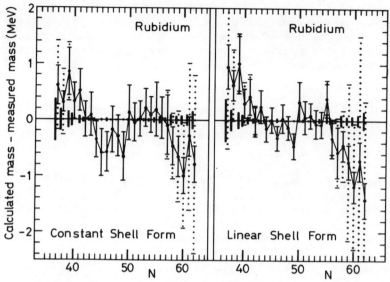

Fig.7. Differences between calculated masses and experimental masses[4] for rubidium isotopes. The errors attached to the points are those of the calculated masses. The errors indicated by solid lines are those given by Eq.(12), and the errors indicated by dotted lines are those including additional errors due to extrapolation as explained in the text. The thick vertical lines along the zero line indicate the experimental errors.

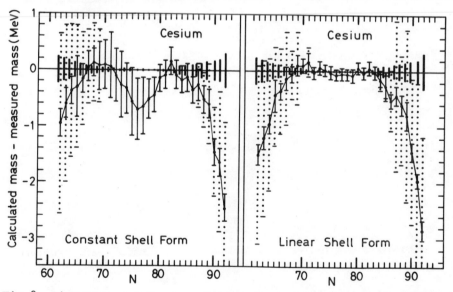

Fig.8. Differences between calculated masses and experimental masses[4] for cesium isotopes. See caption of Fig.7.

In conclusion, we have obtained two empirical mass formulas
which give us mass values and their errors. They will be helpful
in estimating the masses of nuclei far from the β-stability line.
In the course of the analysis we used a new statistical treatment,
which will be useful in other studies which have to employ statistical analyses.

The authors express their thanks to Mr. Y. Ando and Mr. T.
Tachibana for their helpful assistance.

REFERENCES

1. M. Uno and M. Yamada, Prog. Theor. Phys. 53, 987 (1975)
2. M. Yamada, M. Uno and Y. Sakamoto, Proc. of 3rd International
   Conference on Nuclei Far from Stability, Cargese 1976, [report
   CERN 76-13], p.154
3. A. H. Wapstra and K. Bos, At. Data Nucl. Data Tables 19, 175(1977)
4. M. Epherre, G. Audi, C. Thibault, R. Klapisch, G. Huber,
   F. Touchard and H. Wollnik, Phys. Rev. C 19, 1504 (1979)

NUCLEAR MASS RELATIONS AND EQUATIONS

J. E. Monahan and F. J. D. Serduke

Argonne National Laboratory

Argonne, Illinois 60439

INTRODUCTION

Relations among the masses of neighboring nuclei provide a convenient and reasonably accurate method for the estimation of unknown masses. A set of such relations is presented. Each member of this set gives a prediction of roughly equal statistical accuracy for the unknown mass of a given nuclide. The average and dispersion of the set give a "most probable" mass value and a measure of its uncertainty. The structure of these relations (considered as partial difference equations) lead to another set of difference equations the solutions of which can be ordered such that each successive member is expected to be a more nearly accurate representation of ground-state energies as functions of N and Z. The reliability of such solutions for extrapolation far from the region of known nuclei is discussed.

A SET OF MASS RELATIONS

One approach[1] to the prediction of nuclear ground-state energies is based on relations of the form

$$\sum_{ij} a_{ij} M(N+i, Z+j) \simeq 0 \qquad (1)$$

where $M(N+i, Z+j)$ is the mass, or mass excess, of the nuclide with $N+i$ neutrons and $Z+j$ protons. The coefficients a_{ij} in Eq. (1) are to be chosen such that all interactions between nucleons cancel to first order. The minimum condition is clearly that the numbers of neutron-neutron, proton-proton, and neutron-proton interactions must cancel in Eq. (1).

To investigate the structure of equations of the form (1) it is convenient to introduce the displacement operators E_N^i and E_Z^i where

$$E_N^i M(N,Z) \equiv M(N+i,Z) \ , \tag{2a}$$

$$E_Z^i M(N,Z) \equiv M(N,Z+i) \ . \tag{2b}$$

Here i is a positive or negative integer. Equation (1) can be written as a polynomial in these operators acting on the function $M(N,Z)$, i.e.,

$$\sum_{ij} a_{ij} M(N+i,Z+j) = P(E_N,E_Z)M(N,Z) \approx 0 \ , \tag{3}$$

where

$$P(E_N,E_Z) = \sum_{ij} a_{ij} E_N^i E_Z^j \ . \tag{4}$$

The minimum condition that the numbers of nucleon–nucleon interactions cancel in Eq. (3) now takes the form

$$P(E_N,E_Z)N^2 = P(E_N,E_Z)Z^2 = P(E_N,E_Z)NZ = 0 \ . \tag{5}$$

Conditions (5) are satisfied if, and only if, the polynomial $P(E_N,E_Z)$ contains a factor of the form

$$(E_N^\ell E_Z^m - 1)^\gamma (E_Z-1)^\beta (E_N-1)^\alpha \tag{6}$$

with $\alpha+\beta+\gamma \geqslant 3$. Here α, β, and γ are non-negative integers; ℓ and m are arbitrary integers.

The simplest mass equations, i.e. those which refer to the fewest nuclei, are obtained by choosing the polynomial $P(E_N,E_Z)$ to be equal to the factor (6) with the smallest admissible value of $\alpha+\beta+\gamma$. This leads to the mass relations

$$(E_N^\ell E_Z^m - 1)^\gamma (E_Z-1)^\beta (E_N-1)^\alpha M(N,Z) \approx 0 \ , \tag{7}$$

with $\alpha+\beta+\gamma = 3$. For fixed values of ℓ and m there are ten distinct mass relations defined by Eqs. (7). These were evaluated over the known masses for a range of values of ℓ and m and the relations (7) with $\alpha=\beta=\gamma=1$ and $\ell+m$ an even integer were found to have values considerably closer to zero than any of the others. The resulting set of mass relations, namely

$$O^{\ell m} M(N,Z) \equiv (E_N^\ell E_Z^m - 1)(E_Z-1)(E_N-1)M(N,Z) \approx 0 \ , \tag{8}$$

$\ell+m$ = even, has been discussed in Refs. 2 and 3.

An investigation of the distribution of the values of Eqs. (8) over the known masses was carried out as follows. For fixed ℓ, m and $N \gtrsim 10$, $Z \gtrsim 10$ the quantity $O^{\ell m}M(N,Z)$ was calculated for all combinations of mass excesses that are listed with an accuracy of 100 keV or better in the 1975 Wapstra-Bos compilation.[4] No odd-odd N = Z nuclei were included and, for $\ell \neq m$, only nuclei with $N \gtrsim Z$ were considered. Let $R_{\ell m}$ denote the average of these results, i.e.,

$$R_{\ell m} = \frac{1}{n} \sum_{(N,Z)} O^{\ell m}M(N,Z) \ , \tag{9}$$

where n is the number of terms in the sum, and let $W_{\ell m}$ measure the dispersion:

$$W_{\ell m}^2 = \frac{1}{n} \sum_{(N,Z)} (O^{\ell m}M(N,Z)-R_{\ell m})^2 \ . \tag{10}$$

Values of $R_{\ell m}$ and $W_{\ell m}$ are listed in Table I for a few values of ℓ and m.

From the results shown in Table I (and from the more extensive calculations described in Ref. 2) it follows that, for all practical purposes, the quantities $O^{\ell m}M(N,Z)$ are distributed over the known masses with zero mean ($R_{\ell m} = 0$) and equal width ($W_{\ell m} = W$) for a range of values of ℓ and m. Further, over the same range, the correlation coefficients

$$\rho(\ell,m;\ell',m') = \frac{1}{W^2 n} \sum_{(N,Z)} [O^{\ell m}M(N,Z)][O^{\ell'm'}M(N,Z)] \tag{11}$$

are very nearly independent of ℓ, m and ℓ', m' ($\rho(\ell,m;\ell',m')=\rho$).

Consider now the case where there is one unknown excess, say M(N,Z), contained in the relations $O^{\ell m}M(N,Z) = 0$ and let $M_{\ell m}(N,Z)$ denote the estimates of this excess obtained as solutions of these relations, i.e.

$$M_{\ell m}(N,Z) = [E_N^\ell E_Z^m(E_Z-1)(E_N-1) - E_N E_Z + E_N + E_Z]M(N,Z) \ . \tag{12}$$

The excesses on the right-hand side of Eqs. (12) are assumed to be known with negligible uncertainty. (For values of N and Z far from the region of known masses this assumption cannot be satisfied. The use of relations (8) to estimate mass excesses in such cases is described in Ref. 2.) From the discussion given in the previous paragraph it follows that the $M_{\ell m}(N,Z)$ are distributed about the "true" value of M(N,Z) with dispersion and correlation roughly independent of ℓ and m. From the theory of generalized least squares it then

Table I. The mean $R_{\ell m}$ and width $W_{\ell m}$ of the distribution
of $0^{\ell m}M(N,Z)$ over the known masses.

ℓ	m	$R_{\ell m}$ (keV)	$W_{\ell m}$ (keV)	n
1	1	8	198	632
-1	1	-17	211	518
2	0	-7	196	545
0	2	26	214	451
2	2	17	218	595
3	1	7	227	554
4	2	15	229	573
4	0	-4	229	391
6	2	12	226	463

follows that the least squares estimate $\overline{M}(N,Z)$ of the excess $M(N,Z)$ is

$$\overline{M}(N,Z) = \frac{1}{K} \sum_{(\ell m)} M_{\ell m}(N,Z) , \qquad \ell+m = \text{even}, \qquad (13)$$

where K is the number of terms in the sum. Table II contains some
predictions $\overline{M}(N,Z)$ of the mass excess of the nuclide with N neutrons
and Z protons obtained in this manner. Relations (8) with $0 \leqslant \ell \leqslant 6$,
$-4 \leqslant m \leqslant 6$ were used to obtain these results. Also shown in Table II
are the error estimates $\pm$ E, where

$$E^2 = \frac{1}{K} \sum_{(\ell m)} [M_{\ell m}(N,Z) - \overline{M}(N,Z)]^2 . \qquad (14)$$

It is possible to obtain another estimate of the overall accuracy
of the least-squares estimates $\overline{M}(N,Z)$. Since Eq. (13) is linear in
the $M_{\ell m}(N,Z)$, the variance of $\overline{M}(N,Z)$ can be written as

$$\text{var}\{\overline{M}\} = \frac{1}{K^2} \sum_{(\ell m)} \text{var}\{M_{\ell m}\} + \frac{1}{K^2} \sum_{(\ell m) \neq (\ell' m')} \sum \text{cov}\{M_{\ell m}, M_{\ell' m'}\}, \qquad (15)$$

where the arguments (N,Z) have been omitted. As noted previously,
$\text{var}\{M_{\ell m}\} \simeq W^2$ and $\text{cov}\{M_{\ell m}, M_{\ell' m'}\} \simeq \rho W^2$ so that Eq. (15) becomes

$$\text{var}\{\overline{M}\} \simeq [\rho+(1-\rho)/K]W^2 . \qquad (16)$$

Preliminary calculations indicate that ρ is approximately 0.5 and from
Table I, $W \simeq 220$ keV, so that for large values of K

$$(\text{var}\{\overline{M}\})^{\frac{1}{2}} \simeq (\rho W^2)^{\frac{1}{2}} \simeq \pm 156 \text{ keV}.$$

This is indeed about the average of the estimates $\pm$ E shown in
Table II.

Table II. Predictions for the mass excess of a few nuclides.

Nuclide (N,Z)	K (number of relations used)	$\overline{M}(N,Z)$ (excess in keV)
^{26}Ne (16,10)	12	-318 ± 299
^{48}K (29,19)	23	-32517 ± 229
^{76}Ga (45,31)	31	-66439 ± 227
^{86}Zr (46,40)	36	-77933 ± 180
^{109}Sn (59,50)	30	-82698 ± 98
^{153}Nd (93,60)	27	-67356 ± 163
^{179}Yb (109,70)	26	-46744 ± 116
^{189}Hg (109,80)	3	-29508 ± 123
^{207}Hg (127,80)	28	-16133 ± 177
^{214}Ac (125,89)	18	6144 ± 82
^{232}Ac (143,89)	24	39301 ± 102
^{246}Es (147,99)	13	67943 ± 46
^{255}Es (156,99)	15	83910 ± 144

An empirical observation[5] that underlies the set of mass relations (8) involves the so-called residual n-p interaction energy[6] $I_{np}(N+1,Z+1)$, where

$$I_{np}(N+1,Z+1) \equiv (E_Z-1)(E_N-1)M(N,Z). \tag{17}$$

These quantities are generally small and, more importantly, they are strongly correlated separately for neighboring even-A and odd-A nuclei. Thus, for $\ell+m$ an even integer,

$$O^{\ell m}M(N,Z) = I_{np}(N+\ell+1,Z+m+1) - I_{np}(N+1,Z+1) \simeq 0 \tag{18}$$

is the difference between small quantities of comparable magnitude.

Equation (18) is in a form that is convenient for numerical calculations. The quantities I_{np}, Eq. (17), refer to four adjacent nuclei $(N+1,Z+1)$, $(N+1,Z)$, $(N,Z+1)$ and (N,Z). Suppose (N,Z) is the nuclide whose excess is to be estimated and choose $I_{np}(N+1,Z+1)$ such that it contains (N,Z) and three other nuclei whose excesses are known. Then $I_{np}(N+1,Z+1)$ is of the form

$$I_{np}(N+1,Z+1) = M(N,Z) + c , \tag{19}$$

where c is determined from the three known excesses, i.e.,

$$c = M(N+1,Z+1) - M(N+1,Z) - M(N,Z+1) . \tag{20}$$

The value of $I_{np}(N+\ell+1,Z+m+1)$, $\ell+m = $ even, is then calculated in terms of four known excesses for a range of values of ℓ and m. A set of

estimates $M_{\ell m}(N,Z)$ for the excess $M(N,Z)$ is then obtained from
Eq. (18) as

$$M_{\ell m}(N,Z) = I_{np}(N+\ell+1,Z+m+1) - c \; . \tag{21}$$

The set of mass relations (8) exhausts all mass relations of the
form of Eq. (1), which satisfy the condition that the numbers of
n-n, p-p, and n-p interactions cancel, in the following sense. All
such relations (1) give predictions that are linear combinations of
those obtained from the set (8). We have shown that the polynomial
$P(E_N,E_Z)$, Eq. (4), must contain the operator $0^{\ell m}$ as a factor so that

$$\sum_{ij} a_{ij} E_N^i E_Z^j = \sum_{ij} b_{ij} E_N^i E_Z^j 0^{\ell m} \; . \tag{22}$$

From the definition (8) of the operator $0^{\ell m}$ it is easily shown that,
for $i+j$ = even,

$$E_N^i E_Z^j 0^{\ell m} M(N,Z) = (0^{\ell+i,m+j} - 0^{ij}) M(N,Z) \; , \tag{23a}$$

and, for $i+j$ = odd,

$$E_N^i E_Z^j 0^{\ell m} M(N,Z) = (0^{\ell+i-1,m+j} - 0^{i-1,j}) M(N+1,Z)$$

$$= (0^{\ell+i,m+j-1} - 0^{i,j-1}) M(N,Z+1) \; . \tag{23b}$$

From Eqs. (23) it follows that the right-hand side of Eq. (22) is a
linear combination of the $0^{\ell m}$ operating on $M(N,Z)$, $M(N+1,Z)$, or
$M(N,Z+1)$. All of these possibilities are contained in the set of
relations (8).

This result implies that there is small likelihood that a mass
relation of the form of Eq. (1) exists which is distributed over the
known excesses with dispersion significantly less than W, the dispersion associated with each of the relations (8).

A SET OF MASS EQUATIONS

Mass relations, of the type discussed above, are partial difference equations, the solutions of which are mass equations. For
example, relation (8) with $\ell = -m = 1$ is the partial difference
equation

$$(E_N E_Z^{-1} - 1)(E_Z - 1)(E_N - 1) M(N,Z) = 0 \; , \tag{24}$$

and the general solution is the mass equation

$$M(N,Z) = M_0 + f_1(N) + f_2(Z) + f_3(N+Z) \; . \tag{25}$$

Here the f_i are arbitrary functions of their arguments and M_o is a constant independent of N and Z.

To make use of a mass equation of this kind, the point functions f_i are evaluated by a least-squares fit of the equation to the known mass excesses. The resulting equation is then used to evaluate the excesses of nuclei that are not included in the fit. The global nature of the least-squares fit reduces the accuracy of a mass equation compared with that of the set of mass relations (8); however, the calculational convenience of such equations makes it worthwhile to consider their limitations and possible extensions.

Mass equations that arise as solutions of several of the relations (8) were discussed briefly in Ref. 2. However, the members of Eq. (8) are very nearly equivalent statistically and there is no a priori reason to believe that the more complex solutions of the higher-order difference equations are significantly more accurate than the lowest-order ones ($\ell = \pm m = 1$) considered originally by Garvey et al.[1] In Ref. 3 is described an attempt to formulate a set of mass equations that can be ordered such that each successive member gives a more nearly accurate representation of nuclear ground-state energies as functions of N and Z. This work was based on an observation by Jänecke and Behrens[5] that there is a smooth long-range dependence of I_{np}, Eq. (17), on A = N+Z and/or $T_Z = \frac{1}{2}(N-Z)$. The resulting mass equation is

$$M(N,Z) = M_0 + h_1(N) + h_2(Z) + \sum_{\kappa=0}^{k-1} f_\kappa(A)E^\kappa + \sum_{\sigma=0}^{s-1} g_\sigma(E)A^\sigma . \qquad (26)$$

Here h_i, f_κ, and g_σ are arbitrary functions of their arguments, M_0 is a constant independent of N and Z, and E = N−Z. If the set of equations (26) are ordered according to the value of (k,s), the successive members are expected to be more nearly accurate representations of the excesses as functions of N and Z. The first two equations, (k,s) = (0,1) and (1,0), are those considered in Ref. 1.

Some insight into the extrapolation properties of Eqs. (26) can be obtained by consideration of the first three members of these equations, namely

$$M(N,Z) = M_0 + g_1(N) + g_2(Z) + g_3(E), \qquad (27a)$$

$$M(N,Z) = M_0 + f_1(N) + f_2(Z) + f_3(A), \qquad (27b)$$

$$M(N,Z) = M_0 + h_1(N) + h_2(Z) + h_3(E) + h_4(A) , \qquad (27c)$$

where the notation for the point functions has been changed. In addition, consider the mass equations

$$M(N,Z) = M_0 + q_1(N) + q_2(Z) + q_3(E)/A , \qquad (27d)$$

$$M(N,Z) = M_0 + p_1(N) + p_2(Z) + p_3(A) + p_4(E)/A , \qquad (27e)$$

which take into account the long-range dependence of the symmetry energy. Least-squares values for the point functions were obtained by fitting the known excesses to each of these equations and the quantity σ was calculated, where

$$\sigma^2 = \sum_{(N,Z)} [m(N,Z) - M(N,Z)]^2/(n-t) . \qquad (28)$$

Here $m(N,Z)$ is a known excess, n is the number of terms in the sum, and t is the number of free parameters in the mass equation $M(N,Z)$. The results are shown in Table III. Note that Eq. (27c) does not fit the known excesses any better than does Eq. (27b). This would seem to imply that the set of equations (26) is of little or no interest since the expectation was that successive members of this set would be more accurate representations of the excesses as functions of N and Z. However, the value of a mass equation is its usefulness as an extra-polation formula and it is far from clear that the statistic σ is a valid measure of this property.[7]

With the exception of the (1,0) member all of the mass equations (26) probably fit the known excesses with about the same accuracy so that additional criteria are necessary to judge their relative reli-ability as extrapolation formulas. One such criterion is the follow-ing. Except for small Coulomb contributions, the charge symmetry of nuclear forces implies that the n-p interaction energy is invariant under interchange of N and Z, i.e.,

$$(E_Z-1)(E_N-1)M(N,Z) = (E_Z-1)(E_N-1)M(Z,N). \qquad (29)$$

The mass equations (27) satisfy this condition provided that the point functions of E are symmetric about $E = 0$, i.e., if

$$f(E) = f(-E) . \qquad (30)$$

Table III. A comparison of the fits of Eqs. (27) to the known excesses.

	n	t	σ(MeV)
Eq. (27a)	1251	300	0.658
Eq. (27b)	1199	476	0.180
Eq. (27c)	1251	535	0.187
Eq. (27d)	1251	300	0.371
Eq. (27e)	1251	535	0.186

Table IV. The fitted values in MeV of the function of
E = N–Z that occurs in each of Eqs. (27a),
(27c, (27d) and (27e).

E	Eq.(27a) $g_3(E)$	Eq.(27c) $h_3(E)$	Eq.(27d) $q_3(E)$	Eq.(27e) $p_4(E)$
0	0	0	0	0
1	2.800	40.899	66.0	2737.8
–1	2.379	40.800	62.3	2737.7
2	5.652	3.044	148.1	–73.1
–2	5.697	3.008	140.0	–73.4
3	9.920	44.416	277.2	2507.7
–3	9.963	44.410	263.6	2507.7
4	14.537	7.082	421.5	–459.4
–4	14.790	7.128	405.9	–458.5

In Table IV are listed the values of the functions of E that are obtained in the least-squares fits of Eqs. (27) to the known excesses. It is obvious from these results that Eqs. (27c) and (27e) satisfy the symmetry condition (30) to a better approximation than do Eqs. (27a) and (27d). Consequently, it seems reasonable to expect that Eqs. (27c) and (27e) are more reliable extrapolation formulas.

A tabulation of the point functions obtained by fitting Eq. (27e) to the known masses is given in Ref. 3.

This work was supported by the U. S. Department of Energy.

1. G. T. Garvey, W. J. Gerace, R. L. Jaffe, I. Talmi, and I. Kelson, Set of Nuclear-Mass Relations and a Resultant Mass Table, Rev. Mod. Phys. 41: S1 (1969).
2. J. E. Monahan and F. J. D. Serduke, Family of Nuclear-Mass Relations, Phys. Rev. C 15: 1080 (1977).
3. J. E. Monahan and F. J. D. Serduke, Nuclear-Mass Relations and Equations, Phys. Rev. C 17: 1196 (1978).
4. A. H. Wapstra and K. Bos, A 1975 Midstream Atomic Mass Evaluation, At. Data Nucl. Data Tables 17: 474 (1976).
5. J. Jänecke and H. Behrens, Nuclidic Mass Relationships and Mass Equations, Phys. Rev. C 9: 1276 (1974).
6. M. K. Basu and D. Banerjee, Study of the Neutron-Proton Interaction, Phys. Rev. C 3: 992 (1971).
7. E. Comay and I. Kelson, Ensemble Averaging of Mass Values, At. Data and Nucl. Data Tables 17: 463 (1976).

CORRECTIONS TO THE FARADAY AS DETERMINED

BY MEANS OF THE SILVER COULOMETER*

Richard S. Davis and Vincent E. Bower†

Center for Absolute Physical Quantities
National Bureau of Standards
Washington, D. C.   20234

INTRODUCTION

An enduring metrological discrepancy has been the failure of
the Faraday constant as measured electrochemically to agree with
the calculation of that constant from other precisely known
physical constants [1]. At the last meeting of this conference, we
presented results of a measurement of the Faraday constant by
means of a silver coulometer [2,3]. At that time, we were able
to report a high experimental precision for measurements of this
type but had not yet given serious consideration to effects of
impurities. Impurities have now been accounted for but their
effect does not alter the inconsistency described above. A brief
description of the silver coulometer experiment, an analysis of
effects of the impurities in the silver sample, and a discussion
of possible remaining systematic errors follows.

EXPERIMENTAL

The silver coulometer is an uncomplicated device. An anode,
in the form of a pure polycrystalline silver rod is dipped into an
electrolyte which does not dissolve silver. A platinum cathode at
considerable remove from the anode completes the coulometer. In
the first phase of the experiment, a constant current, i, is made
to flow through the coulometer for a known amount of time. During
this time, the anode loses mass according to the reaction

$$Ag \rightarrow Ag^+ + e,$$ (1)

---

*Contribution of the National Bureau of Standards. Not subject to
 copyright.
†Retired from NBS.

The anode also loses a small amount of mass because metallic silver particles simply fall off during the electrolysis. After a time, t, the current is turned off and the total loss in mass, m_s, of the anode is found by weighing the previously tared silver on a sensitive balance.

The second phase of the experiment consists of taking account of the mass loss of the anode due to the residue of metallic silver mechanically separated from the anode. The analysis of this residue has been described in considerable detail elsewhere [4]. It is sufficient to point out that during the analysis: a) the silver never leaves the anode beaker of the coulometer; and b) that the silver is converted to silver ion and plated onto a platinum electrode at constant voltage. The charge, q, consumed during this plating process is found by electronic integration of the current. The electrochemical equivalent of the silver sample, E'_{Ag}, is found from the relation

$$E'_{Ag} = \frac{m_s}{it + q} \, .$$ (2)

CALCULATION OF ELECTROCHEMICAL EQUIVALENT

To find E_{Ag}, the electrochemical equivalent of silver, account must be taken of impurities in the anodes. E_{Ag} is defined as m_{Ag}/q_{Ag} where m_{Ag} is the mass of pure silver dissolved by the passage of q_{Ag} coulombs. If $q_s \equiv it + q$, m_j is the mass of the impurity, j, in the sample, and q_j is the charge required to oxidize the impurity, j, we have

$$E_{Ag} = \frac{m_{Ag}}{q_{Ag}} = \frac{m_s - \Sigma m_j}{q_s - \Sigma q_j} \, .$$ (3)

Because the impurities occur at very low concentrations, equation (3) may be expanded and only the first order terms in the correction retained:

$$E_{Ag} = E'_{Ag}[1 + \Sigma \frac{m_j}{m_s} (\frac{A_r(Ag)}{A_r(j)} \nu_j - 1) + ...] \, ,$$ (4)

where $A_r(j)$ is the atomic weight of the impurity, $A_r(Ag)$ is the atomic weight of silver, and ν_j is the oxidation number of the jth impurity.

The acid in the coulometer was sufficiently concentrated and the concentration of ions formed from any metallic impurities was so tenuous (10^{-6} molar) that in the calculation of the electrochemical equivalent of pure silver the electrochemical oxidation

Table 1.   Estimate of Corrections and Uncertainties
           from Impurity Analyses

Source	Correction to E'_{Ag}	Uncertainty
SSMS-detected (16 elements)	+0.29 ppm	0.3 ppm
IDMS (iron)	+0.79	0.03
IDMS (19 elements)	+0.89	0.5
SSMS-undetected (46 elements)	--	0.3
Total	+1.97	RSS 0.66

was assumed to be the maximum permissible under conditions imposed
by the oxidation-reduction potentials of the ions [5] and by the
actual potential of the silver electrode.

Techniques of spark source mass spectrometry (SSMS) and spark
source isotope dilution mass spectrometry (IDMS) were used to analyze
the impurities in the silver samples. The data from IDMS (with the
single exception of iron impurities) were all upper limits. Elements
detected by SSMS were uncorrected for the mass-spectrometric sensitivity factors. Elements undetected by SSMS might be present at
the resolution level of the apparatus. These diverse kinds of information required specialized statistical analyses which are explained
in detail elsewhere [3]. The results of the impurity analysis are
displayed in Table 1.

One impurity, oxygen, requires special mention. At temperatures near the melting point of silver, the solubility and rate
of diffusion of oxygen in silver are high [6]. If oxygen trapped
in the silver matrix forms the oxide Ag_2O, which is soluble in the
electrolyte used in the silver coulometer, then 1 ppm by weight of
oxygen impurity will cause an error in E_{Ag} of +14.5 ppm. To avoid
significant contamination by oxygen, the silver was annealed in
vacuum at a temperature just below its melting point. It was then
allowed to cool to room temperature under vacuum. When not being
used, the annealed silver was stored in a solution of electrolyte
used in the silver coulometer. One electrode (not used in the
Faraday determinations) was intentionally heated in air at a
temperature of 490°C. Subsequently, a coulometric determination of
E'_{Ag} conducted with the specially treated anode produced a result

15 ppm higher than measurements with samples annealed in vacuum.
Agreement with predictions based on the data of [6] is thought to
be satisfactory.

CALCULATION OF FARADAY CONSTANT

The Faraday constant, F, may be derived from the relation

$$F = \frac{A_r(Ag)}{E_{Ag}} .$$ (5)

Since silver has two stable isotopes, their ratio in our anodes, when
combined with their atomic weights [7], yields the value of $A_r(Ag)$
for our samples. The last measurement we have to date assigns an
uncertainty to $A_r(Ag)$ (one standard deviation) of 2.1 ppm. A
second measurement, now well along, should reduce this uncertainty
significantly.

We advance a value for the Faraday constant of:

$$F = 96486.33(24)A_{NBS} \cdot s \cdot mol^{-1}   (2.5 \ ppm)$$

in terms of the electric units as maintained at NBS in the spring of
1975. The quantities in parenthesis are our estimate of the uncertainty whose components are: systematic, $0.23 \ A \cdot s \cdot mol^{-1}$, and random,
$0.08 \ A \cdot s \cdot mol^{-1}$. The sytematic uncertainty represents a root sum
square of all known systematic uncertainties at the one standard
deviation level. This number is dominated by our uncertainty of the
atomic weight of silver. The random uncertainty is the standard
deviation of the mean of eight measurements of E_{Ag}'.

DISCUSSION

Our value for F remains some 20 ppm higher than that derived
from other than electrochemical means [1]. Obviously an ignored
or underestimated effect plagues at least one of the measurements
involved. Playing devil's advocate by assuming the error to be in
our determination of F, it may be useful to explore where such an
error might or might not occur. If the difference is due to a
systematic error in our measurements, the error must have the effect
of dissolving too little silver for the passage of a given charge.
Ruling out parasitic leakage resistances in the circuit, which were
carefully checked, the first possibility to be considered is that
of less than perfect efficiency of the coulometer anode in generating
silver ion. Unfortunately, electrode efficiency cannot be directly
measured; (the most precise measurements have established the rever-
sibility of the silver coulometer to only 60 ppm [8]). Indirectly,
we may look for correlations of E_{Ag} with current density or maximum
overvoltage of the silver anodes. At equilibrium, thermodynamic
arguments suggest the silver anode should be 100 percent efficient
[8]. These arguments may not, however, be used far from equilibrium.

By increasing the current density of the coulometer, one moves the state of the silver anode further from equilibrium. Increasing the current density by a factor of two had no detectable effect on our measurements of E_{Ag}.

We are left, finally, with the unsatisfying conclusion that the discrepancy between directly measured and indirectly calculated values of F remains unresolved.

ACKNOWLEDGMENTS

The authors are grateful to E. L. Garner, L. P. Dunstan, T. J. Murphy, I. L. Barnes and P. J. Paulsen of the NBS Center for Analytical Chemistry for providing analyses of the samples and for helpful discussions. Useful consultations with W. F.Koch, G. Marinenko, C. Eisenhart and B. N. Taylor are also gratefully acknowledged.

REFERENCES

1. E. R. Cohen and B. N. Taylor, J. Phys. Chem. Ref. Data 2:663 (1973).
2. V. E. Bower and R. S. Davis, in "Atomic Masses and Fundamental Constants 5," J. H. Sanders and A. H. Wapstra, eds., Plenum Press, New York, 578 (1976).
3. V. E. Bower and R. S. Davis (in preparation).
4. R. S. Davis and V. E. Bower, J. Res. NBS (U.S.A.) 84:157 (1979).
5. W. M. Latimer, "Oxidation Potentials," Prentice Hall, Englewood Cliffs (1952).
6. W. E. Eichenaur and G. Muller, Z. Metallkunde 53: 321, 700 (1962).
7. A. H. Wapstra and K. Bos, Atomic and Nuclear Data Tables 19: 175 (1977).
8. D. N. Craig, J. I. Hoffman, C. A. Law, W. J. Hamer, J. Res. NBS (U.S.A.) 64A:381 (1960).

THE VALUE OF THE FARADAY VIA 4-AMINOPYRIDINE[*]

William F. Koch

Center for Analytical Chemistry
National Bureau of Standards
Washington, D.C. 20234

INTRODUCTION

In 1833, Michael Faraday of the Royal Institution in London formulated the laws of electrolysis revealing the relationship between electric current, time, and chemical equivalent weight (mole). Translated into mathematical terms, this relationship can be expressed:

$$\text{mole} = \frac{1}{F} \int_0^t i \, dt. \tag{1}$$

The constant of proportionality, F, has of course been named the Faraday constant and its value is the subject of this presentation.

At the last meeting (AMCO-5), preliminary results were reported on the determination of the Faraday constant by way of coulometric titration of 4-aminopyridine performed at Iowa State University [1]. Additional titrations of the same material have since been conducted at the National Bureau of Standards, taking advantage of improved standards of mass, time, voltage and resistance [2]. A reappraisal of the error analysis has also been undertaken in light of further evidence and is reported herein.

Detailed accounts of experimental procedures, instrumentation and data treatment are well documented [2-5] and need not be repeated here. However a brief review of the experiment may prove helpful to the following discussion.

[*]Contribution of the National Bureau of Standards. Not subject to copyright.

EXPERIMENTAL

4-Aminopyridine is a white, crystalline, non-hygroscopic, mono-protic organic base. It can be purified by a sublimation process thus obviating the most common and persistent impurity, occluded solvent. Its melting point (c. 159 °C) facilitates use of the freez-ing point depression method as a means of establishing its purity. Its chemical composition ($C_5H_6N_2$) yields an equivalent weight of approximately 94g, the value of which can be determined exactly in a particular sample by measuring the isotopic abundance ratios of the three elements, carbon, hydrogen and nitrogen.

The principle of the experiment is to titrate (i.e., react stoichiometric amounts of acid and base) a weighed sample of 4-amino-pyridine with electrogenerated reagent, monitoring the current and time required for complete neutralization. The experiment consisted of two sets of titrations. One set involved the direct titration of 4-aminopyridine with electrogenerated hydrogen ion produced at the hydrazine-platinum anode [5]. The other set involved the determina-tion of the titer of a dilute perchloric acid solution by coulometric titration with cathodically generated hydroxide ion, the addition of a slight excess of this solution to a weighed sample of 4-aminopyridine and the return to the equivalence (neutral) point, again by cathodic generation of hydroxide ion. The results of the titrations conducted at NBS are shown in Table 1, where σ represents standard deviation, σ_m — standard deviation of the mean, and ppm — parts per million.

ERROR ANALYSIS

The estimates of the systematic uncertainties in the experiment at the 1σ level are outlined in Table 2. The Weston cells and standard resistors were calibrated versus the national working standards at NBS; the electronic counter used in the measurement of time utilized the NBS in-house 10 kHz standard frequency. The uncertainties associated with these three components approach their minimum realizable limits and contribute negligibly to the total uncertainty.

The single-piece, stainless steel weights used in the experiment were calibrated with NBS working standards, as well as with a specially constructed set of platinum weights (for use in an absolute volt experiment) which had been calibrated at the International Bureau of Weights and Measures. Hence, the uncertainty attributed to the value of the weights is quite small (0.5 ppm). The predominant factor in the error estimate in the mass measurement is the uncertainty in the density of 4-aminopyridine (2.7 parts per thousand) which adversely affects the correction to mass for air-buoyancy.

Table 1. Electrochemical Equivalent of 4-Aminopyridine

	Cathodic Titrations mg/C_{NBS}	Anodic Titrations mg/C_{NBS}
	0.9754436	0.9754330
	0.9754353	0.9754442
	0.9754523	0.9754414
	0.9754363	0.9754473
	0.9754471	0.9754368
	0.9754379	0.9754467
Mean	0.9754421	0.9754416
σ	0.0000068	0.0000057
σ_m	0.0000028 (2.8 ppm)	0.0000023 (2.4 ppm)

Table 2. Error Analysis

Uncertainties Common to Both Sets (ppm)

Volt	0.2
Resistance	0.2
Time	0.2
Mass	3
Purity	10
Subtotal (RSS)	10.4

Uncertainties Unique to Each Set (ppm)

	Cathodic	Anodic
Equivalence Point	3	7
Titer of Acid	2.8	–
Random (Table 1)	8	2.4
Subtotal (RSS)	5.0	7.4

The uncertainty in the purity of 4-aminopyridine is based on the freezing-point-depression work of Kroeger et. al. [6]. The reported value (10 ppm) is, in fact, a lower limit and should not be revised downward by "probable" corrections as had been done previously [1].

The uncertainty listed as "equivalence point" is related to the ability to effect and detect the completion of the acid-base neutralization reaction. The titer-of-acid uncertainty is the random error (standard deviation of the mean) of six coulometric titrations of the dilute perchloric acid solution used subsequently in the "cathodic" titrations of 4-aminopyridine.

The average of the cathodic and anodic titrations (weighted according to the "unique" uncertainties) yields as the electrochemical equivalent of 4-aminopyridine 0.9754419 mg/C_{NBS}. Combining the "unique" and "common" uncertainties by the root-sum-square method results in a total estimated uncertainty of 11.2 parts per million.

MOLECULAR WEIGHT

In order to calculate an accurate value of the Faraday constant from the electrochemical equivalent of 4-aminopyridine, an accurate value of the compound's molecular weight is required. Since the isotopic abundance ratios of the elements comprising the compound can vary considerably in nature, it was deemed insufficient for work of this caliber to rely on compiled atomic weight tables, the values of which reflect this variability. A sample of the same lot of 4-aminopyridine as used in the experiment has recently been isotopically analyzed for carbon and hydrogen by I. Friedman and J. Gleason of the United States Geological Survey, and for nitrogen by I. Kaplan at the University of California, Los Angeles. Their results have been compiled by I. L. Barnes of the National Bureau of Standards and are summarized in Table 3 [7]. The values of the individual isotopes used in the calculations are those tabulated by Wapstra and Bos [8]. Based on these analyses, the molecular weight of 4-aminopyridine is 94.11504 with an estimated uncertainty of less than 0.1 ppm. This represents a substantial decrease (21 ppm) in the value of the molecular weight previously used [1,2,3], 94.11702, which was based on the 1961 atomic weight tables.

Table 3.   Isotopic Analysis of 4-Aminopyridine

Hydrogen

Absolute value of $^2H/^1H$ in SMOW = 0.00015576 [9]

$^2H/^1H$ 4-AP vs SMOW = -14.2 ± 0.1% = 0.00013365

^1H 99.986637 x 1.007825037

^2H 0.013363 x 2.014101787

Atomic Weight of hydrogen in 4-AP = 1.0079595

Carbon

Value of $^{13}C/^{12}C$ in PDB = 0.0112372 [10]

$^{13}C/^{12}C$ 4-AP vs PDB = -3.44 ± 0.01% = 0.0108506

^{12}C 98.92658 x 12.

^{13}C 1.07342 x 13.003354839

Atomic Weight of carbon in 4-AP = 12.01077

Nitrogen

Absolute Value of $^{15}N/^{14}N$ in air = 0.00367344 [11]

$^{15}N/^{14}N$ 4-AP vs air = -0.196 ± 0.025% = 0.00366624

^{14}N 99.6347 x 14.003074008

^{15}N 0.3653 x 15.000108978

Atomic Weight of nitrogen in 4-AP = 14.006718

CONCLUSION

Combining the electrochemcial equivalent of 4-aminopyridine, its molecular weight, and the estimate of uncertainty, the following value for the electrochemical Faraday constant is advanced,

$$F = 96,484.52 ± 1.08\ A_{NBS} \cdot s \cdot mol^{-1},$$

where A_{NBS} is the NBS as-maintained unit of current as it existed at the time of the experiment (May, 1975).

ACKNOWLEDGEMENTS

The author wishes to acknowledge and express grateful appreciation to V. E. Bower, R. S. Davis, B. N. Taylor, G. Marinenko, and I. L. Barnes for many meaningful discussions; to I. Friedman, J. Gleason, and I. Kaplan for determining the isotopic ratios in the sample; and finally to H. Diehl whose perseverance led this experiment to its conclusion.

REFERENCES

[1] H. Diehl, in Atomic Masses and Fundamental Constants 5, Ed. by
    J. H. Sanders and A. H. Wapstra (Plenum Press, N. Y., 1976),
    p. 584.
[2] W. F. Koch and H. Diehl, Talanta 23, 509 (1976).
[3] W. F. Koch, W. C. Hoyle and H. Diehl, Talanta 22, 717 (1975).
[4] W. F. Koch, D. P. Poe and H. Diehl, Talanta 22, 609 (1975).
[5] W. C. Hoyle, W. F. Koch and H. Diehl, Talanta 22, 649 (1975).
[6] F. R. Kroeger, C. A. Swenson, W. C. Hoyle and H. Diehl,
    Talanta 22, 641 (1975).
[7] I. L. Barnes, NBS, Private Communication.
[8] A. H. Wapstra, and K. Bos, Atomic Data and Nuclear Data Tables
    19-3, 175 (1977).
[9] R. Hagemann, G. Nief, and E. Roth, Tellus 22, 712 (1970).
[10] H. Craig, Geochim. et Cosmochim Acta 12, 133 (1957).
[11] G. Junk, and H. J. Svec, Geochim et Cosmochim. Acta 14, 234
    (1958).

HIGH RESOLUTION PENNING TRAP AS A PRECISION MASS-RATIO SPECTROMETER[*]

Robert S. Van Dyck, Jr., Paul B. Schwinberg and
Samuel H. Bailey

Physics Department
University of Washington
Seattle, Washington 98195

INTRODUCTION

The compensated Penning trap[1] with characteristic dimensions
R_O and Z_O is a D.C. electric potential cage located in a strong
axial magnetic field and as such is an ideal apparatus for isolat-
ing long lived charged particles, nearly at rest (thermalized by a
cryogenic environment) and essentially free of other-particle interactions (due to the ultra-high vacuum used, $<<10^{-10}$ Torr). The
ability to determine mass arises from the inverse mass dependence
of the various motions in the trap. The first such motion is the
axial electric resonance (motion along axis of symmetry) at

$$\nu_z = (1/2\pi)(eV_o/mZ_o^2)^{\frac{1}{2}}$$

which is often used to monitor the presence of the trapped charge,
but is only appropriate for very low precision mass measurements
since the effective applied D.C. potential, V_o, will often depend on
parameters which are difficult to control (i.e. thermal emf's, contact potentials, patch effects, time varying dimensions, dielectric
polarization, unstable power supply, etc.). In contrast to the axial
resonance, the magnetic cyclotron frequency at

$$\nu_c = (1/2\pi)(eB_o/mc)$$

is ideally suited for high precision measurements of mass since the
effective magnetic field at the charge site, B_o, can be controlled
to a large extent by using electrode material that is highly non-
magnetic and shaped with rotational symmetry to minimize any residual
diamagnetism. In addition, modern superconducting technology can

yield magnets which are highly uniform (1 part in 10^7 over 1-cm) and very stable (drift rate less than 1 part in 10^9/hr).

The main advantage of this device is that ion-masses can be compared to the mass of a single electron whose cyclotron frequency can be measured quite accurately (a few parts in 10^9). Accordingly, both ions and electrons will be alternately measured in essentially the same fields and in practically the same small volume of space ($\lesssim 10^{-6}$ cm^3).

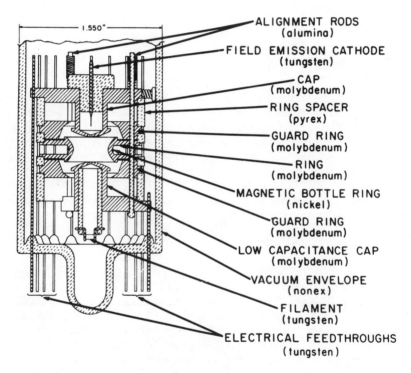

Fig. 1.   Five element compensated Penning trap. The endcaps and
ring form the basic hyperbolic electrode surfaces with
guards of less critical surface geometry to compensate
for truncation of the main electrodes. Other features
shown include a field emission point for electron emission
and a nickel ring to produce a weak magnetic bottle.

EXPERIMENTAL APPARATUS AND MOTION OF CHARGES

   A typical 5-element compensated Penning trap is shown in Fig.
1. Hyperboloids of revolution about the axis of symmetry form the
the trapping electrodes. This geometry produces a quadratic potential distribution when a D.C. potential difference is applied
between endcaps and ring. However, machining and alignment constraints, in addition to the truncation of the surfaces, give rise
to higher order terms. Thus, by allowing some variation of the
curvature of the trapping potential, fourth and fifth electrodes
are able to reduce the influence of these realistic limitations.
Indeed, by monitoring linewidths of the driven axial (electron)
resonance, the guards have led to better than a hundred fold reduction in the higher order terms.[1] The importance of this reduction cannot be emphasized enough, since initial identification
of a trapped ion species requires that the majority of the charges
respond to the same axial frequency, either in absorption or
emission, within the available detection bandpass.

   The basic experiment is shown schematically in Fig. 2. For
negative charges, the ring is positive relative to the grounded
endcaps and reversed for positive charges. This choice allows only
one sign of charge to be harmonically bound axially, with radial
confinement arising from a strong axial magnetic field which then
generates a fast cyclotron rotation in the radial plane. However,
the axial electric field requires the existence of a radial electric
field which is then crossed with the applied axial magnetic field.
The result is the third basic motion in a Penning trap: a slow magnetron drift of the center of the cyclotron orbit around the axis
of symmetry. The effect of this slow rotation is to yield an observed cyclotron frequency ν_c' shifted down by the magnetron frequency ν_m:

$$\nu_c' = \nu_c - \nu_m$$

where ν_c is the free cyclotron frequency. Fortunately, the equation
of motion of a charge in such a trap[2] requires that

$$\nu_m = \nu_z^2/2\nu_c'$$

where ν_z is the observed axial frequency of oscillation. Thus, a
correction frequency $\nu_c - \nu_c'$ can be computed from observed axial and
cyclotron frequencies to yield the free cyclotron resonate frequency.

   A direct measurement of this correction frequency on a single
stored electron can be made[2] and agrees with the calculated magnetron
frequency to an accuracy of 1×10^{-4}. This deviation poses a
possible degree of uncertainty of approximately 1 part in 10^{11} for
the electron and 2 parts in 10^8 for an ion, assuming identical (but

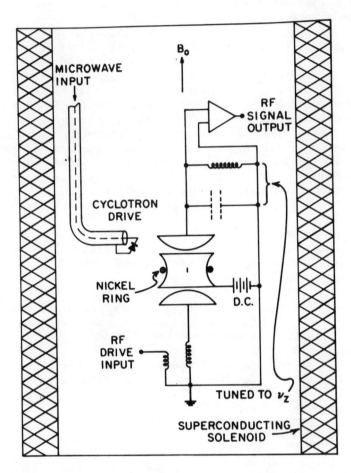

Fig. 2.   Schematic of Experiment. The ring electrode is D.C. biased
           to trap axially and the magnetic field provided by the
           solenoid produces radial binding. R.F. excitation is
           established from a transformer on one endcap and micro-
           wave excitation is obtained via radiation from a Schottky
           diode. Axial detection is accomplished by tuning out the
           intrinsic trap capacity with an external inductor, thus
           producing a high impedance load for the following
           amplifier.

typical) trap parameters. Actually, an investigation of the causes
of this discrepancy leads one to believe that a better alignment of
magnetic and electric axes of symmetry could improve the degree of
agreement. In fact, with this alignment in mind, a recently built
positron/electron trap reduces the discrepancy to $\sim$ 1 part in 10^5.

MODES OF DETECTION

     The axial detection of stored charge arises from image currents
induced in the endcaps by the driven motion which subsequently
develop a small R.F. voltage across a parallel LC circuit, tuned
to the axial frequency (see Fig. 2). The present noise limitations
will allow a few nanovolts of signal to be observed using narrow
band synchronous detection. Direct radial detection of the
rotating current will be achieved by splitting the ring (along perpendicular axial planes) into four equal quadrants, as shown in Fig.
3. Induced current will be synchronously detected in one pair of
opposite quadrants, while driving the harmonic rotation using
balanced drive applied to the second pair of opposite quadrants.

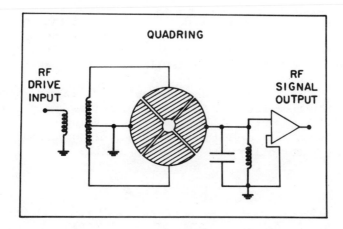

Fig. 3.   Radial detection of ion-cyclotron motion. One pair of
           opposite quadrants will be used for (symmetrical) balanced
           drive and the other pair will be used for synchronous de-
           tection of the driven cyclotron motion (in analogy to
           axial detection).

However, in the case of the electron, the cyclotron frequency
is far too high for direct R.F. detection (>100 GHz). Because of
this, a symmetrically placed loop of nickel wire is put around the
ring electrode (see Fig. 2) in order to produce a weak magnetic
bottle.[2] The effect of this bottle is to establish an axial magnetic field of the form

$$B_z = B_o + B_2 z^2 - B_2 r^2/2$$

which has the effect of producing a weak coupling to the axial
electric resonance. The magnitude of the second order gradient,
$2B_2$, is chosen such that a single quantum of change in cyclotron
energy will give rise to a 1-Hz shift in the axial frequency. This
perturbation is on the order of 2 parts in 10^8 and demonstrates the
importance of improving axial resolution to this order by means of
guard-compensation and improved R.F. amplification techniques.

MAGNETIC RESOLUTION

An example of the potential resolution of the Penning trap
spectrometer is shown in Fig. 4. An electron is stored at 4°K
ambient in a 51 kG magnetic field and its axial resonance (in a 5-
volt deep well) is closely monitored to detect frequency shifts.
Actually, a feedback loop is used in conjunction with synchronous
detection to keep the electron locked to a stable oscillator. Correction voltage is thus monitored to see the effect of accumulated
axial shifts induced by applying an appropriate microwave signal
to the multiplier diode shown in Fig. 2. The result (shown in Fig.
4) is a driven cyclotron resonance with an exponential Boltzmann
tail. The sharp onset corresponds to $Z_{rms} = 0$ in the bottle broadened magnetic field. The tail corresponds to the thermal distribution of axial states in equilibrium with a finite temperature
reservoir (which is the axial preamplifier in this case).

At present, using a somewhat unstable magnet, the sharp feature
is reproducible to a few parts in 10^9 over a time span of a few
minutes. Better magnets will improve this resolution in time.
Linewidths, taken as 1/e of the peak value, are governed primarily
by the bottle interacting with the axial motion:

$$\Delta \nu_c / \nu_c = (B_2 Z_o^2/B_o)(kT/eV_o) \quad .$$

Thus, for a typical bottle with $B_2 = 120$ G/cm^2, $B_o = 51$ kG, $T = 4^{\circ}$K,
$V_o = 10$ volts, and $Z_o = 0.335$ cm, it follows that $\Delta \nu_c / \nu_c = 1 \times 10^{-8}$.
As one can see in Fig. 4 for the electron cyclotron resonance,
$\Delta \nu_c \simeq 6$ kHz yielding $\Delta \nu_c / \nu_c \sim 4 \times 10^{-8}$, which suggests that the
electron is possibly thermalized to a warm amplifier input effect-
ively at $\sim 16^{\circ}$K. Such resonances have recently been obtained alternately with one positron and one electron[3] to yield the pre-

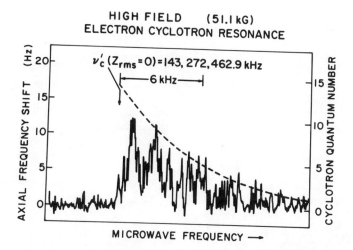

Fig. 4.   Electron cyclotron resonance. The presence of the magnetic bottle is manifested in the shape of this magnetic
resonance. The sharp edge arises from Z_{rms} = 0 and the
tail is determined by the thermal Boltzmann distribution
of axial states. The dashed line is an exponential fit
to the tail with a 1/e linewidth of 6 kHz.

liminary mass ratio (assuming identical charges) $m(e^+)/m(e^-)$ =
$1(\pm1.3\times10^{-7})$ where field drift accounts for the present uncertainty.

As indicated, linewidths of magnetic resonances are governed
by the persistent magnetic bottle, required only for detection of
the electron cyclotron resonance. However, a recent proposal[4] made
by our group suggests that it may be possible to turn off the bottle
during excitations and turn it on only during detection. Such an
alternating scheme may be made possible because of the use of a
small superconducting loop of wire (in place of the nickel loop),
which, as the secondary of a current transformer, will exclude
changes in the total flux through its loop area, thus producing the
usual bottle (dipole) field. Estimates suggest that a multiturn
primary outside the trap's vacuum envelope could achieve $\sim$ 10 Amps

in the superconducting loop, which would produce a bottle comparable
to the present ones. With the use of such a variable bottle, the
relative magnetic linewidths should be less than 10^{-9}, limited by
the residual inhomogeneity and instability of the superconducting
solenoid.

SEEKING A MAGNETIC SADDLE POINT

For the present time, a fixed bottle remains a reality which
may give rise to a possible systematic error associated with doing
separate electron and ion experiments. In particular, the bottle
produces a positional dependence. However, the need to verify
that each charge type "sees" the same magnetic field is clear from
the required cancellation of B in the cyclotron ratio. In order
to achieve this requirement, the special quadring trap is utilized
again. It is proposed that a small D.C. offset bias voltage, ΔV,
be applied asymmetrically to either a pair of opposite quadrants
or the two endcaps in order to provide a first order constant
electric field which effectively shifts the electric saddle point
by Δx:

$$\Delta x = (D/4)(\Delta V/V_o)$$

where D is either the minimum ring diameter, $2R_O$, or the endcap
spacing, $2Z_O$, depending on which pair of electrodes receive $\pm \Delta V/2$
offset voltage.

It is always possible that some residual (versus common) contact potential, on the endcaps for instance, might give rise to a
systematic positional offset, Δx, that then yields a shift in the
cyclotron frequency of each of the charges in the presence of the
magnetic bottle. If the magnetic bottle is exactly centered at the
geometric center of the trap, such shifts will be identical in
magnitude and sign for both charge types (assuming a simple reversal
of applied potential), thus not yielding a systematic error. However, if one assumes that the magnetic bottle is offset axially
by Z_m, then a net cyclotron frequency shift, $\delta\nu_c$, can occur:

$$\delta\nu_c/\nu_c = 4B_2 Z_m \Delta x/B_o .$$

The magnitude of the residual contact potential must then be on the
order of

$$\Delta V/V_o = (B_o/2B_2 Z_o Z_m)(\delta\nu_c/\nu_c)$$

to account for this shift. Due to careful construction of the magnetic bottle in the ring electrode, a worst case uncertainty in the
location of the magnetic center is believed to be ± 0.05 mm. Thus,
using the typical trap parameters given previously, the residual

contact potential needs to be 125 mV to account for a relative
cyclotron frequency shift of 10^{-7}. Such a large residual contact
potential is unlikely in view of experimental data[2] that has determined the common contact potential to be 60 mV or less. Hence, this
possible systematic error is not important in the preliminary e^+/e^-
mass ratio.

Nevertheless, to achieve greater accuracy, the 8-element trap
will be used to plot cyclotron resonances versus D.C. offset in the
three independent directions. This plot effectively yields the
saddle point in the axial magnetic field, which will be the same
for each charge type (i.e., function only of geometry). Thus, the
ratio $\nu_c(\text{ion})/\nu_c(e^-)$ can be taken at this saddle point to guarantee
exact cancellation of B (assuming the sharp edge features are used
as a measure of ν_c). Note, this technique does require the charges
to be dynamically centered in the trap radially via $\nu_z + \nu_m$ side-
band excitation as described elsewhere.[2] Hence, using centered
offset adjust conditions, a magnetron orbit whose radius is as
large as 0.03 mm can give rise to no more than a 1×10^{-8} absolute
shift in ν_c from center to r_m (again with trap parameters typical
of those used in the e^+/e^- mass ratio experiment).

SEARCH FOR PROTONS

As of this writing, results have been limited primarily to the
e^+/e^- mass ratio as a representative experiment of things to come.
The search for single protons has proven experimentally to be more
difficult than expected due to a design flaw in the present quadring
trap. This flaw prevents the guards from compensating the anharmonic
terms in a three times smaller version of the present Penning traps.
Thus, due to their extremely narrow resonances (resulting from their
heavier mass) and some difficulty achieving the desired resolution
in this anharmonic trap, driven proton signals have been observed
only for short times during the loading process. Upon converting
the old single electron experiment with an amplifier tuned to the
proton axial resonance, large numbers of stored protons ($\sim 10^4$-10^5)
have been synchronously detected and centered using $\nu_z + \nu_m$ side-
band excitation for more than 10 hours at a temperature $\geq 200^\circ$K.
Thus, with this success and the knowledge gained from the old quadring trap, work is now proceeding onto a new version of the quadring trap, and it is expected that the goal of obtaining single ions
nearly at rest in free space will be achieved very soon.

*Research supported in part by a U.S. National Bureau of Standards
Precision Measurement Grant and in part by the National Science
Foundation.

REFERENCES

1.   R. S. Van Dyck, Jr., D. J. Wineland, P. A. Ekstrom, and
        H. G. Dehmelt, High Resolution with a New Variable
        Anharmonicity Penning Trap, Appl. Phys. Letters 28,
        446 (1976).

2.   Robert S. Van Dyck, Jr., Paul B. Schwinberg, and Hans G.
        Dehmelt, Electron Magnetic Moment from Geonium Spectra,
        in "New Frontiers in High-Energy Physics," B. Kursunoglu,
        A. Perlmutter, and L. F. Scott, eds., Plenum Press,
        New York (1978).

3.   P. B. Schwinberg, R. S. Van Dyck, Jr., and H. G. Dehmelt,
        Preliminary Positron/Electron Mass Ratio Measurement
        in a Compensated Penning Trap, to be published in Bull.
        Am. Phys. Soc.

4.   P. B. Schwinberg, R. S. Van Dyck, Jr., and H. G. Dehmelt, A
        Variable Magnetic Bottle for Geonium Measurements, to be
        published in Bull. Am. Phys. Soc.

MAGNETIC MOMENTS OF ELECTRONS AND POSITRONS[*]

David E. Newman, Eric Sweetman, Ralph S. Conti, and
Arthur Rich

Randall Laboratory of Physics
University of Michigan
Ann Arbor, Michigan 48109

INTRODUCTION

In this report we describe a number of new features in our precision experiments currently in progress to measure the anomalous
magnetic moments of the electron and positron, and we will point out
the impact these improvements will have in testing several fundamental theories in physics. In the electron experiment, reduction
in both statistical and systematic errors will result in a comparison
with QED predictions at the level of the eighth-order term in the
expansion of the anomaly in powers of α. In addition, by varying
the energy of the electrons we can test special relativity to an
accuracy of 2×10^{-11}. The positron experiment will test CPT by
comparing the electron and positron magnetic moments.

THEORY

The g-factor is defined as the proportionality constant relating
the spin and the magnetic moment:

$$\vec{\mu} = (g/2)(e/mc)\vec{S}$$

The Dirac theory, with g=2, is corrected by quantum electrodynamics
which predicts that $g = 2(1+a)$. The anomaly "a", of order 10^{-3}, is
given as a power series in the fine structure constant α:

---

[*]These experiments are supported by the National Science Foundation
under grant PHY77-26037.

$$a \equiv \frac{g-2}{2} = A(\frac{\alpha}{\pi}) + B(\frac{\alpha}{\pi})^2 + C(\frac{\alpha}{\pi})^3 + D(\frac{\alpha}{\pi})^4 + \ldots.$$

+ weak interaction effects + muonic corrections

+ hadronic corrections.

The coefficients A, B and C are respectively 0.5, -0.328478, and 1.181(10). The prediction is currently limited to an accuracy of 108 ppb (parts in 10^9) by the uncertainty in C.[1] The coefficient D is completely unknown, but if it is comparable to C then this term will contribute about 30 ppb. The value of α is known[2] to 110 ppb. Weak, muonic, and hadronic effects are expected to be 3 ppb or less with an uncertainty of about 0.3 ppb.

We measure the anomaly by observing polarized electrons or positrons trapped in a magnetic well.[3] The trapped particles exe-cute cyclotron orbits at $\omega_C = eB/\gamma mc$, and their spins precess at $\omega_S = (eB/mc)(a + 1/\gamma)$. By observing the difference frequency ω_D between the spin and cyclotron frequencies, we obtain the anomaly directly: $\omega_D = aeB/mc$. These measurements also test special relativity[4] by precisely measuring Thomas precession, as can be seen by rewriting the spin equation as $\omega_S = (g/2)(eB/mc)+(1-\gamma)(eB/\gamma mc)$. The second term represents Thomas precession, a purely relativistic kinematic effect resulting from the particle's acceleration. Comparison of data taken at dif-ferent energies thus permits us to test special relativity at an accuracy 1000 times better than the measurement of ω_D, independently of all QED calculations or measurements of α! The results (if null) may also be interpreted as confirming to high precision the form of the interactions of particles and their spins with magnetic fields, and will rule out any effects of high acceleration. We also expect to set limits on a possible periodic potential in the structure of space.[4]

Finally, the T-violating electric dipole moment of the electron or positron could be measured in these experiments by comparing data taken at different magnetic fields. The magnetic field appears as an electric field in the frame of the electron. The precession resulting from the interaction between a possible electric dipole moment and this motional electric field would thus affect the observed value of ω_D.

THE POSITRON EXPERIMENT

The positron g-factor experiment is constructed and we are expecting to take data soon. As shown in Fig. 1, 1 MeV polarized positrons from ^{68}Ga decay pass through a collimator and a novel "pitch filter", consisting of thin, helical tungsten windings in the source collimator. This can be pulsed to -1 kV to pass positrons

with the right energy and angle to be trapped, or +1 kV to stop
essentially all of the positrons, thus virtually eliminating the
primary backgrounds seen in previous experiments. They are admitted
in short pulses into a shallow (10 to 100 ppm) magnetic well and are
trapped by a retarding electrostatic pulse applied to one of the
cylinders. While in the trap, they are subjected to a longitudinal
rf electric field which transforms, in part, into a radial magnetic
field in the frame of the positrons. When $\omega_{rf} = \omega_D$, this signal
resonantly rotates the positron spins into the $\pm Z$ direction.[5] They
are then ejected from the trap and the Z-component of their polarization is measured in the polarimeter.[6] The phase of the driving rf,
and thus the final polarization direction, can be alternated auto-
matically, thereby cancelling out numerous systematic drifts.

In the current run, using a relatively weak initial source of
3 mCi of ^{68}Ga, we expect to be limited by statistics to about 20 ppm
accuracy. However, we have developed a technique for electroplating
sources of up to 100 mCi of ^{68}Ga on a 3 mm^2 spot. This could permit
us to measure the positron anomaly to 3 ppm or better. Such a
measurement would constitute a test of CPT to an accuracy of 3×10^{-9}
and will push the relativity test to $\beta = 0.9$ with this same accuracy.

THE ELECTRON EXPERIMENT

The electron g-factor experiment (Fig. 2) will use a new pulsed
source of polarized electrons based on the Fano effect[7] in a Cs
atomic beam. In this source, electrons with 90% polarization are
produced from atomic Cs by a short laser pulse and are then accelerated to 100 keV. We plan to use a novel injection scheme in which
the electrons are focused into the central well by the fringing
field of the solenoid, with trapping at the well by an electrostatic
pulse. This will permit the use of shallower wells (1 ppm) with a
much higher trapping efficiency (10^{-3}) than previously possible.
The wells will consist of superposed electric and magnetic fields
to minimize systematic errors arising from such effects as stray
electric field the longitudinal velocity of the trapped particles,
and the different time-averaged magnetic fields seen by different
particles in the well. The trapping time of 10 to 100 msec is expected to be limited by vacuum and magnetic field inhomogeneities.
After the trapping period, the electrons will be ejected from the
well and the solenoid, spin rotated by 90° using a Wien filter, and
then polarization analyzed by Mott scattering.

Anticipated data are shown in Fig. 3. Here ω_D is the frequency
at which the spin rotates relative to the electron velocity, and
thus appears directly as the frequency of the final polarization
oscillations. Note that ω_D is measured by fitting a sinusoidal
curve to the data taken with short and long trapping times; thus
many systematic errors associated with injection, ejection, source

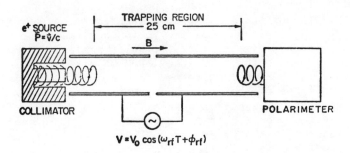

Fig. 1. Apparatus for the positron g-2 experiment.

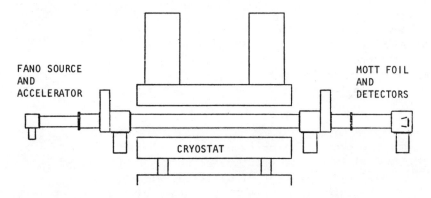

Fig. 2. Apparatus for the electron g-2 experiment.

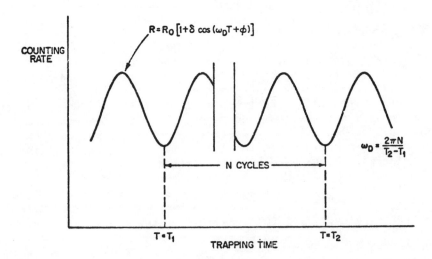

Fig. 3. Expected electron g-2 data.

characteristics, and detection efficiencies cancel out. Further cancellations can be obtained by using the same time base for the trapping interval and for the NMR measurement of the magnetic field.

The experiment is being designed to achieve an accuracy of 20 ppb or better, with the largest error (11 ppb) coming from the calibration[8] which relates NMR frequency to absolute magnetic field. With this result we expect to test QED at a level comparable to the eighth-order term, and to test special relativity with an accuracy of 1 to 2×10^{-11} at $\beta = 0.5$ to 0.7. This is two orders of magnitude more precise than previous tests[4] of special relativity at significant velocities. In addition, it will allow a limit of order 10^{-18} e-cm to be placed on the electric dipole moment of the electron. This is less precise than electric dipole moment measurements using atomic electrons, which have reached a limit[9] of 10^{-24} e-cm but it has the esthetic advantage of being performed directly on a free, rather than bound, particle.

REFERENCES

1. M.J. Levine, E. Remeddi, and R. Roskies, to be published in Phys. Rev.
2. P.T. Olsen and E.R. Williams, Proc. of the Fifth International Conference on Atomic Masses and Fundamental Constants, J.H. Sanders and A.H. Wapstra, eds. (Plenum Press, New York, 1976).
3. J. Wesley and A. Rich, Rev. Mod. Phys. 44, 250 (1972).
4. D. Newman, G.W. Ford, A. Rich and E. Sweetman, Phys. Rev. Lett. 40, 1355 (1978).
5. G.W. Ford, J. Luxon, A. Rich, J. Wesley, and V. Telegdi, Phys. Rev. Lett. 29, 1691 (1972).
6. A complete description of the latest version of our positron polarimeter may be found in G. Gerber, D. Newman, A. Rich, and E. Sweetman, Phys. Rev. D15, 1189 (1977).
7. W.V. Drachenfels, U.T. Koch, R.D. Leppner, T.M. Müller, and W. Paul, Z. Physik 269, 387 (1974).
8. W.D. Phillips, W.E. Cooke, and W. Kleppner, Phys. Rev. Lett. 35, 1619 (1975).
9. M.A. Player and P.G.H. Sandars, J. Phys. B 3, 1620 (1970).

COMPARISON OF THE K X-RAY ENERGY RATIOS OF HIGH Z AND LOW Z ELEMENTS WITH RELATIVISTIC SCF DF CALCULATIONS

G.L. Borchert, P.G. Hansen, B. Jonson and H.L. Ravn

The ISOLDE Collaboration, CERN
Geneva, Switzerland

J.P. Desclaux

Centre d'Etudes Nucléaires de Grenoble, LIH
Grenoble, France

INTRODUCTION

Todays atomic theory allows the transition energies of atomic K X-rays to be calculated with high precision. As they have sizeable contributions from terms of fundamental importance precise experimental data in particular for the heavy elements [1,2,3] are very much in demand. In the present contribution we report an experimental and theoretical study of ratios of K X-ray energies of high Z and low Z elements. With the method used the results become independent of the absolute energy scale.

EXPERIMENTAL TECHNIQUES

The only experimental method to determine X-ray energies between 20 keV and 120 keV with a precision in the ppm range is the use of a crystal spectrometer. With such an instrument usually the largest contribution to the uncertainty is caused by the reproducibility of the angular setting and its linearity within the angular range required for the independent measurement of the reference line. The reported measurements are based on a new technique [4] in which the energies of pairs of overlapping lines, measured with a bent crystal spectrometer are compared with a precision of 10^{-3} of the natural line width or better. This is possible because of the special source arrangement which allows a measurement of 2 independent sources at essentially the same time for each spectrometer setting. Its principle is shown in fig. 1. The technique was originally developed in order to permit comparisons of X-rays of the same element but following different excitation mechanisms [5,6,7]. In order to apply the method to different elements we have exploited the fact that

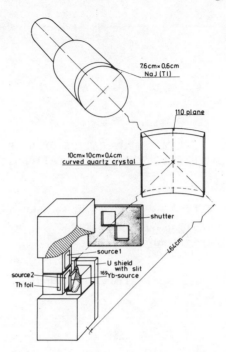

Fig. 1.   Schematic view of the source arrangement of the crystal
          spectrometer. The photo-ionization sources are mounted on
          Al cubes which can be adjusted in the dispersion direction.
          Two sources of 60 Ci ^{169}Yb are used for the excitation and
          a uranium shutter allows only the X-rays from one source at
          a time to reach the crystal. A Th foil placed behind the
          source absorbs some of the gamma rays that have passed
          through the source and the re-emitted Th X-rays serve to
          increase the fluorescence efficiency.

certain pairs of K X-rays from a light and a heavy element by
accident overlap almost exactly when measured in different orders
n of Bragg reflection. Thus our experiment determines the small
difference

$$\Delta = E_1/n_1 - E_2/n_2$$

almost independently of the accuracy of the angular setting and
linearity. The main contribution to the uncertainty is then given
only by counting statistics.

THE THEORETICAL CALCULATION

     We performed the calculations with a relativistic Dirac-Fock
programme [8]. The X-ray energies are determined as the difference
of the total energies of the final and the initial state thus inclu-

ding full relaxation. To achieve self consistency by minimizing the
total energy the electron-electron interaction is restricted to the
instantaneous Coulomb repulsion while the kinematics and the interaction with the nuclear field are represented by the single particle
Dirac Hamiltonian. After self consistency is obtained the expectation
value of the Breit operator, estimates for the Lamb shift, the vacuum
polarization and the self energy are added as first order corrections.
The vacuum polarization is estimated as the expectation value of the
Uehling potential [9]. The self energy is approximated by an interpolation of the hydrogenic results of Mohr [10,11]. Instead of using
the bare nuclear charge we assign to each orbital an effective charge
that is determined such that the expectation values of the orbital
radii for this atom are equal to those of an hydrogenic ion with this
charge. This approximation introduces only a minor error [12]. For the
charge density of the nucleus we use a Fermi distribution. The radius
and skin parameters c and t are taken from standard tables [13]. The
total uncertainty of the results is estimated to be smaller than 1 eV.

MEASUREMENTS AND RESULTS

The pairs of elements studied here are listed in the first
column of table 1. To avoid energy shifts due to chemical structure [14]
we used pure metallic foils as sources. As the creation of K holes by
means of EC or IC can cause energy shifts which are very specific for
the element [5,7,15,16] we induced the X ray transitions by photoexcitation with the photon flux from a 60 Ci ^{169}Yb source. To eliminate
the uncertainties of the different position of the sources we always
measured the positive and negative Bragg reflections. The intensity of
each reflection is recorded at 15 to 20 points across the line. A
typical pair of reflections is shown in fig. 2. These scans are repeated
at least 10 times; then we interchanged the positions of the sources
and repeated the measurement in the same way once more. By averaging
these two data sets we could eliminate a shift due to a small asymetry
in the effective source position with respect to the central crystal
plane [4]. The data are analysed with our standard procedure [4,17] and
the results Δm and the experimental uncertainties are given in col.5 & 6.

As the photoexcitation with significant probability creates
multihole states we expect an energy shift of the reflection due to
unresolved satellites [6]. These shifts were estimated in the same way
as given elsewhere [7] using a frozen orbital calculation for the hole
production probability [18] and a relaxed orbital HF calculation for
the satellite energy [19]. The resultant shift of the center of gravity
of the line depends of course on the range of the reflection that is
used in the analysis.

In a few experiments the metallic exciter foils were composed
of the stable isotopes of an element in the natural abundance. In
these cases we renormalized the measured energy difference to the
isotope that has been calculated by using standard tables [20].

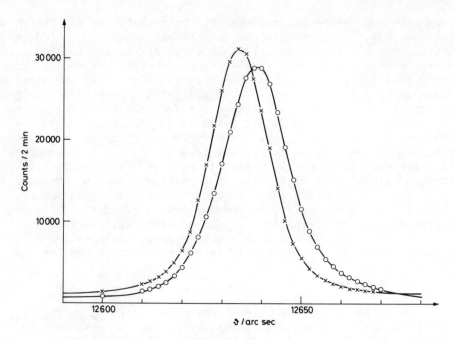

Fig. 2.   Typical scan across the positive Bragg reflection from the
Th-Ho experiment. The second order reflection of the Th $K\alpha_1$
line almost coincides with the first order reflection of
the Ho $K\alpha_2$ line. o means contents of counter 1 measuring
the intensity of the Th $K\alpha_1$. x means contents of counter
2 measuring the intensity of the Ho $K\alpha_2$. The individual
sources were tilted so that the overlapping reflections
had about the same line widths.

In the spectrometer the radiation is affected by various energy
dependent absorption processes. To estimate the effect of the absorption in the source material and in the reflection crystal we use a
linear approximation. The shift due to the energy dependence of the
reflectivity of the quartz crystal is determined by means of a extrapolation of experimental data [21]. Finally the effect of the energy
dependent efficiency of the NaJ detector is found by means of a
linear approximation from standard curves [22].

The sum of these corrections is listed in column 7 of table 1.
After application to the measured data we obtain the final results
given in column 8. They can be compared directly with the theoretical
calculations the results of which are shown in the last column. The
differences between experiment and theory are plotted in fig. 3 as
a function of the charge of the heavier element.

Table 1.

El.	Trans.	Order	Energy[23] keV	Δm meV	Uncert. meV	Corr. meV	Δ exp meV	Δ theo meV
Pt	$K_{\beta 1}$	3	75.750	-22 040	100	180	-22 060	-23 770
Sn	$K_{\alpha 1}$	1	25.271			40		
Pt	$K_{\alpha 1}$	2	66.832	-26 450	80	100	-26 450	-26 000
La	$K_{\alpha 1}$	1	33.442			50		
Bi	$K_{\alpha 2}$	3	74.815	- 3 750	50	100	- 3 680	- 4 080
Ag	$K_{\beta 1}$	1	24.943			100		
Th	$K_{\alpha 1}$	2	93.350	-26 300	80	120	-26 290	-26 375
Ho	$K_{\alpha 2}$	1	46.700			70		
U	$K_{\beta 1}$	2	111.300	-23 530	400	200	-23 490	-24 250
Er	$K_{\beta 1}$	1	55.674			140		

The first two results correspond to the Δ values of the 3rd order reflection of the $K_{\beta 1}$ of Pt compared to the first order of the $K_{\alpha 1}$ of Sn and the 2nd order of the $K_{\alpha 1}$ of Pt compared to the first order of the $K_{\alpha 1}$ of La. For the pair Bi-Ag we measure the 3rd order reflection of the $K_{\alpha 2}$ of Bi and the 1st order of the $K_{\beta 1}$ of Ag. Here the shift Δ has the smallest value, therefore the precision is highest in spite of an additional correction for the $K_{\beta 3}$ line of Ag. In the 4th experiment we measured the difference between the 2nd order reflection of the $K_{\alpha 1}$ of Th and the 1st order of the $K_{\alpha 2}$ of Ho. Finally in the last case we use the 2nd order reflection of the $K_{\beta 1}$ of U overlapping with the 1st order of the $K_{\beta 1}$ of Er. As the $K_{\beta 1}$ line is about a factor of 5 weaker than the $K_{\alpha 1}$ we obtain in this case an uncertainty of about 400 meV.

As can be seen from the table the sum of the corrections is of the order of the experimental uncertainties but it cancels out to a good deal when taking the appropriate differences. If one compares our $(\Delta \text{exp} - \Delta \text{theo})/\Delta Z$ ratios for the individual pairs of X-ray transitions with those of Deslattes et al. [2] our values tend to be smaller. If we try to group the results it seems that the deviations between experiment and theory are larger for the pairs where transitions from the M shell are involved, which is to be expected, as the correlation terms are less well known for the higher electronic shells.

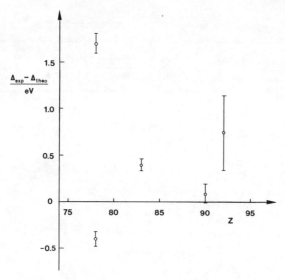

Fig. 3. Difference between the experimental values Δexp and the
        calculated ones Δtheo as a function of Z of the heavier
        element. The errors include only the experimental uncer-
        tainties. The accuracy of the calculated values Δtheo is
        estimated to be less than 1 eV.

        Further investigations of the corrections, especially of the
multihole excitation and the atomic structure effect, as well as
additional measurements are carried out at present. Probably the
final uncertainties will will decrease considerably. Therefore
the presented results should be regarded as preliminary. Especially
the atomic structure effect [7] may be of importance for the present
work. Although it seems to be qualitatively understood for the 4f
metals [7] the situation with respect to 5d metals is less satisfactory [16],[24]). As for the calculation, we assume a free atom whereas
the experiment uses atoms in a metallic lattice. To simulate this
effect we calculated the energy of the $K_{\alpha 1}$ line of Th in the case
where we added an electron to different outer orbitals. The energy
shifts we found were an order of magnitude smaller than the experimental errors. However, the calculation is quite sensitive to
changes of the nuclear shape parameters. In the case of Th a 1%
change of the radius parameter shifts the K-binding energy by about
2 eV. Therefore, further progress must be made before it becomes
possible to exploit fully the precision of our method.

Acknowledgements: We would like to thank Dr. Benenson and the
                   AMCO 6 Conference Committee who made the
                   presentation of this contribution possible
                   by their generous financial support.

REFERENCES

1. M.S. Freedman, F.T. Porter and J.B. Mann, Phys. Rev. Lett. 28, 711 (1972).
2. R.D. Deslattes, E.G. Kessler, L. Jacobs, W. Schwitz, Phys. Lett. 71A, 411 (1979).
3. M.O. Krause and C.W. Nestor, Phys. Scri. 16, 285 (1977).
4. G.L. Borchert, P.G. Hansen, B. Jonson, H.L. Ravn, O.W.B. Schult and P. Tidemand-Petersson, A Crystal Spectrometer for Precision Measurements of Small Energy Shifts, to be published.
5. G.L. Borchert, P.G. Hansen, B. Jonson, H.L. Ravn, O.W.B. Schult and P. Tidemand-Petersson, Phys. Lett. 63A, 15 (1977).
6. G.L. Borchert, P.G. Hansen, B. Jonson, I. Lindgren, H.L. Ravn, O.W.B. Schult and P. Tidemand-Petersson, Phys. Lett. 65A, 297 (1978).
7. G.L. Borchert, P.G. Hansen, B. Jonson, I. Lindgren, H.L. Ravn, O.W.B. Schult and P. Tidemand-Petersson, Phys. Lett. 66A, 374 (1978).
8. J.P. Desclaux, Comp. Phys. Comm. 9, 31 (1975).
9. E.A. Uehling, Phys. Rev. 48, 55 (1935).
10. P.J. Mohr, Ann. Phys. (New York) 88, 52 (1974).
11. P.J. Mohr, Phys. Rev. Lett. 34, 1050 (1975).
12. K.T. Cheng and W.R. Johnson, Phys. Rev. A14, 1943 (1976).
13. R. Engfer, H. Schneuwly, J.L. Vuilleumier, H.K. Walter and A. Zehnder, At. Data Nucl. Data Tables 14, 509 (1974).
14. O.I. Sumbaev, Modern Phys. and Chem. (Academic, London, 1976), Vol. I, p. 31.
15. A.I. Yegorov, A.A. Rodinov, A.S. Rylnikov, A.E. Sovestnoye, O.I. Sumbaev, V.A. Shaburov, Pisma ZhETF 27, 9, 514 (1978).
16. K.C. Wang, A.A. Hahn, F. Boehm, P. Vogel, Phys. Rev. A18, 2580 (1978).
17. P.G. Hansen, Nucl. Instr. Meth. 154, 321 (1978).
18. C.W. Nestor et al., ORNL-4027 (1966).
19. A. Rosén and I. Lindgren, Phys. Rev. 176, 114 (1978).
20. K. Heilig and A. Steudel, ADNDT 14, 614 (1974).
21. J. Cl. Dousse, thesis, Fribourg 1978.
22. Harshaw tables.
23. C.M. Lederer and V.S. Shirley, Table of Isotopes (J. Wiley, New York) 1978.
24. G.L. Borchert, P.G. Hansen, B. Jonson, H.L. Ravn and O.W.B. Schult, to be published.

PRECISION MEASUREMENTS OF THE TRIPLET AND SINGLET POSITRONIUM

DECAY RATES*

D.W. Gidley, A. Rich, and P.W. Zitzewitz

Randall Laboratory of Physics
University of Michigan
Ann Arbor, Michigan 48109

INTRODUCTION

Positronium (Ps) is the hydrogen-like bound state of the electron and its antiparticle, the positron. Positronium is an attractive testing ground for the theory of quantum electrodynamics (QED) since its constituent particles interact to high order only through the electromagnetic interaction. Fundamental tests of QED involving Ps may be classified as measurements of either fine structure in the n=1 or n=2 levels or of annihilation decay properties. After a brief overview, the precision measurement of the triplet and singlet ground state decay rates will be discussed.

With a history of nearly 30 years the measurement of the ground state fine structure has become the most precise Ps test of QED, reaching an accuracy of 6 ppm in 1977.[1] The advantage of a purely leptonic system is evident when this experiment is compared with the hydrogen Lamb shift measurements,[2] which are of comparable precision. The Lamb shift theory[3] must include a 100 ppm correction and a corresponding 6 ppm uncertainty due to the effects of the finite charge distribution of the proton. With the recent observation of the n=2 state of Ps[4] an excited state fine structure measurement[5] (2^3S_1-2^3P_2) has been performed at the 300 ppm level. As tests of charge conjugation invariance in the electromagnetic interactions experiments have searched for the C parity violating decay modes of the singlet state[6] ($^1S_0 \rightarrow 3\gamma$) and of the triplet state[7]

---
*Research carried out in collaboration with D.A.L. Paul and D.H.D. West, and supported by the National Science Foundation.

$(^3S_1 \rightarrow 4\gamma)$. Charge conjugation invariance requires that singlet Ps decay into an even number of photons and triplet Ps into an odd number.

DECAY RATE OF TRIPLET POSITRONIUM

Measurements of the annihilation decay rate, λ_T, of the triplet (1^3S_1) state attracted attention when a 2% discrepancy arose between new measurements[8] and the existing theoretical calculation. In addition the calculation was supported by two independent measurements of λ_T, each at the 0.2% level (see ref. 8). This discrepancy prompted Caswell, Lepage, and Sapirstein[9] to perform a new calculation of the radiative corrections to λ_T which resulted in a 3% decrease in the decay rate. Their most recent result[10] is

$$\lambda_T = \frac{2(\pi^2-9)}{9\pi} \frac{\alpha^6 mc^2}{\hbar} \left[1 - \frac{\alpha}{\pi}(10.266\pm0.011) - \frac{\alpha^2}{3}\ell n(\alpha^{-1}) \right]$$

$$= 7.0386\pm0.0002 \ \mu sec^{-1} \qquad\qquad\qquad Eq.1$$

Even more recently the resultant 1% discrepancy has considerably narrowed as systematic problems in the experiments have been resolved. The presently accepted experimental values are summarized in Table I. The measurements are characterized by the medium in which orthopositronium (o-Ps, the triplet ground state) is formed since the interaction of o-Ps with the surrounding atoms is the major systematic effect.

TABLE I

Summary of the most recent measurements of λ_T

Reference	Ps Medium	$\lambda_T (\mu sec^{-1})$
(11) Michigan	SiO_2 powder	7.067±0.021
(11) Michigan	Gas	7.056±0.007
(12) London	Gas	7.045±0.006
(13) Michigan	Vacuum	7.050±0.013

In the two gas experiments listed in Table I, as well as the earlier measurements of λ_T, positrons from a radioactive source (^{68}Ga or ^{22}Na) are injected into a gas chamber containing typically Freon 12, isobutane, or various gas mixtures. After moderating to an energy of roughly 10 eV the positrons either form Ps or freely annihilate with electrons in the gas or chamber walls. The time

interval between positron emission and the subsequent detection of
the annihilation γ-rays is measured using a time-to-amplitude
converter (TAC) and the resulting time spectrum is stored in a
multichannel analyzer. The time spectrum beyond the prompt peak of
freely annihilating positrons is fitted to a flat background of
random coincidence events and a single exponential with fitted decay
constant taken to be the decay rate of orthopositronium at the given
gas density. The decay rate, increased by collisional quenching
with the gas atoms, is assumed to be linear in the gas density (for
low enough pressures). Extrapolation to zero gas density yields the
vacuum decay rate, λ_T.

    In an attempt to avoid the (1-5)% extrapolation in λ_T of the
gas experiments, decay rates were measured in different samples of
SiO_2 powder.[2] Paulin and Ambrosino[14] had reported decay rates in
such powders within 1-2% of the vacuum value. Although we found
that a 1% extrapolation of λ_T in the density was required for the
lightest powder we could obtain, the high positron stopping power
of the powder yielded time spectra with much higher signal-to-chance
noise ratios than the gas. However, new systematics at the level of
0.1-0.2% arose due to uncertainties in measuring and guaranteeing
the uniformity of the powder density as well as possible physical
effects (such as Stark shifts due to possible charging of the powder
grains).

    The "vacuum" experiment (see Table I) is designed to minimize
the interaction of the o-Ps with its environment. A low energy
($\approx$400 eV) beam of positrons is focused onto the MgO-coated cone of
a channel electron multiplier (CEM). The CEM detects the secondary
electrons expelled by the incident positron as the start signal to
the TAC. The o-Ps formed in the 0.1 mm layer of MgO diffuses into
an evacuated MgO-lined cavity and the annihilation γ-rays are detected as the stop signal. The cavity confines the o-Ps to a region of
uniform γ-ray detection efficiency and the MgO coating minimizes
the probability of collisional pick-off with the cavity walls. The
wall collision rate is estimated to be less than 10 per lifetime
compared with a typical grain collision rate in the powder experiment
of over 10^4 per lifetime. The major systematic uncertainty (0.010
μsec^{-1}) in this experiment is associated with the possibility that
some fraction of the o-Ps may remain trapped or bound in the MgO
layer of the CEM cone.

    The agreement between theory and any one of the experimental
values of λ_T in Table 1 is satisfactory (typically 1σ-2σ). However,
it is disturbing that all the results are uniformly high. A new
measurement of λ_T at the level of 0.002 μsec^{-1} should definitively
resolve any possible remaining discrepancy. Such a measurement is
discussed at the end of the next section.

NEW EXPERIMENT TO MEASURE THE SINGLET DECAY RATE

Since the theoretical and experimental values for the triplet
decay rate have changed so radically in the last four years it is
only natural that interest in the singlet decay rate, λ_S, would be
renewed. The first radiative corrections to λ_S were first calculated
in 1957 by Harris and Brown[15] and have recently been independently
verified by Cung et al.,[16] Tomozawa,[17] and Freeling.[18] The theoretical value is

$$\lambda_S = \frac{\alpha^5}{2} \frac{mc^2}{\hbar} \left[1 - \frac{\alpha}{\pi} \left(5 - \frac{\pi^2}{4} \right) \right]$$

$$= 7.985 \times 10^3 \ \mu sec^{-1}. \qquad\qquad\qquad\qquad\qquad\qquad\qquad Eq.2$$

The experimental situation is much worse. A direct timing
measurement of the 1/8 nsec lifetime would, at present, not be
feasible to the accuracy of interest. The best measurement of λ_S
to date is $\lambda_S = (7.99 \pm 0.11) \times 10^3 \ \mu sec^{-1}$, obtained indirectly from
measurements of the linewidth obtained in the fine structure experi-
ment.[19] This experiment is not sufficiently accurate to check λ_S
to order α.

We are just beginning to acquire data in a new experiment
designed to measure λ_S to 0.2%. The technique is based on the fact
that in a magnetic field, B, the singlet and triplet m=0 states are
mixed to yield perturbed states. The decay rate, λ_T', of the perturbed
triplet state is

$$\lambda_T' = \frac{1}{1+y^2} \lambda_T + \frac{y^2}{1+y^2} \lambda_S , \qquad\qquad\qquad\qquad\qquad Eq.3$$

where $y = x/(1+\sqrt{1+x^2})$ and $x = 0.0276B(kG)$. The experiment, performed
in a uniform magnetic field (≈ 4 kG), involves measuring the perturbed
(m=0) and the unperturbed (m = ± 1) rates in isobutane gas at various
densities. By extrapolating the measured decay rates to zero den-
sity λ_T' and λ_T can be obtained and thus λ_S can be determined from
Eq.3. If, as expected, the collisional quenching rates for the
perturbed and unperturbed triplet states are equal then the decay
rates in Eq.2 may be interpreted as the observed decay rates. Since
collisional quenching is negligible for the singlet state, λ_S could
then be determined without extrapolation over the gas density!
Decay rates will be measured versus density to check these expectations.

While measuring λ_S to 0.2% this experiment should also yield a
new measurement of λ_T at the level of 0.05%. The principle advantage

of this experiment is the use of a magnetic field that confines the positrons and, because of its small diffusion length, the positronium, to a narrow region along the field axis. This significantly improves the annihilation γ-ray detection efficiency and thus the signal-to-noise ratio in the lifetime spectrum. An accuracy of 0.02% in λ_T should be feasible by replacing the TAC with a crystal controlled digital timer and by increasing the magnetic field strength so as to enhance the separation of the two triplet components in the lifetime spectrum.

REFERENCES

1.   P.O. Egan, V.W. Hughes, and M.H. Yam, Phys. Rev. A15, 251 (1977).
2.   S.R. Lundeen and F.M. Pipken, Phys. Rev. Lett. 34, 1368 (1975) and D.A. Andrews and G. Newton, Phys. Rev. Lett. 37, 1254 (1976).
3.   G.W. Erickson, Phys. Rev. Lett. 27, 780 (1971) and P.J. Mohr, Phys. Rev. Lett. 34, 1050 (1975).
4.   K.F. Canter, A.P. Mills, Jr., and S. Berko, Phys. Rev. Lett. 34, 177 (1975).
5.   A.P. Mills, Jr., S. Berko, and K.F. Canter, Phys. Rev. Lett. 34, 1541 (1975).
6.   A.P. Mills, Jr. and S. Berko, Phys. Rev. Lett. 18, 420 (1967).
7.   K. Marko and A. Rich, Phys. Rev. Lett. 33, 980 (1974).
8.   D.W. Gidley, P.W. Zitzewitz, K.A. Marko, and A. Rich, Phys. Rev. Lett. 37, 729 (1976).
9.   W.E. Caswell, G.P. Lepage, and J. Sapirstein, Phys. Rev. Lett. 38, 488 (1977).
10. W.E. Caswell and G.P. Lepage, Phys. Rev. A20, 36 (1979).
11. D.W. Gidley, A. Rich, P.W. Zitzewitz, and D.A.L. Paul, Phys. Rev. Lett. 40, 737 (1978).
12. T.C. Griffith, G.R. Heyland, K.S. Lines, and T.R. Twomey, J. Phys. B 11, L743 (1978).
13. D.W. Gidley and P.W. Zitzewitz, Phys. Letters A69, 97 (1978).
14. R. Paulin and G. Ambrosino, J. Phys. (Paris) 29, 263 (1968).
15. I. Harris and L. Brown, Phys. Rev. 105, 1656 (1957).
16. K. Cung, A. Devoto, T. Fulton, and W. Repko, Michigan State University preprint, March, 1978.
17. Y. Tomozawa, submitted to Annals of Physics.
18. J. R. Freeling, Ph.D. Thesis, University of Michigan, 1979.
19. E.P. Theriot, Jr., R.H. Beers, and V.W. Hughes, Phys. Rev. Lett. 18, 767 (1967).

PRECISION GAMMA- AND X-RAY ENERGIES

R.D. Deslattes and E.G. Kessler, Jr.

Center for Absolute Physical Quantities
National Bureau of Standards
Washington, D.C. 20234

INTRODUCTION

Several goals encourage efforts to improve internal consistency
of γ-ray and X-ray transition energy scales. These include but are
not confined to: detector calibration, rationalization of nuclear
level schemes and chemical analysis. Fewer and less obvious goals
motivate efforts to relate these scales to external "standards".
Although our work has predominantly addressed the latter class of
applications, the results obtained do exhibit increased measurement
precision thereby also contributing to the former.

Our main efforts have been to improve determination of certain
X-ray characteristic lines and relatively low energy (0.04 < E < 1.1
MeV, to date) γ-ray lines relative to the Rydberg constant, R_∞.
Clearly, the point of such measured connections is, in the case of
electronic X-ray spectra, to sharpen tests of (atomic) theory. The
corresponding interest in nuclear γ-rays is evidently not for
comparison with (nuclear) theory since *ab initio* calculations of
nuclear level schemes do not yet tax the reliability of (absolute)
measurement technology. On the other hand, several nuclear γ-lines
have served (and will continue to serve) as reference points for
muonic and pionic atom spectra which are (loosely) lumped together
under the appellation "mesic X-ray spectra". Interest here follows
from the fact that certain of these mesic X-ray spectra are substantially those of single particles bound in a Coulomb potential.
This simplification occurs for angular momenta $\ell > 1$, a condition
which reduces the sensitivity of calculated term values to internal
nuclear details. Also, for principal quantum numbers, n, not too
large, screening of the potential seen by the "meson" due to atomic
electrons can be treated as a small correction. Energy term systems

or transition arrays calculated for these "exotic" atoms have had different significances at different times depending on knowledge available from other sources about elementary particle properties especially their masses.

In all cases, expressions for energy terms appear as modifications of hydrogenic calculations multiplied by an effective Rydberg constant. The modified hydrogenic expressions contain values for magnetic moments, g-factors and possible couplings to nuclear parameters. The effective Rydberg constant is $R_{eff} = \frac{m_x}{m_e} R_\infty$ where m_x is the mass of the "exotic" particle, m_e the mass of the electron and R_∞ the electronic Rydberg. If the mass, m_x, is not otherwise well known, such an experiment can be considered as a mass determination. This was originally the case for muons and remains the case for pions and kaons. When the exotic particle mass is otherwise well determined (i.e., the muon), such spectra become testing grounds for the theoretical models, in this case, quantum electrodynamics especially the vacuum polarization terms.

One other type of external reference procedure for γ-ray spectra is well known and is especially significant for this conference. The process of building up an overall nuclidic mass scheme for both ground state and excited state nuclei proceeds, as reviewed elsewhere in this volume, by combining mass decrement measurements with γ-ray transition arrays. This process requires, in an obvious way, that the scale for mass decrements be related to that for γ-rays by $\Delta E = \Delta M c^2$. To the extent that γ-rays are anchored to visible wavelengths and mass decrements (in atomic mass units) are converted to the macroscopic scale by $\Delta M = \Delta M * N_A^{-1}$, (the (*) signifies units of the atomic mass scale), it is not clear that the scales are thereby connected with higher energy γ-ray lines ($E \gtrsim 2$ MeV). This problem is presently resolved by using ΔM intervals (for which a corresponding ΔE can be established) to fix the energies of the cascade components. As will be clear at the end, we hope to extend the optically based measurement chain into this same energy region with a resulting over-determination which may fix a value for Faraday's constant in the future.

OUTLINE OF MEASUREMENT CHAIN

Numerical estimates of the Rydberg constant, R_∞, follow from Doppler-free spectroscopy of atomic hydrogen [1]. This is nowadays carried out with respect to 633 nm HeNe lasers stabilized by locking to Doppler-free saturation signals in molecular I_2 [2]. It transpires that the laser wavelengths are well determined with respect to the ^{86}Kr standard and are consistent with a proposed redefinition of the meter fixing $c = 299\ 792\ 458$ m sec^{-1} [3]. We used this same laser system to anchor the measurement chain from the visible to the γ-ray region.

The procedures we have used involve three separate measurement steps; the output of the first is the input to the second, and so on. In the first step, the spatial periodicity of a particular specimen of Si was established in terms of the I_2 stabilized laser. In the second step, the calibration embodied in the particular Si specimen used in the first step was transferred (without significant loss of accuracy) to other specimens of Si and Ge. Resulting from this "dissemination" step were several pairs of "calibrated" crystals suitable for use in various photon energy regions in the third step. In the third, and last, step, crystals obtained from the second step have been used in transmission double-crystal instruments equipped with an absolute angle measurement capability. Gamma-ray and X-ray diffraction angles obtained with these instruments thus yield numerical results for the corresponding transitions in terms of the I_2 stabilized 633 HeNe laser. If the laser is assigned an energy, a frequency or a wavelength, then the X-ray or γ-ray transition becomes known, via our measurement chain, in the corresponding units.

Since these procedures have been described in several previous conferences [4], [5], [6], [7], and [8] and a detailed report of the entire experimental program submitted for publication [9], it seems appropriate to comment only very briefly on these steps. A general indication of the accuracy levels achieved thus far will, however, be given. Thereafter, we turn to a synopsis of the results and a brief survey of their applications thus far.

a) X-ray/optical interferometry

The process of determining the 0.2 nm repeat distance in Si in terms of a visible laser is by now well known. The diagram of Fig. 1 reminds the reader of its essential features. A separated-crystal, symmetric Laue-case X-ray interferometer produces "fringes" as a function of relative translation of the crystals along the 110 direction as indicated. These fringes are essentially independent of X-ray wavelength and can be measured ultimately with an imprecision determined by crystal non-uniformity (∿ 0.01 ppm).
In our work, mirrors of a high finesse hemispherical Fabry-Perot interferometer were attached to the separated crystals. What is suggested in the figure is that one can imagine carrying a simultaneous determination of a common displacement in terms of both fringe systems thereby establishing the ratio of the optical period, λ/2 to the X-ray period, d(220). In practice the exercise is easier than the implied fringe counting exercise and somewhat more complex because of diffraction phase shifts on the optical side and residual curvature of the measurement trajectory.

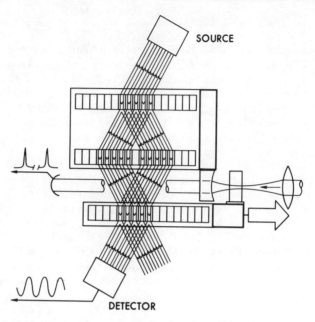

Fig. 1.   Schematic diagram showing the principle of operation of
the optical and X-ray interferometer system.

        The actual measurement strategy used, required corrections and
final results can be briefly summarized. Rather than operate in a
scanning mode we chose to secure an optical signal derived lock to
various Fabry-Perot orders at each of which the X-ray intensity in
the interfering beam could be noted. From observed intensities at
successive orders, a set of phases could be determined which, in
turn, yielded cumulative X-ray phase versus optical order. The
slope of this function is simply the desired ratio, modulo the
separately determined integer part, namely 1648. Further measurements were made at intervals where the X-ray phase had almost
returned to its original value, modulo 2π, thereby effecting
precise determination of the fringe fraction in a very short
sequence of observations. The data shown in Fig. 2 give an impression
of the precision obtained in this exercise. The two histograms
reflect different weightings of the observations while the indicated
partial means show the effect of grouping data according to time of
observation. In preparing Fig. 2, all measurements have been
corrected to a common temperature and the effect of wavefront
curvature removed from the data (diffraction phase shift correction).
A correction of 0.27 ± 0.03 ppm was also required for the measured
path curvature. The final result for d(220) stated at 25°C is
192.01707 picometers (0.1 ppm). This result follows from data near
22.5°C using $\alpha = 2.56 \pm 0.03 \times 10^{-6}$ K^{-1} Discussion of the details
of wavefront curvature and path curvature determinations can be
found in Ref. 5 while a more thorough data analysis is given in Ref. 9.

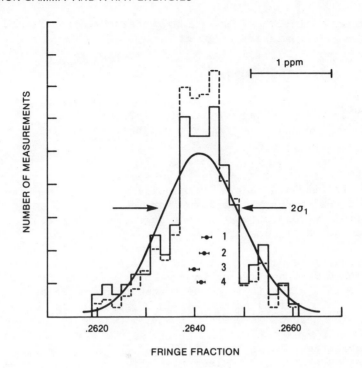

Fig. 2.   Histograms and common fitted Gaussian for two possible weightings of the interferometer results.   The partial means shown adjacent to the numbers 1-4 represent a time ordered disection of the data.

b)   Lattice parameter transfer

What one needs at this point are some relatively simple methods to transfer the information contained in the calibration, (a), to other specimens suited, for example, to γ-ray diffraction. The procedures used are mentioned in our previous reports, [5-8], and are more thoroughly described in Ref. 9.

What leads one to look for special procedures is that direct Bragg angle measurements yield results no better than the angle measurement while also being limited by the great width and complex shape of X-ray lines.   The methods which emerge are thus contrived to have very small effective dispersion yielding a small angle or shift in angle corresponding to a difference in lattice parameter.

In order to make such methods work, one needs to assume that (1) adjacent parts of the same boule of crystal have equal lattice parameters and that (2) for the (cubic) lattices of Si and Ge used

Table I. Lattice Spacing of Crystals Used for X- and
          Gamma-ray Diffraction

Crystal	Diffraction Planes	a_o (nm) at 22.5°C
Si I	220	0.54310271 (0.14 ppm)
Si II	111	0.54310278 (0.12 ppm)
Ge	400	0.56578216 (0.14 ppm)

here, $d(hk\ell) = a_o (h^2 + k^2 + \ell^2)^{-1/2}$. The second condition is what
permits contriving a situation of closely equal spacing between
crystals of the different species. For instance, following Baker
and Hart [10], we exploited the near degeneracy of Si(355) and
Ge(800) in the main transfer.

The kind of results we obtained for the principal pairs of
crystals in current use are shown in Table I. These data are
quoted at 22.5°C since it is close to the temperature at which all
measurements are carried out. This is a particularly serious
problem in the case of Ge where $\alpha = 5.95 \pm 0.11 \times 10^{-6}$ K^{-1} [5,9].

Although the accuracy levels claimed in Table I are sufficient
for present purposes, it is clear that the transfer measurements
have perceptable effects on the initial calibration accuracy of 0.1
ppm. In assessing transfer errors, an attempt has been made to
estimate current experimental limits on homogeneity and symmetry
[9]. On the other hand, each specimen has, in fact, its own spacing
which is far better defined (whether or not measured) than limits
set by transfer imprecision, possible inhomogeneity and asymmetry.
A possible conclusion is that, at some point in the future, one
will have to turn to direct γ-ray diffraction to make crystal-to-
crystal comparisons.

c)   Transmission two-crystal instrument

Crystals calibrated in b) are used in transmission two-crystal
instruments provided with high accuracy interferometric angle
measurement systems. By now, two such instruments are in operation
which, while sharing many basic features, have distinct capabilities.
Such instruments have a long history [11] and the first of ours has
been described elsewhere [5,9]. This section therefore aims merely
to remind the reader of the general features, summarize the γ-ray
results and comment on the newer instrument's features.

As suggested in Fig. 3, diffraction angles are determined by
noting the interval between two orders of diffraction. As shown,
these might be equal and symmetric (n, ±n) but one may have, in
general (n, n'') and (n, n'). For the equal order case, the angular

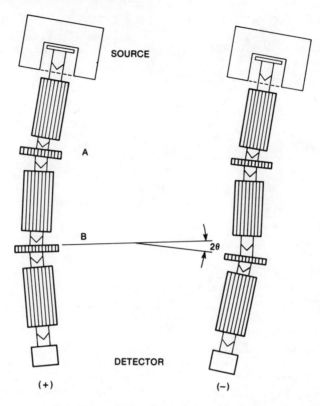

Fig. 3. Schematic diagram indicating the principle of the transmission geometry double crystal wavelength measurement. The figure
illustrates the case of equal diffraction orders in the dispersive
and non-dispersive configurations for which Bragg's equation holds
exactly.

interval is twice the Bragg angle, θ; i.e., $\lambda = 2d\sin\theta$ holds exactly
for rays parallel to the plane of dispersion. For finite divergence
out of the plane of dispersion, corrections are needed which can
generally be kept to the order of a few ppm. Thus geometrical
determination of the out-of-plane collimation to a few percent
suffices for sub ppm measurements.

The instruments are equipped with sensitive angle measuring
laser interferometers. These are of the polarization encoded
variety where the output plane of polarization rotates by 90° for
a crystal rotation of the order of 0.07 arc sec. We use Faraday
modulators, polarizing prisms at extinction and null-seeking
servoes to measure and control angular motions well below 10^{-4} arc
sec. Since typical γ-ray diffraction angles studied thus far are
in the range $0.3 < \theta < 3°$, the angular precision is adequate for
sub ppm work. These angle interferometers require calibration.

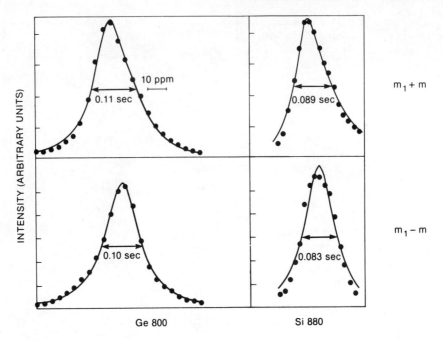

Fig. 4. Examples of diffraction profiles obtained with Si 880 and Ge 800 crystal pairs at 411 keV. The anti-parallel curves labeled n,+n show asymmetry due to the presence of finite vertical divergence whose effect is included in the model curves shown by the solid lines.

This has been obtained by mounting optical polygons on the crystal tables. A space-fixed, sensitive autocollimator is used to locate each face normal and all polygon interfacial angles determined in terms of the interferometric measuring systems. The closure condition on these interfacial angles, α_i, namely $\Sigma\alpha_i = 2\pi$ holds independent of polygon errors thereby fixing the calibration factor. We have found that these calibration exercises can be carried out to better than 0.1 ppm. With the first instrument, slow drifts in the calibration occurred over time requiring that an interpolation be made between calibrations carried out before and after a set of γ-ray measurements. This process permitted, in recent years, absolute angle determinations with an accuracy of 0.1 ppm.

Typical diffraction profiles are indicated in Fig. 4. The asymmetry in the (n, +n) curves is that expected from vertical divergence which has, in fact, been modelled in the solid curves. Widths in these cases correspond to spectroscopic resolving powers of 4×10^4 and 7.5×10^4. Evidently, the precision with which one locates these peak positions is limited by counting statistics. If these are adequate, there is evidently no difficulty in securing 0.1 ppm in this exercise. In passing, it should be noted that we

use a model line shape which is Lorentzian for (n, -n) and the
convolution of a Lorentzian with the (asymmetric) vertical divergence function for (n, +n). Data are fit to these model functions
and the parameters of the Lorentzians used to locate the peak
positions.

In summary, we have a three step process which goes from the
visible to γ-rays. Each step is associated with error contributions
near the 0.1 ppm level except that the γ-ray measurements may be
additionally limited by counting statistics. With a typical
resolving power of 2×10^4, accumulation of 10^4 counts in the peak
permits location of a peak to 0.5 ppm. Since the Bragg angle
follows from two such measurements, drifts must be accounted for
and backgrounds determined; more typically appreciably more events
are required to reach this level of accuracy.

RESULTS AND DISCUSSION

Gamma-ray results obtained to date [12,13] are summarized in
Table II. Only final average values are shown here, however, in
most cases, several repetitions were involved often with different
crystals and diffraction orders. Also, as indicated in the Table,
certain γ-ray energies were obtained by summation of cascade transitions with appropriate corrections for recoil before and after
summation. The entries in Table II are intended to represent
observed γ-ray energies, i.e., they should be increased by the
recoil factor before being treated as energy level differences.

We feel that these results represent a considerably higher
level of accuracy than was previously available and contribute
also, to a lesser extent, to improvements in internal consistency.
The reader will note the appearance of two "error" columns in Table
II. The first is intended to reflect the quality of our connection
to visible standards. The energy results are assumed to be expressed
in "conventional" eV which are obtained from the wavelength (or
wavenumber) determinations taking the present V-λ conversion factor,
1.2398520×10^{-6} eV-m, as exact. These smaller uncertainties are
appropriate for use in connection with detector calibration, level
scheme determination, mass values for elementary particles, and so
forth. On the other hand, when a connection is required to an
electrically determined energy, then the additional 2.6 ppm uncertainty in the voltage wavelength conversion factor needs to be
included as has been done in the last column. Where closure tests
are available from cascade-crossover relations, consistency is
obtained within expected limits as is discussed in more detail in
the original publications [12,13]. Also, where measurements have
been repeated over a long period with several crystals and orders,
an estimate of overall uncontrolled systematic error has been
obtained as 0.13 ppm. A term reflecting this contribution has been
included in the error estimates of Table II.

Table II. Measured energies and uncertainties. The uncertainty, σ_T, includes contributions from the angle measurements, X-ray/ optical interferometer measurement, crystal-to-crystal transfers, angle calibration, and the vertical divergence correction. The uncertainty σ_E, combines σ_T with the 2.6 ppm uncertainty in the V-λ product (V·λ = 1.2398520 x 10^{-6} eV·m) which was used to convert wavelengths to energies. Entries indicated by a † are obtained by summation and those with a * are obtained by combining direct measurement and summation results.

| Source | Energy (keV) | Uncertainties (ppm) | |
		σ_t	σ_E
^{198}Au	411.80441	0.30	2.62
^{198}Au	675.88743	1.00	2.79
^{198}Au†	1087.69033	0.64	2.68
^{192}Ir†	136.34347	1.86	3.20
^{192}Ir	205.79549	0.24	2.61
^{192}Ir	295.95825	0.38	2.63
^{192}Ir	308.45689	0.42	2.63
^{192}Ir	316.50789	0.52	2.65
^{192}Ir†	416.47136	1.76	3.14
^{192}Ir	468.07147	0.52	2.65
^{192}Ir	484.57797	0.82	2.73
^{192}Ir	588.58446	1.21	2.87
^{192}Ir*	604.41459	0.32	2.62
^{192}Ir*	612.46559	0.36	2.62
^{192}Ir†	884.54174	0.82	2.73
^{169}Yb	63.12077	1.34	2.93
^{169}Yb	93.61514	1.25	2.88
^{169}Yb	109.77987	0.47	2.64
^{169}Yb	118.19018	1.53	3.02
^{169}Yb	130.52368	0.25	2.61
^{169}Yb	177.21402	0.30	2.62
^{169}Yb	197.95788	0.30	2.62
^{169}Yb*	261.07851	1.23	2.88
^{169}Yb*	307.73757	0.29	2.62
^{170}Tm	84.25523	0.88	2.75
W Kα_1	59.319233	0.90	2.75

The effect of these new measurements as a whole on previous knowledge appears consistent with the change noted in the case of the 411 keV transition of the Hg daughter of ^{198}Au. Where the previous estimate obtained in an elegant experiment by Murray, Graham, and Geiger [14] was 411794 eV (17 ppm), our value is

411804.41 eV (0.35 ppm) [12]. The Murray, Graham, and Geiger
result when reanalyzed putting in more current values of physical
constants and the value of U Kα_1 suggested by our X-ray work (see
below) becomes 411804 eV (17 ppm) [9].

Some of the γ-ray lines listed in Table II have served as
reference values for determination of muonic and pionic X-ray
spectra. In the case of muonic spectra, use of the new scale
together with improvements in the muonic data have effectively
removed [15] previously significant discrepancies which loosely
speaking constituted the vacuum polarization puzzle [16]. Also,
the scale revision in the context of pionic X-ray spectra has
effectively removed a small discrepancy between two different
routes to the pion mass [9]. Overall the new scale appears to be
significantly improved in accuracy, perceptibly more consistent and
generally beneficial in the resolution of historical quandries.

In addition to γ-ray measurements, we remeasured a few characteristic X-ray lines which have been of significance in various
definitions of X-ray wavelength scales. This work included a
relatively recent measurement of W Kα_1 [17] and earlier determinations
of Cu Kα_1 and Mo Kα_1 [18]. If the careful measurements by Borchert
[19] of ratios between the Kα lines of U, Th, Pu and Tm and and the
^{198}Au (411 keV) line are revised to account for our new measurement
of the 411 keV line, a small but significantly improved set of X-
ray data is obtained [20]. When these are compared with precise
relativistic SCF calculations that have become available in recent
years [21], a significant pattern of discrepancies emerges as shown
in Fig. 5. To pursue this issue further we have employed a new and
wider range instrument at the NBS 4 MeV electron accelerator and
begun a systematic study of X-ray K series spectra especially in
heavy elements. We expect to combine these results with L and M
series measurements and with selected photoelectron threshold data
to obtain new values for inner-shell term energies. Our objective
is to study the dependence of the above noted discrepancies on
principal and orbital quantum numbers as well as on Z.

NEW INSTRUMENTATION - FUTURE PROGRAMS

At several points in the above report, the existence of a
"second generation" transmission two-crystal instrument has been
noted. Salient characteristics of the new instrument [22] in
comparison with the earlier version are: a wider angular range,
± 15° in place of ± 2.5°; computer control including on-line atmos-
pheric corrections and, the ability to deal with stationary radiation
sources (the first generation required source movement to accommodate
different Bragg angles). The last mentioned capability is now
being exploited at the electron Van de Graaff where a stationary
target chamber is an assured convenience. This facility has a
larger significance as is described below in view of which, we were

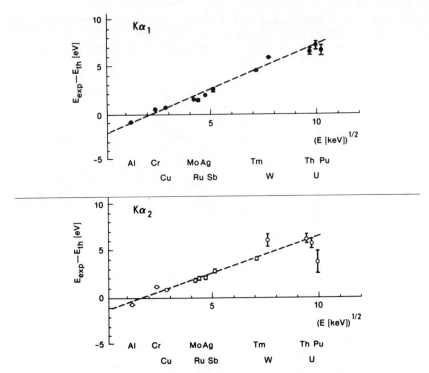

Fig. 5. Differences between relativistic self-consistent field
calculations and selected atomic data for the case of Kα lines in
the indicated elements. A few of these were obtained directly, the
rest by using gamma-ray to X-ray ratios together with our direct
gamma-ray measurements.

led to provide it with a considerable measure of portability.

     As can be noted from Table II, γ-ray measurements obtained
with the earlier instrumentation do not extend beyond 1.1 MeV.
This truncation does not reflect either a limitation of measurement
technology or a failure of interest. Instead it is a consequence
of the fact that the first generation instrumentation required
intense, long-lived sources. These were produced by neutron
activation in the NBS reactor, extracted into casks and transported
to our measurement facility. There the casks were positioned so as
to furnish radiation at an appropriate Bragg angle for the (stationary) two-crystal instrument. Our examination of nuclear data
tables does not reveal any convenient sources which could be
prepared by neutron activation having useful activities in excess
of 10^2 Ci, half lives in excess of 1 day and energies in excess
of 2 MeV. Thus, in order to proceed to higher energies (4 MeV seems

practical from the standpoint of measurement precision), we have to
consider use of relatively prompt γ's from in-pile sources. At the
same time it must be noted that facilities for insertion, cooling
and removal of such sources are not widely available. Therefore in
addition to stationary source capability, our second generation
instrumentation was built with a measure of portability in mind.

The second generation instrumentation project was undertaken
in collaboration with two visiting scientists, L. Jacobs of Leuven
University and W. Schwitz, University of Fribourg. In view of
extensive γ-ray and X-ray programs planned, it was provided with a
high level of computer automation (CAMAC) including the ability to
make real time adjustments to the angle commands to account for
ambient changes. In addition, awareness of potential interest in
the X-ray region led us to provide a wide angular range ($\pm$ 15°).
Ability to deal with stationary sources was obtained by placing the
main instrument support plate (carrying also the detector) on
wheels and a pivot so that an average entrance Bragg angle can
readily be established. On the support plate, the detector and its
10^3 kg shield have also a pivot and wheels permitting access to
both "plus" and "minus" diffracted beams.

While this instrument is presently active on atomic X-ray
problems, plans are underway to bring it to bear on in-pile capture
γ-sources. A particularly suitable facility is available in the
high flux reactor at Institute Laue-Langevin in Grenoble. Among
the problems which could be addressed in such a facility, one
cannot fail to take note of simply extending the optically controlled
scale to energies, ΔE, sufficiently large ($\sim$ 10 MeV) that the
corresponding mass defect, ΔM, could be determined to $\sim$ 1 ppm.
Associated with ΔE there is an effective wavelength, $\hat{\lambda}$, such that
$\Delta E = hc/\hat{\lambda}$. The corresponding mass decrement, ΔM, is, of course,
derived from measurements on the atomic mass scale of ΔM^*, where
$\Delta M = \Delta M^*/N_A$. When these relations are inserted into $\Delta E = \Delta M c^2$,
one obtains a value for the "mass-wavelength" product, namely

$$\Delta M^* \hat{\lambda} = N_A h/c \tag{1}$$

Although this grouping of fundamental constants has not been used
in past least squares adjustments, there is no reason why it
should not be used in future adjustments if data are available. In
the event of sufficient improvement in realization of the Ampere,
equation (1) may be usefully rewritten in terms of Faraday's
constant, $F = N_A e$ and the Josephson effect value of e/h to obtain:

$$\Delta M^* \hat{\lambda} \, c \, (e/h) = F \tag{2}$$

This evidently desirable approach to F is not yet viable as may be
seen by writing the absolute values of (e/h) and F in terms of
local units and the conversion factor $K = A_{NBS}/A_{ABS}$ [23].

SUMMARY

Overall, it appears that a significant improvement has been
made in the determination of γ-ray transition energies. This
improvement has already yielded certain dividends in the area of
muonic and pionic X-ray spectra. Atomic X-ray data, newly acquired
revised seem also to tell a possibly significant story. Finally,
with still newer instrumentation it appears possible to extend the
congruent electromagnetic scale to of the order of 10 MeV where it
conveniently overlaps the scale of nuclidic masses. The synthesis
effected in this work is seen to have impacted tests of quantum
electrodynamics, mass values for elementary particles and the
validity of single configuration estimates of relativistic self-
consistent field calculations.

REFERENCES

1.   T.W. Hänsch, M.H. Nayfeh, S.A. Lee, S.M. Curry, and I.S.
        Shahin, Precision Measurement of the Rydberg Constant by
        Laser Saturation Spectroscopy of the Balmer α Line in
        Hydrogen and Deuterium, Phys. Rev. Lett. 32:1336 (1974).
        See also, J.E. Goldsmith, E.W. Weber, and T.W. Hänsch,
        New Measurement of the Rydberg Constant Using Polarization
        Spectroscopy of Hα, Phys. Rev. Lett. 41:1525 (1978).
2.   W.G. Schweitzer, Jr., E.G. Kessler, Jr., R.D. Deslattes, H.P.
        Layer, and J.R. Whetstone, Description, Performance, and
        Wavelengths of Iodine Stabilized Lasers, Appl. Opt.
        12:2927 (1973).
3.   Recommendation of CCDM, June 1979, K.G. Kessler, private
        communication.
4.   A preliminary result appeared in, R.D. Deslattes, E.G. Kessler,
        Jr., W.C. Sauder, and A. Henins, Visible to Gamma-ray
        Wavelength Ratio in: "Atomic Masses and Fundamental
        Constants 5," J.H. Sanders and A.H. Wapstra, Eds., Plenum
        Press, New York (1976).
5.   R.D. Deslattes, Reference Wavelengths – Infrared to Gamma-
        rays, Avogadro's Constant, Mass and Density in:
        "Proceedings of the International School of Physics
        'Enrico Fermi'," Course LXVIII 12-14 July 1976 (to be
        published).
6.   E.G. Kessler, Jr., R.D. Deslattes, W.C. Sauder, and A. Henins,
        Precise γ-ray Energy Standards in: "Proceedings of
        Neutron Capture Gamma-ray Spectroscopy Symposium," R.E.
        Chrien and W.R. Kane, Eds., Plenum Press, New York (1979).
7.   R.D. Deslattes, Rydberg Values for X- and γ-rays, Jap. Jour.
        Appl. Phys. 17:1 (1978).
8.   R.D. Deslattes, X-ray Interferometry: The Optical to Gamma-
        ray Connection in: "Proceedings of the Workshop on
        Neutron Interferometry," 1979 (to be published).

9. R.D. Deslattes, E.G. Kessler, Jr., W.C. Sauder, and A. Henins,
    Remeasurement of Gamma-ray Reference Lines, submitted to
    Annals of Physics.

10. J.F.C. Baker and M. Hart, An Absolute Measurement of the
    Lattice Parameter of Germanium Using Multiple Beam X-ray
    Diffractometry, Acta Cryst. A31:364 (1975).

11. T.R. Cuykendall and M.T. Jones, A Two-Crystal Spectrometer for
    X-rays of Wave-Length 0.030 < λ < 0.215 Å, Rev. Sci.
    Instr. 6:356 (1935); J.W. Knowles, A High Resolution Flat
    Crystal Spectrometer for Neutron Capture γ-ray Studies,
    Can. J. Phys. 37:203 (1959); J.W. Knowles, Measurement of
    γ-ray Diffraction Angles to ± 0.02 Second of Arc with a
    Double Flat Crystal Spectrometer, Can. J. Phys. 40:237
    (1962); J.W. Knowles, and H.M.B. Bird, A Computer Con-
    trolled Interferometer System for Precision Relative
    Angle Measurements, Rev. Sci. Instr. 42:1513 (1971); V.L.
    Alexeyev, V.A. Shaburov, D.M. Kaminker, O.I. Sumbaev, and
    A.I. Smirnov, A Double Crystal Diffraction Spectrometer
    for Studies of High Energy Gamma Rays Resulting from
    Thermal Neutron Capture, Nucl. Instr. and Meth. 58:77
    (1968).

12. E.G. Kessler, Jr., R.D. Deslattes, A. Henins, and W.C. Sauder,
    Redetermination of ^{198}Au and ^{192}Ir γ-ray Standards
    between 0.1 and 1.0 MeV, Phys. Rev. Lett. 40:171 (1978).

13. E.G. Kessler, Jr., L. Jacobs, W. Schwitz, and R.D. Deslattes,
    Precise γ-ray Energies from the Radioactive Decay of
    ^{170}Tm and ^{169}Yb, Nucl. Instr. and Meth. 160:435 (1979).

14. G. Murray, R.L. Graham, and J.S. Geiger, A Determination of
    the Absolute Energy on the Hg^{198} 412 keV γ-radiation,
    Nucl. Phys. 45:177 (1963); G. Murray, R.L. Graham, and
    J.S. Geiger, The Precision Determination of Some γ-ray
    Energies Using a β-Spectrometer, Nucl. Phys. 63:353
    (1965).

15. C.K. Hargrove, E.P. Hincks, R.J. McKee, H. Mes, A.L. Carter,
    M.S. Dixit, D. Kessler, J.S. Wadden, H.L. Anderson, and
    A. Zehnder, Further Muonic-Atom Test of Vacuum Polariza-
    tion, Phys. Rev. Lett. 39:307 (1977); T. Dubler, K.
    Kaeser, B. Robert-Trissot, L.A. Schaller, L. Schellenberg,
    and H. Schneuwly, Precision Test of Vacuum Polarization
    in Heavy Muonic Atoms, Nucl. Phys. A294:397 (1978).

16. P.J.S. Watson and M.K. Sundareson, Discrepancy Between Theory
    and Experiments in Muonic X-rays - A Critical Discussion,
    Can. J. Phys. 52:2037 (1974); J. Rafelski, B. Müller, G.
    Soffard, and W. Greiner, Critical Discussion of the
    Vacuum Polarization Measurements in Muonic Atoms, Annals
    of Phys. 88:412 (1974).

17. E.G. Kessler, Jr., R.D. Deslattes, and A. Henins, Wavelength
    of the W Kα$_1$ X-ray Line, Phys. Rev. A 19:215 (1979).

18. R.D. Deslattes and A. Henins, X-ray to Visible Wavelength
    Ratios, Phys. Rev. Lett. 31:972 (1973).

19.   G.L. Borchert, Precise Energies of the K-Röntgen-Lines of Tm,
      Th, U and Pu, Z. Naturforsch. 31a:102 (1976).

20.   R.D. Deslattes, E.G. Kessler, Jr., L. Jacobs, and W. Schwitz,
      Selected X-ray Data for Comparison with Theory, Phys.
      Lett. 71A:411 (1979).

21.   K-N. Huang, M. Aoyagi, M.H. Chen, B. Crasemann, and H. Mark,
      Neutral-Atom Electron Binding Energies from Relaxed-
      Orbital Relativistic Hartree-Fock-Slater Calculations
      $2 \leq Z \leq 106$, Atomic Data and Nuclear Data Tables 18:243
      (1976).

22.   The second generation instrumentation project was undertaken
      in collaboration with two visiting scientists, L. Jacobs
      of Leuven University, and W. Schwitz, University of
      Fribourg.

23.   We are indebted to Barry N. Taylor for several discussions
      regarding interpretations of these measurements.

PRECISE γ-RAY ENERGIES FROM RADIONUCLIDE DECAY AND THE (n,γ) REACTION: REVISED VALUES FOR THE NEUTRON MASS AND SELECTED NEUTRON BINDING ENERGIES†

R. C. Greenwood*, R. G. Helmer*, R. J. Gehrke*
and R. E. Chrien**

*Idaho National Engineering Laboratory, EG&G Idaho, Inc.
Idaho Falls, Idaho 83401, U. S. A.
**Brookhaven National Laboratory, Upton, New York, 11973,
U. S. A.

ABSTRACT

Precise measurements of γ-ray energies up to 3.5 MeV from several radionuclides have been made using Ge(Li) detectors. Also, the ^{1}H(n,γ) reaction γ-ray energy was measured and from it a value was obtained for the deuteron binding energy. Combining this $S_n(^2$H) with published mass differences, values were obtained for the neutron binding energies of ^{3}H, ^{13}C, ^{14}C and ^{15}N, and for the neutron mass excess. This value of $S_n(^{15}$N) was used in conjunction with measured γ-ray energy differences to obtain energies of selected γ rays from the ^{15}N(n,γ) reaction up to 10.8 MeV.

INTRODUCTION

In this paper, we review the status of our continuing effort to develop a consistent set of precisely measured γ-ray energies. There has been significant progress in this work since our report at the last Mass Conference.[1] A new energy scale based on a measurement by Kessler et al.[2] of the absolute γ-ray wavelength of the 411-keV line from ^{198}Au together with recent precise γ-ray energy measurements made with crystal-diffraction spectrometers have provided a firm base for our measurements made with Ge-semiconductor detectors. Thus, with this base of crystal-diffraction spectrometer data, we have been able to extend our measurements of the energies of γ rays from several radionuclides up to 3.5 MeV.[3,4]

---

†Work performed under auspices of the U. S. Department of Energy.

In other experiments, we have measured the $^1H(n,\gamma)^2H$ reaction γ-ray energy using a number of these radionuclides as energy standards and have thus obtained an improved value for the binding energy of the deuteron $S_n(^2H)$[5]. By combining this binding energy value with published mass differences[6-8] of doublets involving neighboring isotopes and ^1H and ^2H we have obtained revised neutron binding energies of ^3H, ^{13}C, ^{14}C and ^{15}N together with a value for the neutron mass excess. Also, based on this $S_n(^{15}N)$ and γ-ray energy difference measurements we have obtained selected γ-ray energies from the $^{14}N(n,\gamma)^{15}N$ reaction up to an energy of 10.8 MeV.

For all of the uncertainties quoted in this paper a distinction is made between the contribution from the energy scale σ_r and that from all subsequent measurements σ_m. The total uncertainty value σ_t is obtained by summing in quadrature σ_r and σ_m. These quoted uncertainties are intended to be estimates of standard deviations.

WAVELENGTH BASED ENERGIES

Definition of Energy Scale

The recent article of Kessler et al.[2] reports a series of measurements to determine a very precise absolute wavelength for the 411 keV γ ray from the decay of the ^{198}Au. The resulting 411 - ^{198}Au value is 3.0107788 pm with an uncertainty of 0.37 ppm. Based on a voltage-wavelength conversion factor of 1.2398520×10^{-6} eV.m (+2.6 ppm) from Cohen and Taylor[9] the corresponding γ-ray energy is 411.80441 ± 0.00108 keV. In the present work all of the wavelength-based energies are referenced to this value. Our choice of 411 - ^{198}Au to define the energy scale may be considered somewhat arbitrary since Kessler et al.[2] have also reported wavelength values for several ^{192}Ir decay γ rays. Our choice, however, was based upon the fact that more extensive measurements were made for ^{198}Au, and the fact that all other recent crystal-diffraction spectrometer results have been referenced to the 411 - ^{198}Au line. The uncertainty in the 411 - ^{198}Au energy includes error contributions of 0.37 ppm from the wavelength determination and 2.6 ppm from the conversion to the keV unit. The combined error of 2.63 ppm is therefore taken as the σ_r for the wavelength-based energy scale.

Energies from Crystal Diffraction Spectrometers

In recent years, several sets of precise γ-ray energies measured with crystal-diffraction spectrometers have been published. Radionuclides for which such measurements have been made include ^{51}Cr[10], ^{57}Co[11], ^{60}Co[11], ^{109}Cd[12], ^{110m}Ag[13,14], ^{137}Cs[11], ^{139}Ce[12], ^{141}Ce[12], ^{152}Eu[11], $^{153}Sm(^{153}Gd)$[11], ^{160}Tb[15], ^{169}Yb[10,16-18], ^{170}Tm[10,16,18], ^{182}Ta[19,20], ^{183}Ta[20,21], ^{192}Ir[2,10,12,16], ^{198}Au[2,11], ^{199}Au[20], ^{203}Hg[10,11], and ^{241}Am[22]. As can be seen, these measurements resulted

primarily from four groups; Reidy, Kern and coworkers, Borchert
et al., and Kessler et al. For ^{192}Ir measurements have been made
by all four groups and in Ref. 3 their data are compared. This
comparison indicates that the four sets of data are quite consistent: which we accept then as a validation of the individual measurement techniques and have therefore used measurements from each of
the groups in developing our energy set.

These crystal-diffraction spectrometer measurements provide
the set of basic calibration energies for our γ-ray energy measurements with Ge-semiconductor detectors. The necessity for such Ge-
detector measurements results from the fact that precise crystal
diffraction spectrometer measurements have to date been limited to
energies below ~1.0 MeV, with cascade sums in ^{110m}Ag increasing
this range to 1.5 MeV. Also, γ-ray energies from several radioisotopes of interest have not been measured with crystal diffraction
spectrometers.

Energies from Ge Semiconductor Detectors

Since Ge-semiconductor systems are generally nonlinear, we have
not considered it satisfactory to rely on an energy interpolation
over a large energy range, even with the availability of nonlinearity corrections. Instead, we have generally preferred to use
the existing crystal-diffraction spectrometer data and measure
energy differences between closely spaced γ rays with simultaneous
counting of the unknown and calibration radioisotopes, as discussed
in Refs. 3 and 4. In order to build up to higher γ-ray energies,
we have then made use of cascade-crossover relationships with the
energies of the cascade γ rays being measured by energy differences.
As discussed in Ref. 1, two chains of energy combinations were used
to get us up to an energy of 1.3 MeV; with one chain consisting of
γ rays from ^{192}Ir and ^{160}Tb and the other consisting of γ rays from
^{198}Au, ^{59}Fe and ^{182}Ta. As an internal check on the consistency of
this procedure it was possible to compare, by difference measurements, energies of γ rays >1 MeV derived separately through each
chain. Such a comparison is shown in Table I where the results
indicate good agreement between the two chains with no evidence for
a systematic bias. Thus, the proposed measurement uncertainties,
typically 4 eV at an energy of 1 MeV seem reasonable.

The ^{110m}Ag decay provides an intermediate step for achieving
precise γ-ray energies up to 1.5 MeV; from cascade-crossover relationships. Furthermore, curved-crystal diffraction spectrometer
measurements also exist[13,14] for this decay. These data, together
with our own Ge(Li) detector measurements, were included in a least-
squares fit involving seven excited states of ^{110}Cd in Ref. 4 and
shown to be consistent. The output of this fit gave γ-ray energies
>1 MeV with σ_m of 1.7 - 2.7 ppm.

TABLE I. Comparisons of energy chains <1.3 MeV

Transitions in difference	Energy difference (keV)		Discrepancy (eV)
	Computed	Measured	
1177(^{160}Tb)-1189(^{182}Ta)	11.088 8(50)	11.085 0(24)	+3.8(55)
1271(^{160}Tb)-1291(^{59}Fe)	19.715 9(89)	19.716 6(30)	-0.7(94)
		Average	2.7(47)

The series of experiments designed to extend the region of precise γ-ray energies from 1.3 MeV up to 3.5 MeV, using energy difference measurements and cascade-crossover relationships required several interrelated steps which are indicated schematically in Fig. 1 and are discussed in more detail in Ref. 4. Because ^{56}Co is such a useful calibration standard, emitting strong γ rays approximately evenly spaced in the energy region 0.8 - 3.5 MeV, step 6 might be considered as the culmination of this measurement effort and a listing of the ^{56}Co γ-ray energy values is shown in Table II. The three 3.2 MeV γ-ray energies in ^{56}Co result from 2-MeV γ rays summed with the 1238-keV γ ray. Because of the importance this places on the 2-MeV region in this building process, an internal check of the consistency on the energies and uncertainties in this region is desirable. Such a consistency check is shown in Table III and the results indicate that the uncertainty estimates are quite reasonable. Thus, typical measurement uncertainties in the energies range from 5 to 15 eV between 1.3 and 3.5 MeV, i.e., ~4 ppm compared to 2.63 ppm for the reference scale uncertainty.

Measurement of $S_n(^2H)$

The value of the deuteron binding energy $S_n(^2H)$ is, at the present time, most accurately obtained from measurement of the energy of the prompt γ ray emitted by the $^1H(n,\gamma)^2H$ reaction with slow neutrons. Currently, an uncertainty of 30 eV is quoted on the adjusted value of $S_n(^2H)$. Furthermore, even this large uncertainty estimate may be unrealistic since the energy values with the smallest errors[23-25] included in the adjustment involved analyses of γ-ray spectra measured with Ge(Li) detectors containing mixed full-energy and escape peaks without any corrections being made for the shifts induced between the different types of peaks by the electric fields in the detectors[26]. For example, we would estimate that the electric field correction to the $S_n(^2H)$ value quoted in Ref. 24 should be ~40 eV, with the sign of the correction now indeterminate. Thus, a redetermination of $S_n(^2H)$ was considered desirable.

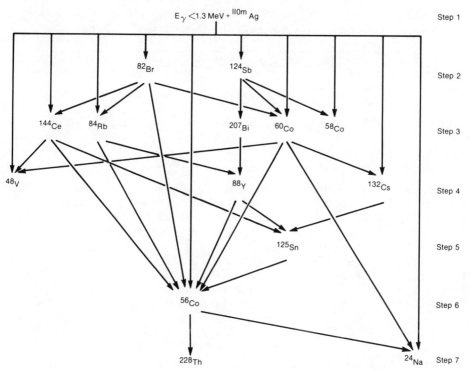

FIG. 1 Schematic diagram of measurement steps to extend region of precise γ-ray energies from 1.3 MeV up to 3.5 MeV

TABLE II. γ-RAY ENERGIES FOR ^{56}Co

γ-ray energy	Uncertainty (eV)		
(keV)	σ_m	σ_r	σ_t
846.764	6	2.2	6
1037.844	3	2.7	4
1175.099	8	3.1	8
1238.287	5	3.3	6
1360.206	5	3.6	6
1771.350	14	4.7	15
1810.722	17	4.8	17
1963.714	11	5.2	12
2015.179	10	5.3	11
2034.759	10	5.4	11
2113.107	11	5.6	12
2212.921	8	5.8	10
2598.460	7	6.8	10
3009.596	15	7.9	17
3201.954	12	8.4	14
3253.417	11	8.6	14
3272.998	11	8.6	14
3451.154	10	9.1	13

TABLE III.   Comparison of crossover γ-ray energies derived
from cascade sums in the region ~2.2 MeV

Transitions in difference	Energy Difference(keV)		Discrepancy (eV)
	Computed	Measured	
$2185(^{144}Ce)-2240(^{48}V)$	54.734(8)	54.729(8)	5 ± 11
$2185(^{144}Ce)-2212(^{56}Co)$	27.257(10)	27.260(16)	-4 ± 19
		Average	3 ± 10

As described in Ref. 5, our recent measurements were conducted
using an external beam of thermal neutrons from the HFBR at Brookhaven with γ rays being detected in either of two Ge-semiconductor
detectors. In these experiments, an energy difference value of
37585+15 eV was measured between the 2223 - $^1H(n,\gamma)$ and the 2185 -
^{144}Ce decay γ rays. Thus, we obtained a value for the energy of
the $^1H(n,\gamma)$ reaction γ ray of 2223.247 keV(σ_m=16 eV, σ_t=17 eV) and
using a value of 1.317 keV for the nuclear recoil correction we
obtain the following value for the binding energy of the deuteron.

$$S_n(^2H) = 2224.564 \text{ keV } (\sigma_m = 16 \text{ eV}, \sigma_r = 5.8 \text{ eV}, \sigma_t = 17 \text{ eV})$$

This value is significantly lower, i.e., by 64 eV, than the adjusted
$S_n(^2H)$ value given in Ref. 8 but for the reasons cited above it
should be used in preference to that value.

Concurrent with the present work Vylov et al.[27] also remeasured
the energy of the $^1H(n,\gamma)$ reaction γ ray and reported a value of
2224.628+0.016 keV for $S_n(^2H)$, corrected by us to the current 411 -
^{198}Au reference[2]. This is 64 eV (29 ppm) higher than our value for
$S_n(^2H)$. However, comparison of the calibration energy values used
by Vylov et al.[27], also corrected to the 411 - ^{198}Au reference
standard[2], to those used in our measurement, from Refs. 3 and 4,
shows that ~19 ppm of this discrepancy can be accounted for by the
different calibration energy values used in the two works. In the
report of Vylov et al.[27] it is not clear how, or if, they have
corrected for shifts in peak positions that might have occurred as
a result of the electric field effect[26] due to the extended nature
of their $^1H(n,\gamma)$ source as compared to the presumably point radio-
nuclide calibration sources.

MASS BASED ENERGIES

Definition of Energy Scale

The existance of this second reference scale for γ-ray energies
results from the improved accuracy achieved in recent years for mass

difference measurements[6-8], which permits neutron binding of a
number of light-mass nuclides to be determined with sufficient
precision to be of value as calibration energies in (n,γ) spectroscopy. Neutron binding energies are obtained in this manner from
mass differences of doublets involving neighboring isotopes together
with ^1H and ^2H by using the following relationships:

$$^A_Z{}^2H - {}^{A+1}_Z{}^1H = S_n(^{A+1}Z) - S_n(^2H) \qquad (1)$$

or

$$^A_Z{}^1H - {}^{A+1}_Z = S_n(^{A+1}Z) - [n - {}^1H] \qquad (2)$$

where

$$n - {}^1H = S_n(^2H) - [{}^1H_2 - {}^2H] \qquad (3)$$

In these relationships, the mass denoted by ^AZ is for the isotope
with atomic number Z and mass number A in its nuclear and atomic
ground states.

It is immediately obvious from inspection of these relationships that the values of $S_n(^{A+1}Z)$ determined in this way result from
linear combinations of mass-based numbers (the mass differences) and
411 - ^{198}Au wavelength-based numbers [$S_n(^2H)$]. In order to compute
correctly the reference uncertainties associated with such mixed-
scale values it is necessary to utilize the variance-covariance
matrix terms associated with the 1973 adjustment of the Fundamental
Constants[9], recognizing that the relevant scale conversion factors
can be expressed as follows:

Wavelength Conversion: $\qquad 1 \ m^{-1} = \left\{\dfrac{he}{c}\right\} eV$

Mass Conversion: $\qquad 1 \ u \quad = \left\{\dfrac{c^2}{10^3 F}\right\} eV$

where following Ref. 9 the braces indicate numerical values only.
From these scale conversion relationships, and using the procedure
outlined by Cohen and Taylor[9], we have computed a variance-
covariance matrix which is applicable for computing realistic
estimates of the reference uncertainties in mixed wavelength-energy
and mass-energy relationships. This matrix is shown in Table IV,
where C_λ is the conversion factor from m^{-1} to eV and C_M is the conversion factor from u to eV. The value of 0.811 for the correlation
coefficient between these two conversion factors indicates a high
degree of correlation. Thus, the reference uncertainty associated
with an energy value, E, determined from a linear combination of
wavelength (E_λ) and mass (E_M)-based energies is obtained as
follows:

$$E = E_\lambda + E_M = (f_\lambda + f_M) E$$

and

$$\sigma_r^2 = \frac{\partial E}{\partial C_\lambda} \frac{\partial E}{\partial C_\lambda} \nu_{11} + 2 \frac{\partial E}{\partial C_\lambda} \frac{\partial E}{\partial C_M} \nu_{12} + \frac{\partial E}{\partial C_M} \frac{\partial E}{\partial C_M} \nu_{22}$$

where the ν_{ij} are the covariances. If σ_r is in ppm and the ν_{ij} in $(ppm)^2$ this expression reduces to

$$\sigma_r = f_\lambda^2 \nu_{ii} + 2f_\lambda f_M \nu_{12} + f_M^2 \nu_{22}$$

TABLE IV.   Variance-covariance and correlation coefficient matrix used to compute the reference uncertainties of energies (in eV) for mixed 411- ^{198}Au wavelength-based and mass-based scales. Following Table 33.4 of Ref. 9, the variances and covariances are in $(ppm)^2$ and the correlation coefficient is given below the diagonal.

	C_λ	C_M
C_λ	6.835	6.014
C_M	0.811	8.045

Neutron Binding Energies

For several of the light-mass stable nuclei, for which mass differences of doublets involving neighboring isotopes together with ^{1}H and ^{2}H have been precisely measured, the relationships expressed by Eqs. 1 and 2 provide by far the most accurate values for the neutron binding energies of the higher mass isotopes $S_n(^{A+1}Z)$. Such mass differences, having comparable accuracy to the present $S_n(^2H)$ value, have been reported in Refs. 6 and 7 and included in the 1977 Atomic Mass Evaluation of Wapstra and Bos[8] but, for the reasons cited in that evaluation their absolute uncertainties were increased by a factor of 2.5. There were no neutron binding energies, except for $S_n(^2H)$, included in that adjustment with uncertainties comparable to these mass difference values. Thus, the Wapstra and Bos[8] adjusted value of these mass differences depends solely on the adjustment of the total body of mass difference data, and does not include any γ-ray data through Eqs. 1-3. We have therefore chosen to use these adjusted mass difference values to obtain the neutron binding energies shown in Table V. In the case of $S_n(^{14}C)$ we obtained a value of 7937.883± 0.031 μu for the ^{13}C ^{1}H- ^{14}C mass difference based on the adjusted values of the mass excesses.[8]

Each of the nuclei shown in Table V can be produced in an (n,γ) reaction with the ^AZ stable-element targets. Furthermore, there is a strong capture state-to-ground state γ-ray transition in

each case. Thus, by applying the appropriate nuclear recoil
corrections equally precise ground-state γ-ray energies can be
obtained. These primary γ-ray energies are also given in Table V.

The Neutron Mass

The mass of the neutron is currently most accurately determined
from Eq. 3 Using the adjusted value of 1548.287 ± 0.006 μu
$(1442.232 \pm 0.007$ keV)[8] for the ${}^1H_2 - {}^2H$ mass difference we obtain

$$n - {}^1H = 782.332 \text{ keV } (\sigma_m = 17 \text{ eV}, \sigma_r = 3.4 \text{ eV}, \sigma_t = 17 \text{ eV}).$$

Alternatively by rearranging Eq. 3 as follows

$$n-1 = S_n({}^2H) - [{}^1H_2 - {}^2H] + [{}^1H-1] \tag{4}$$

and using the adjusted ${}^1H-1$ value of 7825.037 ± 0.010 μu $(7289.034 \pm 0.023$ keV)[8] we obtain for the neutron mass excess

$$n-1 = 8071.367 \text{ keV } (\sigma_m = 19 \text{ eV}, \sigma_r = 22 \text{ eV}, \sigma_t = 29 \text{ eV})$$

or

$$n-1 = 8664.898 \text{ μu } (\sigma_m = 21 \text{ nu}, \sigma_r = 4.0 \text{ nu}, \sigma_t = 21 \text{ nu}).$$

${}^{14}N(n,\gamma)$ Reaction γ-Ray Energies

The prompt γ rays emitted in the ${}^{14}N(n,\gamma){}^{15}N$ reaction with slow
neutrons have long proven to be a valuable calibration source in
neutron capture γ-ray spectroscopy and in muonic x-ray studies.

TABLE V. Neutron binding energies obtained from mass differences
and the present $S_n({}^2H)$ value. Also shown are the energies
of the capture state-to-ground state γ rays emitted
following slow neutron capture into those nuclei.

(n,γ) Reaction product nucleus	$S_n({}^{A+1}Z)$ (keV)	Nucleus recoil energy (keV)	Ground-state γ-ray energy (keV)	Error (eV)		
				σ_m	σ_r	σ_t
${}^3T$	6257.268	6.953	6250.316	17	16.5	24
${}^{13}C$	4946.329	1.010	4945.319	20	12.9	24
${}^{14}C$	8176.483	2.561	8173.922	33	21.9	40
${}^{15}N$	10833.297	4.196	10829.101	24	29.3	38

Apart from providing an equally accurate value for the capturing
state-to-ground state γ ray, the value for $S_n(^{15}N)$ shown in Table V
can also be used to improve the accuracy for several of the other
prompt γ rays emitted in this reaction. This results directly from
the fact that all cascade paths must sum to equal the capturing-state
energy, $S_n(^{15}N)$. Two of the most prominent cascade relationships
are

$$E_t(5534) + E_t(5298) = E_t(5563) + E_t(5270) = S_n(^{15}N) \qquad (5)$$

where the E_t represent transition energies. Since the 5.2 - 5.5 MeV
differences are small (<6%) they can be measured quite precisely,
and combined with $S_n(^{15}N)$ can be used to obtain improved accuracy
for the energies of the 5.2 and 5.5-MeV γ rays. The energy
difference values shown in Table VI were obtained both as new
measurements (BNL-2 and OR-15) and re-evaluation of earlier
published results[28]. No differences involving the 5534 keV transition were measure because of spectral interference problems
involving full- and double-escape peaks.

Additional input to the determination of these γ-ray cascade
energies was obtained from the cascade-crossover relationship.

$$E_t(1678) + E_t(1999) + E_t(1884) = E_t(5563) \qquad (6)$$

From re-evaluation of earlier data using the calibration energies
of Refs. 3 and 4 we obtained the following revised γ-ray energies:
1678.241(47), 1681.589(228), 1884.808(37) and 1999.566(137) keV.

TABLE VI. Measured energy differences in the $^{14}N(n,\gamma)$ reaction

Transitions in difference	Detector	Number of measurements	γ-ray energy difference (eV)	Average energy difference	
				ε^2	γ-ray(eV)[tran-sition(eV)]
5297-5269	OR-1	6	28708(108)	0.5	28676(28)
	OR-4	11	28680(53)		
	P-11	8	28631(41)		[28687(28)]
	P-37	3	28713(93)		
	BNL-2	2	28547(125)		
	OR-15	11	28756(86)		
5562-5297	OR-1	6	264320(132)	0.6	264283(32)
	OR-4	11	264285(59)		
	P-11	8	264337(53)		[264386(32)]
	P-37	3	264322(96)		
	BNL-2	2	264210(169)		
	OR-15	11	264171(86)		

The ^{14}N(n,γ) reaction γ-ray energies shown in Table VII were obtained from a least-squares fit which included as input $S_n(^{15}N)$, the < 2-MeV γ-ray energies and the transition energy differences. Since the measurement uncertainties associated with the transitions in Eq. 6 are much greater than that of $S_n(^{15}N)$ we choose to normalize the fit to $S_n(^{15}N)$. The energy of the 4.5-MeV γ ray in Table VII was obtained from subsequent analysis using the ^{14}N(n,γ) and ^{12}C(n,γ) reaction γ rays as calibrations, with the 6.3 MeV γ ray energy then being obtained from the cascade-crossover relationship with $S_n(^{15}N)$.

FUTURE DIRECTIONS

In the energy region > 3.5 MeV, there are still only a few precisely measured γ ray energies; principally from the (n,γ) reaction and measured on a mixed wavelength-mass scale. Additional precisely measured γ-ray energies in the region > 3.5 MeV are there-fore clearly essential. Possibilities for such measurements exist using radioisotopes such as ^{88}Rb and ^{66}Ga and nuclear reactions such as (n,γ) and (p,γ). Such new data can therefore be obtained on either the wavelength- or the mass-based energy scales, or a mix-ture of both, and thus it will be essential to address the question of the compatibility of the two separate energy scales. Fortunately, a direct comparison of these scales is available through Eq. 1; by measuring, through cascade sums following the (n,γ) reaction with

TABLE VII.   γ-Ray energies emitted in the ^{14}N(n,γ) reaction with
            thermal neutrons

γ-ray energy (keV)	Uncertainty (eV)	
	σ_m	σ_t[a]
1678.260	59	59
1681.589	228	228
1884.820	47	47
1999.729	77	77
3677.748	56	56
4508.665	52	53
5269.121	32	35
5297.794	32	35
5533.400	32	35
5562.072	32	35
6322.474	57	60
9148.608	229	230
10829.101	24	38

[a]The reference uncertainty was taken to be identical to that of $S_n(^{15}N)$; i.e., as 2.7 ppm.

slow neutrons, a wavelength-based value of $S_n(^{A+1}Z)$ corresponding to one of the precisely measured doublet mass differences[6-8]. Perhaps the most promising reaction for this purpose is $^{12}C(n,\gamma)^{13}C$ since $S_n(^{13}C)$ is only 4.9 MeV and the cascade-crossover relationship

$$E_t(1262) + E_t(3684) = S_n(^{13}C) \tag{7}$$

is the sole decay mode from the capturing state. Possibilities for measuring the energy of the 3684 keV cascade transition on the wavelength scale, using Ge-semiconductor detectors, include calibrations against radioisotopic sources such as ^{56}Co and ^{66}Ga.

REFERENCES

1. R. G. Helmer, R. C. Greenwood and R. J. Gehrke, Atomic Masses and Fundamental Constants, Vol. 5, ed. by J. H. Sanders and A. H. Wapstra, (Plenum, New York, 1976), p. 30.
2. E. G. Kessler, R. D. Deslattes, A. Henins and W. C. Sauder, Phys. Rev. Lett. 40, 171 (1978).
3. R. G. Helmer, R. C. Greenwood and R. J. Gehrke, Nucl. Instrum. Methods 155, 189 (1978).
4. R. C. Greenwood, R. G. Helmer and R. J. Gehrke, Nucl. Instrum. Methods 159, 465 (1979).
5. R. C. Greenwood and R. E. Chrien, Proc. 3rd Intern. Symp. on Neutron Capture Gamma-Ray Spectroscopy and Related Topics (Plenum, New York, 1979), p. 618.
6. L. G. Smith, Phys. Rev. C 4, 22 (1971).
7. L. G. Smith and A. H. Wapstra, Phys. Rev. C 11, 1392 (1975).
8. A. H. Wapstra and K. Bos, At. Data Nucl. Data Tables 19, 177 (1977); ibid 20, 1 (1977).
9. E. R. Cohen and B. N. Taylor, J. Phys. Chem. Ref. Data 2, 663 (1973).
10. G. L. Borchert, W. Scheck and K. P. Wieder, Z. Naturforsch. 30a, 274 (1975).
11. G. L. Borchert, Z. Naturforsch. 31a, 387 (1976).
12. J. J. Reidy, The Electromagnetic Interaction in Nuclear Spectroscopy, ed. by W. D. Hamilton, (North-Holland, Amsterdam, 1975), p. 873.
13. G. L. Borchert, W. Scheck and O. W. B. Schult, Atmic Masses and Fundamental Constants, Vol. 5, ed. by J. H. Sanders and A. H. Wapstra, (Plenum, New York, 1976) p. 42.
14. J. Kern and W. Schwitz, Nucl. Instrum. Methods 151, 549 (1978).
15. M. A. Ludington, J. J. Reidy, M. L. Wiedenbeck, D. J. McMillan, J. H. Hamilton and J. J. Pinajian, Nucl. Phys. A119, 398 (1968): J. J. Reidy, private communication (1961).
16. W. Beer and J. Kern, Nucl. Instrum. Methods 117, 183 (1974).
17. W. Schwitz and J. Kern, Proc. 2nd Intern. Symp. on Neutron Capture Gamma Ray Spectroscopy and Related Topics (RCN, Petten, 1975), p. 697.

18. E. G. Kessler Jr., L. Jacobs, W. Schwitz and R. D. Deslattes,
      Nucl. Instrum. Methods 160, 435 (1979).
19. O. Pillar, W. Beer and J. Kern, Nucl. Instrum. Methods 107, 61
      (1973).
20. G. L. Borchert, W. Scheck and O. W. B. Schult, Nucl. Instrum.
      Methods 124, 107 (1975).
21. U. Gruber, R. Koch, B. P. Maier and O. W. B. Schult;
      Z. Naturforch. 20a, 929 (1965).
22. R. W. Jewell, W. John, R. Massey and B. G. Saunders, Nucl.
      Instrum. Methods 62, 68 (1968).
23. W. J. Prestwich, T. J. Kennett, L. B. Hughes and J. Fiedler,
      Can. J. Phys. 43, 2086 (1965).
24. R. C. Greenwood and W. W. Black, Phys. Lett. 21, 702 (1966).
25. H. W. Taylor, N. Neff and J. D. King, Phys Lett. 24B, 659 (1967).
26. R. G. Helmer, R. J. Gehrke and R. C. Greenwood, Nucl. Instrum.
      Methods 123, 51 (1975).
27. Ts. Vylor, K. Ya Gromov, A. I. Ivanov, B. P. Osipenko, E. A.
      Frolov, V. G. Chumin, A. F. Shchus and M. F. Yudin, Sov. J.
      Nucl. Phys. 28, 585 (1978).
28. R. C. Greenwood, Phys. Lett. 27B, 274 (1968).
29. R. G. Helmer, J. W. Starner and M. E. Bunker, Nucl. Instrum.
      Methods 158, 489 (1979).

THE ABSOLUTE MEASUREMENT OF ACCELERATOR BEAM ENERGIES

P. H. Barker, D. P. Stoker, H. Naylor,
R. E. White and W. B. Wood

Department of Physics
University of Auckland
Auckland, New Zealand

SUMMARY

A method has been developed for the measurement of accelerator beam energies to an accuracy approaching one part in 10^5. As a check, the energy of a resonance in $^{27}Al(p,\gamma)$ at E_p = 0.992 MeV has been determined. In addition, a new determination of a resonance energy in $^{16}O(p,p_0)$ at E_p = 6.50 MeV has been made. Finally, a provisional value for the $^{14}N(p,n)^{14}O$ threshold energy has been obtained.

INTRODUCTION

In recent years, our group in Auckland has made a study of (p,n) reaction Q-values, particularly as they relate to ft values of super-allowed Fermi beta decays. These experiments have consisted of the examination of the appropriate (p,n) yield curves close to threshold and the extraction of the threshold point, all on a nominal energy scale, together with the subsequent absolute calibration of this energy scale. As the theoretical considerations involved in the calculation of the ft values have gradually stabilised, the need for improved accuracy in the experimental data has grown, and consequently both parts of the experimental technique mentioned above have undergone several states of refinement.

Until now, the Auckland energy measurements have been made by comparing the momenta of protons in the incoming beam with that of ^{212}Bi alpha particles from a ^{228}Th source using an Enge split-pole magnetic spectrograph. Originally[1] this was achieved by scattering the proton beam from a thin gold line target and observing the

particles at 80° or 135°. The alpha particles were then focused
to the same point in the spectrograph focal plane as the protons
had been by altering the magnetic field and the proton momentum was
calculated using that of the alpha particles, multiplied by the
ratio of the appropriate nuclear magnetic resonance frequencies.
An improvement in this method was subsequently brought about by
positioning the spectrograph at 0°, thus removing the need to
measure a scattering angle precisely and allowing an improved
method of particle detection in the focal plane[2].

    The chief disadvantages at this stage were two-fold. Firstly,
the use of different magnetic fields for the protons and alpha
particles, with the consequent assumptions relating to differential
hysteresis behaviour, and secondly the use of radioactive alpha
particle sources, which tends to be something of a 'black art'.
Although the former factor has been exhaustively examined, (see
reference [3]), it would nevertheless be desirable to eradicate it
if possible, whilst the latter is a messy business which always
carries a slight tinge of uncertainty. In addition, one relies
for the values of the alpha particle momenta on one set of experimental data by Rytz[4], and whilst these are not seriously in doubt,
a possible discrepancy has been reported[5].

    The present paper reports a method for the determination of
the energy of a charged particle beam, based on a method given
originally by White et al[6] for measuring the energy of ^{210}Po
alpha particles. Ions of a heavy element (either potassium or
thallium) are accelerated through a voltage, V, which is adjusted
until the ions and the beam particles have the same momenta, as
determined in the fixed magnetic field of the spectrograph. The
voltage V is measured relative to an absolute one volt standard.
The energy of the beam particles is then given directly in terms of
V and of the masses of the ions and the beam particles.

    It will be shown that for beam energies which are of immediate
interest, an accuracy approaching 1 in 10^5 is obtainable, although,
of course, limitations on the understanding of the exact processes
taking place in an experiment may well reduce the accuracy which
is ascribable to the energetics of particular features of that
experiment.

    As an example we have chosen to measure the energy of a
resonance in ^{27}Al(p,γ) at E_p = 0.992 MeV. This is not very well
suited to our accelerator, being rather low in energy, but has the
advantage that the accepted value has an error of only ±40 eV[7].
We could find no reaction energies in the range 2 - 10 MeV which
are quoted to comparable accuracy. In addition, we have determined
the energy of a narrow, isolated resonance in ^{16}O(p,p) at
E_p ≃ 6.50 MeV, which could serve as a convenient energy calibration

for low energy accelerators. Lastly, we have examined the $^{14}N(p,n)$ threshold curve, and obtained a Q-value, and hence a maximum energy for superallowed beta decay, which is more accurate than any previous value.

THE HEAVY ION SOURCE SYSTEM

The Ion Beams

The calibration of the proton energies involved in the present work, 0.99 MeV and 6.3 - 6.5 MeV, is conveniently carried out using 26 keV $^{38}K^+$ and 31 - 33 keV $^{205}Tl^+$ ions respectively. For the former, a solution containing alkali salts is applied to a filament and evaporated to dryness. The filament is then placed, together with its high voltage mounting, into the target chamber. A sketch plan of the arrangement is shown in figure 1. Potassium ions are produced by heating the filament to about $1000^\circ K$ with an alternating current. The ions are accelerated by the voltage V applied between the filament and the source cathode. This cathode is a v-shaped electrode at ground potential, having a slit aperture 0.05 mm wide by 3 mm high placed at the spectrograph object position. For discussions of general considerations relating to such ion beams see references [8,9,10].

Beams of thallium ions are produced in essentially the same way as for potassium except that the thallium is not applied directly to the filament as a salt, but is contained instead, in metallic form, in a separately heated ceramic oven held at ground potential. When the oven, which is drawn dotted on figure 1, is heated, a cone of thallium vapour emerges, much of which strikes the filament.

After the ion beam has been deflected by the vertical magnetic field of the spectrograph, it passes through a vertical slot, 0.05 mm wide, in the focal plane and then strikes an insulated collector plate. Scanning over a small range of accelerating voltage allows an ion beam intensity profile to be recorded by observing the variation in the current. Figures 2 and 3 show typical curves for potassium and thallium.

An examination has been made of the dependence of the position of the ion beam image on such factors as the filament position, its physical condition and temperature, the height or orientation of the v-shaped tantalum extraction slot, or the size of the spectrograph entrance aperture (normally $\lesssim \pm \frac{1}{6}^\circ$ wide x $\pm \frac{1}{3}^\circ$ high). All these produce negligibly small uncertainties.

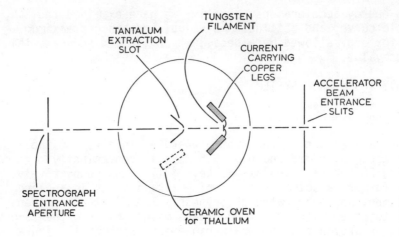

Fig. 1.   A sketch of the plan of the ion source arrangement.

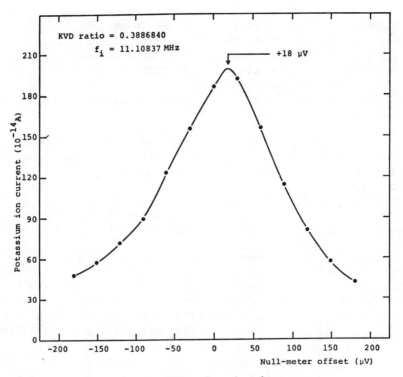

Fig. 2.   A potassium ion beam scan.

Voltage Measurements

The accelerating voltage is obtained from a 0 - 60 kV power supply, having a nominal stability of 0.005% per hour. A diagram of the voltage measuring circuit is presented in figure 4. It consists essentially of a $(10^4 + 1):1$ step-down resistive divider, the output of which is compared with a one volt standard using a precision step-down resistance box and a sensitive microvoltmeter.

The one volt transfer standard is Hewlett-Packard model 735A. It is checked periodically against standard cells at the Physics and Engineering Laboratory, New Zealand Department of Scientific and Industrial Research, and they themselves are checked anually against standards which have been calibrated using the AC Josephson effect. The accumulated uncertainty in the transfer standard is certainly not greater than five microvolts. The value of the standard is 1.000002 volts.

The upper arm of the $(10^4 + 1):1$ divider is made up of one hundred resistors connected in series, each nominally $1 M\Omega$ resistance. These were supplied by General Resistance Corporation, and their values are guaranteed to be distributed within 0.01% of their mean value.

The lower arm is a parallel arrangement of two sub-arms. The first has a 10 kΩ composite resistor (actually nine 10 kΩ resistors in 3 x 3 series-parallel) in series with a 1.105 kΩ resistor and a 10 Ω trim helipot. The second sub-arm is the input impedance, (100 kΩ), of a resistance divider, Julie Research Corporation model VDR307 (Referred to as KVD). Thus the resistance of the bottom arm can be adjusted to have an overall value of 10^{-4} of that of the upper arm. The Julie Corporation resistive divider is of the Kelvin-Varley type incorporating seven decades, with a fixed input impedance. The output ratio is guaranteed to ± 1 in the sixth decade, and the input impedance to ± 0.5 ppm/$^\circ$C.

In order to set the overall step-down ratio to $(10^4 + 1):1$, use may be made of a result attributed to Lord Rayleigh[11]. If a set of N resistors has mean resistance M and individual fractional differences m_i from M, so that $R_i = M (1 + m_i)$ then their total series resistance R_s is related to their total parallel resistance R_p by

$$R_s = N^2 (1 + \frac{1}{N} \sum m_i{}^2) R_p$$

assuming higher powers of m_i are negligible. In the present case, since the m_i are smaller than 10^{-4}, a knowledge of R_p provides R_s to better than 0.01 ppm. Hence an arm ratio of $10^4:1$ was achieved by connecting the one hundred resistors in the top arm in parallel, and, using the top and bottom arms in a bridge arrangement, adjust-

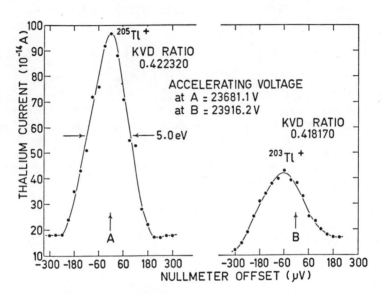

Fig. 3.   Beam scans for two Tℓ isotopes.

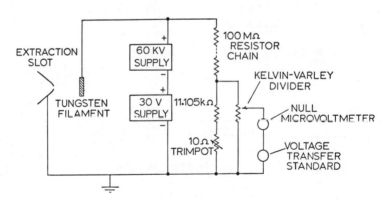

Fig. 4.   Diagram of the voltage measuring circuit.

ing the trim helipot in the bottom arm until the resistances of the top and bottom were equal. The parallel connections on the top arm were then removed, leaving it series connected with a step-down ratio of $(10^4 + 1)$ with the only error contribution coming from the balancing procedure and estimated to be < 3 ppm.

The series 100 MΩ chain is in the form of a square spiral supported on the outside of a perspex former, which is of side 350 mm and height 450 mm. Each resistor is shielded in a copper cylinder, 15 mm diameter which is maintained at the potential of the lower end of the resistor, and the top and bottom resistors are connected to spun aluminimum domes. These measures reduce corona currents. The whole structure is maintained in a thermo-statically controlled dried atmosphere inside a polythene cylinder approximately 1100 mm high and 1100 mm diameter (A commercial garbage container). A photograph of the divider, with the container lid raised, is shown in figure 5. The effects of leakage down the perspex former, of corona leakage, and of temperature variation of the enclosure have been studied and all are negligible at the level of one part in 10^5.

As a check on the balancing procedure and on the time stability of the apparatus, the divider was balanced, the parallel connections removed, and then replaced three months later with the divider having been used in the meantime. It was found that the balance point had moved by 2 parts in 10^6.

Because of Joule self heating, and the Peltier and Seebeck effects, the effective resistance of the 1 MΩ resistors in the upper arm is a function of the applied voltage. At 50 kV (i.e. 500 V across each resistor) effects of the order of 100 ppm are to be expected. (The balancing procedure had, of course, been conducted with a low voltage across each resistor). Accordingly the total series arm resistance was determined as a function of applied voltage using bridge techniques in two independent series of measurements. The resulting curve is shown in figure 6 and is accurate to better than 3 ppm over the voltage range 0-50 kV.

Heavy Ion Kinetic Energy

Sumin has shown that atomic ions produced by surface ionisation and thermionic emission processes from surface ionisation sources have an initial Maxwellian energy distribution characteristic of the ioniser surface temperature, T[12)[13). Thus when the ions are subsequently accelerated through a potential V their mean kinetic energy is given by

$$E_i = eV + \tfrac{3}{2} kT.$$

It is to be noted, however, that the true accelerating voltage V

Fig. 5. Photograph of the divider with container lid raised.

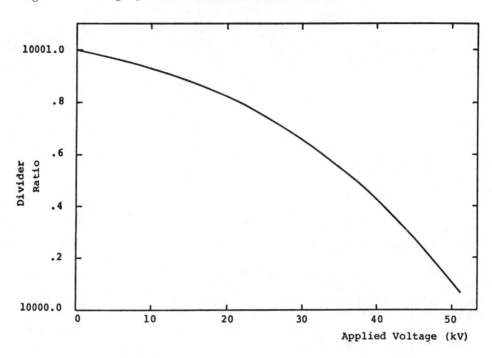

Fig. 6. The divider step-down ratio as a function of the applied
         voltage.

is the voltage measured by the voltage divider, suitably corrected for the effective contact potential between the filament surface and the ground defining point of the spectrograph, i.e.

$$V = V_{meas} - V_c$$

where V_c is the contact potential, which should be empirically determined for each experimental set up.

In the present cases, where the filament surface is tungsten oxide contaminated with either alkali salts or thallium, the appropriate V_c's were found by bringing either caesium and potassium, or thallium and potassium beams, at constant spectrograph field, round through the detector slot. Then, if M_i is ion mass, the requirement that the product $M_i E_i$ be constant for each pair provides the values of V_c, which were found to be for alkali-tungsten oxide, 1.04 ± 0.10 V, and for thallium-tungsten oxide, 2.20 ± 0.10 V. (Potassium is normally present in the metallic tungsten).

Beam Energy Measurements

The procedure for a beam energy measurement during a typical experiment is as follows. The experiment is first performed roughly to determine at what NMR frequency of the accelerator 90° analysing magnet the beam energy needs to be calibrated. The spectrograph magnetic field is then set to such a value that the particle beam passes through the object and image slots.

At this point the heavy ion source system is introduced into the target chamber and a potassium or thallium energy profile taken. As the peaks are ~ 0.02% FWHM this is most easily done by arranging a variable low voltage to act in series with the main accelerating voltage power supply. With the ion beam striking the slot and this low voltage near midrange, the null microvolt-meter is balanced by adjusting the 7-decade divider. Varying the low voltage supply then allows a curve of ion beam intensity versus microvoltmeter reading, such as the one in figure 2, to be obtained easily. After this, and without altering the spectrograph magnetic field, the object slot or the image slot, the ion source is removed, the accelerator beam passes through both slots, and the experiment in question performed. Thus the spectrograph functions as a momentum analyzer of resolution approximately 10^4, for which the magnetic rigidity of the central path is calibrated to approximately 1 in 10^5.

THE $^{27}Al(p,\gamma)$ RESONANCE AT 992 keV

This resonance has been investigated using the methods outlined above. Targets were of aluminium, evaporated in situ on a tantalum backing, and were struck by the momentum analysed beam in the back of the spectrograph magnet box. The thickness of the targets was estimated to be ~ 100 μgm/cm^2, and the yield of the resonance was monitored by observing 1.78 MeV gamma rays in a 2" x 3" NaI detector at ~ 45° to the incident beam.

To observe the resonance, the beam energy was set a few keV above the resonance energy and the gamma ray yield followed as an increasing, positive voltage was applied to the insulated target. Figure 7 shows such a yield curve.

Both before and after each yield curve was taken, the magnetic rigidity of the central path through the spectrograph was calibrated using a beam of ^{39}K$^+$. Table I shows the contributions of the various factors to the calibration number (ME/f^2), where M and E are the mass and kinetic energy of the ion respectively, and f is the spectrograph NMR frequency. The kinetic energy of the proton beam, E_p is then given by

$$E_p = \frac{ME}{f^2} \cdot \frac{1}{m_0} \cdot \left(1 + \frac{E_p}{2m_0 c^2}\right)^{-1} f_p^2$$

where m_0 is the mass of a proton and f_p is the spectrograph NMR frequency at which the resonance experiment was carried out. Normally $(f - f_p)$ is less than 3 parts in 10^5 that is, the spectrograph field is essentially unchanged.

In Table 2 is shown the calculation, using the curve of figure 7 together with the calibration of table 1 of the resonance half height energy. In two separate experiments, values obtained were 991.92 (4) keV and 991.90 (4) keV, giving a mean of 991.91 ± 0.03 keV. This value is uncorrected for non-uniform proton energy loss in the target or for the effect of the natural resonance width. It may be compared with the two currently most accurate values also uncorrected, of 991.91 ± 0.05 keV[14] and 991.88 ± 0.10 keV[15]. The mean of these last two results will probably rise to 991.92± 0.04 keV when updated using the 1973 fundamental constants.

THE ^{16}O(p,p$_0$) RESONANCE AT 6.5 MeV

A resonance in ^{16}O(p,p$_0$) at 6.50 MeV is reported in ref.[16] as being narrow (~ 2 keV), isolated and roughly symmetric at back angles. Accordingly the resonance was investigated using essentially the same technique as discussed in section 4. Targets were of boron oxide of thickness greater than 10 keV to the beam, evaporated on to 20 μgm/cm^2 carbon foils. The scattered protons were observed at (165 ± 5)° to the incident beam in a silicon semiconductor detector and the spectrograph calibrated with ^{205}Tl$^+$ ions of approximately 32 keV.

A yield curve for the resonance is shown in figure 8 from which a resonance half-height was extracted.

The weighted mean of seven such determinations was 6494.36 keV with internal error 0.05 keV, and external scatter error of 0.06 keV. This is again uncorrected for non-uniform proton energy loss or resonance width. Because the thin target resonance has a slight high energy tail, care was taken to define the full height at least 3 - 4 keV above the half-height of the thick target yield.

THE $^{14}N(p,n)^{14}O$ THRESHOLD AT 6.31 MeV

Again the experiment was performed after the spectrograph with a momentum analysed beam and the calibration carried out with $^{205}Tl^+$ ions. Targets of melamine, evaporated on to a gold backing were thicker than 7 keV to the proton beam.

To obtain a yield curve near threshold, a target was bombarded for 100 second and was then moved 40 cm, using a magnetically coupled pneumatic piston, to a shielded counting station. After a 5 second delay period, the yield of ^{14}O was examined by detecting the 2.31 MeV gamma rays which follow the ^{14}O beta decay using a Ge(Li) detector. During this count period the proton beam was stopped on a tantalum shutter 5 meters upstream of the spectrograph.

A yield curve such as figure 9 was obtained by applying increasing positive voltages to the target. This yield curve may be analysed using the well known function

$$Y = A + B (E - E_0)^{\frac{3}{2}}$$

where A is the background and E_0 the threshold energy, and the latter extracted.

The four yield curves taken had background statistically compatible with zero and give a threshold energy of 6353.99 keV, with internal error 0.08 keV and external scatter error of 0.07 keV. This value is uncorrected for non-uniform proton energy loss and for the beam energy width but these will certainly be smaller than 100 eV. It leads to a Q-value for the superallowed beta decay of ^{14}O of 1809.2 ± 0.1 keV.

Unfortunately the above figure must be treated as provisional only, since intermittent malfunctioning of the constant temperature resistive divider enclosure made a completely satisfactory calibration impossible.

Other determinations of this energy to sub-keV accuracy are shown in table 3.

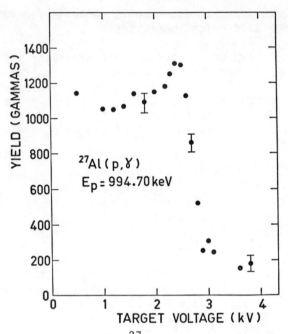

Fig. 7. Yield curve for the ^{27}Al(p,γ) resonance at ep=994.70 keV.

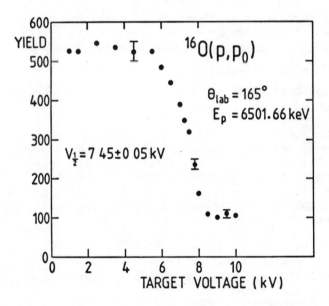

Fig. 8. Yield curve for the ^{16}O(p,p$_0$) resonance at 6.50 MeV.

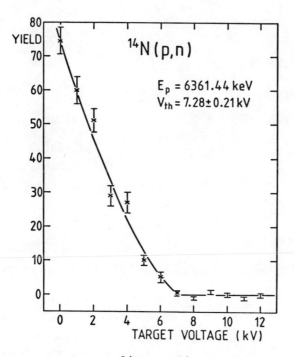

Fig. 9.   Yield curve for the $^{14}N(p,n)^{14}O$ reaction near threshold.

Table 1. Calibration of the magnetic rigidity of the central
            path through the spectrograph.

KVD ratio	0.388684	(1)
Null meter offset for $^{39}K^+$ peak	18	(8) μV
High voltage divider ratio	10,000.74	(3)
Transfer standard voltage	10.00002	(3) V
∴ Voltage applied at $^{39}K^+$ peak	25730.26	(25) V
$\frac{3}{2}$ kT	0.16	(2) eV
Contact potential	- 1.04	(6) eV
∴ $^{39}K^+$ energy, E	25729.38	(26) eV
Spectrograph frequency, f	11.10837	(6) MHz
$^{39}K^+$ mass, M	38.963159	(1) u
∴ $\frac{ME}{f^2}$	8124.24 eV.u.(MHz)$^{-2}$	(12)

Table 2. Calculation of resonance half height energy.

Initial $\frac{ME}{f^2}$	8124.24 eV.u.(MHz)$^{-2}$	(12)
Final $\frac{ME}{f^2}$	8124.01 eV.u.(MHz)$^{-2}$	(12)
Mean $\frac{ME}{f^2}$	8124.12	(12)
Proton mass	1.007276	u
Spectrograph frequency for protons f_p	11.10824	(9) MHz
∴ Proton energy	994.69	(2) keV
Target voltage for resonance half height	2.77	(3) kV
∴ Resonance half height energy	991.92	(4) keV

Table 3. Q-value of the ^{14}O Superallowed Beta Decay.
          (Values taken from ref[21])

Q_β (keV)	Reference
1809.34 ± 0.7	[16]
1810.24 ± 0.5	[17]
1810.54 ± 0.5	[18]
1808.78 ± 0.4	[19]
1807.88 ± 0.8	[20]
1810.37 ± 0.6	[21]
1809.2 ± 0.1	Present provisional value

REFERENCES

1) D.C. Robinson and P.H. Barker, Nucl. Phys. A225 109 (1974).
2) P.H. Barker, R.E. White, H. Naylor and N.S. Wyatt, Nucl. Phys.
       A279 199 (1977).
3) P.H. Barker, H. Naylor and R.E. White, Nucl. Instr. and Meth.
       150 537 (1978)
4) A. Rytz, At. Data and Nucl. Data Tables, 12 479 (1973).
5) P. Glässel, E. Huenges, P. Maier-Komor, H. Rösler, H.J. Scheerer,
       H. Vonach and D. Semrad, Proc. 5th Int. Conf. on Atomic
       Masses and Fundamental Constants, Paris, (Plenum Press,
       New York) 110 (1975).
6) F.A. White, F.M. Rourke, J.C. Sheffield and R.P. Schuman,
       Phys. Rev. 109 437 (1958).
7) P.M. Endt and C. van der Leun, Nucl. Phys. A310 1 (1978).
8) N.I. Ionov, Prog. Surf. Sci. 1 237 (1972).
9) E.Ya. Zandberg and N.I. Ionov, Poverkhnostnaya Ionizatsiya
       (Surface Ionisation), Nauka, Moscow (1969), English
       translation: SFCSI COMM (TT-70-50148) (NTIS).
10) M. Kaminsky, Atomic and Ionic Impact Phenomena on Metal Sur-
       faces (Academic Press Inc., N.Y. 1965).
11) Lord Rayleigh, Phil. Trans. Roy. Soc. (Lond.) 173 697 (1882).
12) L.V. Sumin, Sov. Phys. Tech. Phys. 19 1205 (1975).
13) L.V. Sumin, Sov. Phys. Tech. Phys. 21 303 (1976).
14) M.L. Roush, L.A. West and J.B. Marion, Nucl. Phys. A147 235
       (1970).
15) A. Rytz, H.H. Staub, H. Winkler and W. Zych, Helv. Phys. Acta
       35 341 (1962).
16) J.W. Butler and R.O. Bondelid, Phys. Rev. 121 1770 (1961).

17) R.K. Bardin, C.A. Barnes, W.A. Fowler and P.A. Seeger, Phys.
    Rev. 127 583 (1962).
18) M.L. Roush, L.A. West and J.B. Marion, Nucl. Phys. A147 235
    (1970).
19) R.E. White and H. Naylor, Nucl. Phys. A278 333 (1977).

20) H. Vonach et al., Nucl. Phys. A278 189 (1977).
21) P.H. Barker and J.A. Nolen, Tokyo Conference on Nuclear Physics,
    (1977).

ABSOLUTE MEASUREMENT OF THE ENERGY

OF ALPHA PARTICLES EMITTED BY ^{239}Pu

Albrecht Rytz

Bureau International des Poids et Mesures
F-92310 Sèvres, France

INTRODUCTION

The spectrum of the alpha particles emitted by ^{239}Pu has been investigated in considerable detail by several authors since 1948. The energy values corresponding to the various lines observed were determined with respect to well known standards. Table 1 gives a chronological summary of the reported energy values for the most intense transition.

The long half-life ($T_{1/2} \approx 2.4 \times 10^4$ a) implies a low specific activity and considerable line broadening which renders absolute energy measurements difficult, even with an isotopically pure source. Since ^{239}Pu sources of high quality were available, it seemed worthwhile to try such a measurement. The result obtained consolidates the knowledge of the ^{239}Pu–^{235}U mass difference.

EXPERIMENTAL PROCEDURES

The BIPM spectrometer as described in [9] covers α particle energies from 2 to 11.5 MeV and particle orbits of a radius between 365 and 475 mm. The line shape appearing on the detector (Ilford K2 nuclear track plate) can be predicted from geometrical considerations. It further depends more or less strongly on the properties of the source except for the high-energy edge which contains all the useful information on particle energy. This edge extends usually to a distance corresponding to less than 2 keV from the endpoint. In this interval the observed intensity falls off with the power 3/2 of the distance, thus permitting an extrapolation to zero intensity as described in [9].

249

Table 1. Alpha-particle energy values for the main line of ^{239}Pu

Originally published (keV)	Standard(s) used	Adjusted value (keV)	year	Reference
5 159 ± 5	^{210}Po, ^{226}Ra, ^{222}Rn	5 159	1948	Cranshaw [1]
5 147	^{212}Bi	5 148	1950	Rosenblum [2]
5 150 ± 2	^{210}Po	5 156	1952	Asaro [3]
5 147.4 ± 0.2	^{210}Po	5 153.4	1955	Gol'din [4]
5 150	^{210}Po ?	5 156	1961	Dzhelepov [5]
5 155.7 ± 0.6	^{212}Bi	5 156.3	1962	Leang [6]
5 157	^{240}Pu	5 156.1	1962	Baranov [7]
5 156.77 ± 0.41	^{240}Pu ?	5 156.65	1975	Baranov [8]
5 156.70 ± 0.14	absolute	5 156.70	1979	present work

For the present experiment a new spectrograph was constructed. It represents a length standard connecting the entrance slit with the plate holder by a rigid bar. The plate holder is furnished with twenty-four radio-active markers in two rows. The distance of each marker from the inner edge of the slit was measured on a classical transverse comparator using a recently calibrated standard meter bar. The thermal expansion coefficient was determined between 18 and 25 °C. Each marker exposes a fine tungsten wire to a collimated pencil of α particles from a ^{241}Am source of about 10 kBq (see Fig. 1), covered by 12 μg/cm^2 of aluminum. The wire is mounted perpendicular to the spectrograph axis and almost in contact with the photographic plate onto which it casts a shadow which is clearly visible after development (see Fig. 2). For further details of the length measurements see [9].

The spectrograph is surrounded by a box consisting of 3 mm thick copper sheet. This box is maintained at a constant temperature by a thermoregulated water circuit. It has been verified that the mean temperature of the spectro-

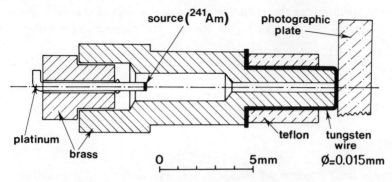

Fig. 1. Longitudinal section of one of the 24 markers for impressing a scale
of reference lines on the photographic plate (see Fig. 2).

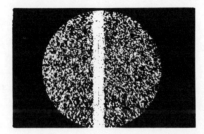

Fig. 2. Micrograph of a reference line impressed on the nuclear track plate
by one of the markers (Fig. 1). The white area is the shadow of
a straight tungsten wire of 15 μm diameter; the black spots are
α-particle tracks in the emulsion.

graph does not deviate by more than 0.1 K from that of the water, for
a temperature difference $\leqslant 3$ K between the water and the vacuum chamber
walls.

With the two sources available six different runs could be completed.
The summed durations were 1 395 h (four runs with source no. 1) and 935 h
(two runs with source no. 2).

In the present experiment, where only very small numbers of tracks
could be obtained, an alternative extrapolation method [10] based on the
principle of maximum likelihood might be expected to give more reliable
results than the usual least-squares adjustments. Both methods were applied
to all six runs. As no systematic difference appeared, the mean of each pair
of values was taken.

The determination of the radius of curvature ρ and of the mean magnetic induction B is described in $[9]$. In the formula for the energy

$$E_\alpha = a\,(B\rho)^2 + b\,(B\rho)^4 + d\,(B\rho)^6$$

the factors a, b and d were derived from the latest adjustment of fundamental constants $[11]$.

SOURCE PREPARATION

For the present experiment it was imperative to use highly purified ^{239}Pu because of the vicinity of the ^{240}Pu lines. Such material was made available by the Central Bureau for Nuclear Measurements (Euratom, Belgium) where two sources were prepared at an interval of 22 months.

The primary ^{239}Pu had an isotopic purity of (99.977 ± 0.003) %, the principal impurity being ^{240}Pu (0.021 %). The PuF$_3$ was placed in a Pt boat, calcined at 500 °C and subsequently transformed to the anhydrous fluoride by interaction with gaseous HF and H$_2$ at 550 to 600 °C. This fluoride was then used for evaporation in high vacuum from a Pt crucible onto a support at 25 cm, made of Pt for the first source and of Metglas for the second. The Pu masses per unit surface were 92 and 57 μg·cm^{-2}, the total masses 6.9 and 4.3 μg, respectively. The technique for obtaining well-defined and uniform fluoride deposits of fissile materials is described in $[12]$.

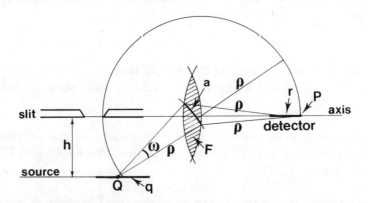

Fig. 3. Schematic view of the spectrograph geometry, with
       q = useful source section,   h = source/slit distance,
       r = useful detector section,   a = arch corresponding to r,
       ρ = radius of particle orbits,   P = endpoint   and
       F = surface containing the centers of all orbits reaching r.

LIMITING CONDITIONS FOR ABSOLUTE ENERGY MEASUREMENTS

Transmission

Fig. 3 explains the calculation of the transmission factor D, making use of simple geometric considerations inspired by Collins et al. [13]. The detector section, useful for energy determination, of length r over which the intensity drops according to a 3/2-power law, determines the surface F containing the centers of all the orbits of radius ρ reaching the detector within r from the endpoint P and originating in the useful source section q which is determined by ρ, h and r. The arch length a which is proportional to the number of particles leaving the source at Q and arriving within r from P subtends the angle ω while the detector width b subtends the angle $\nu = b/(\pi \rho)$. The transmission factor is therefore $D = b \bar{\omega}/(4 \pi^2 \rho)$. With $\rho = 405$, $h = 30$, $r = 0.1$ and $b = 4.5$ (all in mm) one gets $D = 4.6 \times 10^{-7}$. Here $\bar{\omega}$ is the mean angle ω calculated from ρ, h and r.

Rate of α particles arriving within the useful detector section

Knowing the superficial density σ of the ^{239}Pu source, we can calculate the activity on the useful surface of the source. Multiplying this activity by the emission probability and the transmission factor, we get the rate of tracks to be expected in the absence of self absorption. This rate being about sixty times as high as experimentally observed (0.14 α particles per hour), it may be concluded that the effective source thickness is $d_e = \sigma/(60\mu) \approx 10^{-7}$ cm, where $\mu = 9.3$ g $\cdot$ cm^{-3} is the density of PuF$_3$. This thickness corresponds only to two or three layers of atoms.

Alternatively, dividing the energy loss of 0.64 keV, corresponding to a spectral displacement of r/2, by the stopping power of Pu for 5.2 MeV α particles, an activity is obtained which leads to an expected track number three times the experimental one. This result seems quite satisfactory in view of the simple model used.

Limits of observation

Experience has shown that, in order to get enough precision, the number of genuine particles reaching the useful detector section must not be less than fifty and should be at least five times the corresponding number of background particles. With a background rate of $B = 0.01$ h^{-1} the time needed is 1 000 h or 42 d. The limiting half-life $T'_{1/2}$ may thus be calculated using the observed particle rate $N = 0.14$ h^{-1} for a half-life of $T_{1/2} = 24\ 100$ a and an emission probability of $p = 0.73$:

$$T'_{1/2} = N \cdot T_{1/2} / (5 \, p \, B) \approx 95 \, 000 \, a.$$

For practical reasons and in order to get an acceptable precision of the extra-polation, the exposure time should not exceed 20 days, which sets a practical limit for the half-life of 45 000 a (for p = 1).

UNCERTAINTIES

The random uncertainty of the result for each run follows from the two corresponding extrapolations. By summing the inverse squares of these uncertainties, the inverse internal variance of the final result is obtained. Table 2 summarizes a detailed analysis of the systematic uncertainty which is based on experience and a few auxiliary measurements.

RESULTS

The weighted mean result of the six runs is E_α = 5 156.70 keV, with external and internal standard deviations of 0.17 keV and 0.11 keV, respectively. The estimated total systematic uncertainty is 0.25 keV.

The six results plotted versus source age (Fig. 4) may suggest a decrease of the measured value with age. However, a definite answer would only be possible with additional results.

One might also be tempted to discard the two lowest and least precise values as well as the idea of an age effect linear in time. The weighted mean of the four remaining values would then be 5 156.85 keV, whereas the linear extrapolation to zero age of all the six values gives 5 156.90 keV.

As the three results derived above are well within the estimated systematic uncertainty and as the source-age effect is not assured, we recommend   as the final result

$$E_\alpha = (5 \, 156.70 \pm 0.14) \, keV.$$

CONCLUSION

The energy of α particles of the main transition from [239]Pu has been measured by an absolute method using sources of high purity. The value obtained is in excellent agreement with the most recent relative determina-tion [8] and leads to a decay energy of 5 244.6 keV which is 0.9 keV higher than the adjusted value in [14] .

Table 2. Components of systematic uncertainty (in $10^{-6} E_{\alpha}$)

a) <u>Length measurements</u>

Mean distance of markers from slit	3
Distance between markers	5
Alignment of the plate	15
Spectrograph temperature during run	3
Dilation during scanning of plate	5
Transverse displacement during scanning	5
Extrapolation method	42

b) <u>Measurement of magnetic induction</u>

Mean difference before and after run	12
Hartree correction	3
Correction for diamagnetism and shape of probe	2
Frequency measurement	2
Residual fluctuations	3

c) <u>Constants</u>

Combined effect of uncertainties of m_e , e and N_A	3
	—
Sum in quadrature	48×10^{-6}

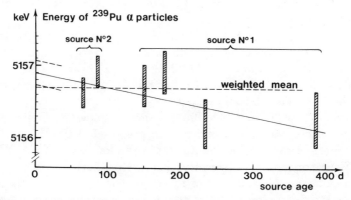

Fig. 4. Graphical representation of results from six runs versus source age.
Tentative extrapolation to zero age by least–squares adjusted straight
line. The broken lines indicate the limits corresponding to one
standard deviation.

The author wishes to express his sincere thanks to Dr. J. Van Audenhove and his coworkers from CBNM for preparing the two sources. He is further indebted to his colleague Dr. J.W. Müller for performing the extrapolation calculations and for stimulating discussions.

REFERENCES

1.   T.E. Cranshaw, and J.A. Harvey, Measurement of the energies of α-particles, Can. J. Res. 26 A : 243 (1948)

2.   S. Rosenblum, M. Valadares, and B. Goldschmidt, Structure fine du spectre magnétique α du Plutonium 239, C.R. Acad. Sci. Paris 230 : 638 (1950)

3.   F. Asaro, and I. Perlman, The Alpha-Spectra of ^{239}Pu and ^{240}Pu, Phys. Rev. 88 : 828 (1952)

4.   L.L. Gol'din, E.F. Tret'yakov, and G.I. Novikova, The alpha-spectra of heavy elements, "Proc. of a Meeting on At. En.", Moscow (1955), UCRL transl. 242

5.   B.S. Dzhelepov, R.B. Ivanov, and V.G. Nedovessov, Alpha decay of ^{239}Pu, Zh. Eksp. Teor. Fiz. (Sov. Phys.-JETP) 41 : 1725 (1961)

6.   C.F. Leang, Détermination des énergies des groupes intenses α émis par les noyaux des plutoniums 238, 239, 240 et les noyaux d'américium 241, C.R. Acad. Sci. Paris 255 : 3155 (1962)

7.   S.A. Baranov, V.M. Kulakov, and S.N. Belen'kii, Fine structure of ^{239}Pu alpha radiation, Zh. Eksp. Teor. Fiz. (Sov. Phys.-JETP) 43 : 1135 (1962)

8.   S.A. Baranov, and V.M. Shatinskii, Energy levels of the ^{241}Pu nucleus, Yad. Fiz. (Sov. J. Nucl. Phys.) 22 : 670 (1975)

9.   B. Grennberg, and A. Rytz, Absolute measurements of α-ray energies, Metrologia 7 : 65 (1971)

10.   D.J. Gorman, and J.W. Müller, Maximum-likelihood fit to points originating from different Poisson distributions, Rapport BIPM-72/6, Recueil de Travaux du BIPM 3 (article 22) (1971-72)

11.   E.R. Cohen, and B.N. Taylor, The 1973 least-squares adjustment of the fundamental constants, J. Phys. Chem. Ref. Data 2 : 663 (1973)

12.   J. Van Audenhove, P. De Bièvre, and J. Pauwels, Fission foils and alloys containing fissile materials prepared at CBNM, preliminary paper presented at the Materials Workshop, 2nd ASTM-EURATOM Symp. on Reactor Dosimetry, Palo Alto (Oct. 1977)

13.   E.R. Collins, C.D. McKenzie, and C.A. Ramm, A precise technique for the determination of some nuclear reaction energies, Proc. Roy. Soc. (London) A 216 : 219 (1953)

14.   A. Wapstra, and K. Bos, The 1977 atomic mass evaluation, ADNDT 20 : 1 (1977).

ATOMIC MASS DETERMINATIONS AT THE UNIVERSITY OF MANITOBA

R.C. Barber, K.S. Kozier*, K.S. Sharma, V.P. Derenchuk,
R.J. Ellis

Department of Physics
University of Manitoba
Winnipeg, Manitoba, Canada   R3T 2N2

V.S. Venkatasubramanian+ and H.E. Duckworth++

University of Winnipeg
Winnipeg, Manitoba, Canada   R3B 2E9

INTRODUCTION

We have, in our laboratory, two large mass spectrometers
which may be operated routinely with a resolving power of 100,000
or more. The older and larger instrument[1] ("Manitoba I") has a
2.7 m radius and is of a modified Dempster geometry. While it
was not used for the work reported here, it is currently in
operation and is being used to extend some of these measurements.

The newer instrument[2] ("Manitoba II") has a 1.0 m radius and
was built according to a second-order double focussing geometry
proposed by Hintenberger and König. Despite its smaller size, the
improved ion optics and mechanical design of this instrument have
made possible a performance generally somewhat superior to that
of Manitoba I.

We have used it to pursue two kinds of work: the first is
the extension of measurements amongst stable isotopes in the upper

* Now at Whiteshell Nuclear Research Establishment, Pinawa,
   Manitoba, Canada.
+ On leave from the Indian Institute of Science, Bangalore 560012,
   India.
++ Adjunct Professor of Physics, University of Manitoba, Winnipeg,
   Canada

half of the mass table, while the second is directed toward the
use of this instrument for the determination of the masses of
unstable nuclides.

In connection with this latter aspect, it should be noted
that the geometry of the instrument has the important feature
that the inclined boundaries of the magnetic field produce a net
focussing in the z-direction so that high transmission may be
achieved.

A small segment of the mass spectrum is scanned across the
collector slit by means of Helmoltz coils located in the final
drift space, and the ion current passing through the slit is
recorded in a 4096 channel multiscalar or viewed live on an
oscilloscope screen. Fig. 1 illustrates an accumulated spectrum
for the favourable material mercury chloride at mass number 274.
The resolution of the instrument is indicated at 5%, 15% and 50%
of the peak height. At the top of the figure the spacings of the
members of the triplet are also shown.

To determine the mass difference between the two members
of a mass doublet, the "peak matching" technique is used.[2] One
peak is chosen as a reference. Then, for every second sweep of
the oscilloscope display, the potentials in the instrument are
changed so as to displace the pattern. In particular, for the
electrostatic analyser, the potential change required to bring
a peak in the displaced spectrum to coincidence with the reference
peak, may be used to calculate the mass difference of the doublet.
The techniques by which the coincidence of the two peaks is
determined have been described elsewhere.[2,3,4]

As in all work reported by us since 1973, a systematic
correction has been made to the doublets reported here.[5] This
correction is necessary because unswitched surface potentials are
present. These may give rise to a systematic bias of perhaps 100
ppm in a doublet determination. The magnitude of the correction
is determined by the measurement of a well known, wide doublet
(that is, 1 or 2 u wide) and then applied as a linear correction
to the raw value obtained for the narrow doublet under study.

RECENT RESULTS

Region Hf to Re

We have been working in the region immediately above the
rare earths and have been attempting to establish mass links
between all of the naturally occurring nuclides. A part of this
work[6]   was incorporated in the 1977 Mass Evaluation[7]   while
one mass difference in Hf[8]   and a set of mass differences and

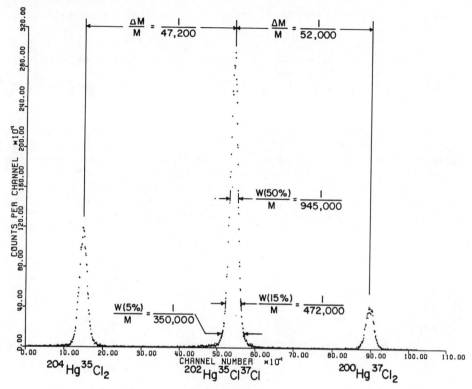

Fig. 1. High Resolution Mass Spectrum at M = 274.

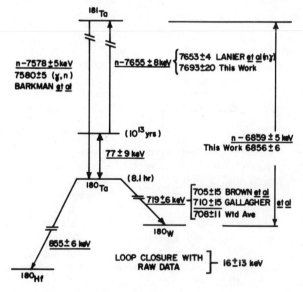

Fig. 2. Detail of Available Data for M = 180. The results of a
least squares evaluation are underlined.

atomic masses for W and Re [9]   appeared too late to be included.

In table 1 we present a set of new mass differences which complement the previous data by providing mass links between Hf, Ta and W. Further, doublet A provides an important independent value for the difference between the very rare isotope ^{180}Ta and ^{181}Ta.

Also in Table 1 a comparison with the 1977 Mass Evaluation is made by showing the difference between the two numbers. Inasmuch as the new values are generally superior in precision, the error reflects mostly the uncertainty assigned in the Mass Evaluation. In the particular case of doublet A, the poor agreement and large uncertainty in the present value reflects the very low natural abundances of ^{17}O(0.04%) and ^{180}Ta(0.01%).

Table 1. New Atomic Mass Difference (μu)

	Doublet	This Work	Mass Evaluation[a] − This Work
A	^{181}Ta^{17}O^{35}Cl$_2$−^{180}Ta^{16}O^{35}Cl^{37}Cl	7572±21	−43±25
B	^{181}Ta^{35}Cl$_2$−^{179}Hf^{35}Cl^{37}Cl	5128.6±2.1	8.5±4.6
C	^{183}W^{35}Cl−^{181}Ta^{37}Cl	5177.2±1.2	3.9±7.5
D	^{183}W^{16}O$_2$^{35}Cl−^{178}Hf^{35}Cl^{37}Cl	30 455.7±5.0	5.9±7.4
E	^{184}W^{16}O$_2$−^{181}Ta^{35}Cl	23 917.5±2.8	−2.0±10.1

[a]Reference 7

In Fig. 2, the data relating to the neutron separation energy of ^{181}Ta and to the ground states at mass number 180 are shown schematically. The ordering of levels in ^{180}Ta, with the naturally occurring 10^{13} year state above the 8.1 hour state, is that suggested by Wapstra and Bos [7].

The difference between ^{181}Ta and ^{180}W is given by our mass doublets. Also shown are the two determinations for Q_β-for the 8.1 hour state and a recent, precise (γ,n) Q-value by Barkman et al at McMaster University [10].

The loop which involves the 8.1 hour state as the ground state closes to 16 ± 13 keV as shown.

Lanier, et al [11]   have reported the value shown for neutron capture by the 10^{13} year state as determined with a natural sample of Ta. Our new value, which also can involve only the 10^{13} year state, is consistent with their value, despite the lower precision. When these data are combined, they place the 10^{13} year state 77 keV above the 81 hour ground state.

The underlined values are the results of a least squares adjustment of the mass differences.

Hg

In Table 2 we present new values for a set of doublets which yield mass differences and atomic masses for the stable isotopes of Hg. The data constitute 2 independent sets which overdetermine separately the odd-A and even-A atomic masses (except for ^{196}Hg). The two sets may be related by the addition of the (n,γ) Q-values on ^{198}Hg, ^{199}Hg, ^{200}Hg and ^{201}Hg 12.

Agreement between the doublet values which give mass difference (F,G,H,J,K,L,M) and the Mass Evaluation is extremely good. Agreement with sums of pairs of (n,γ) Q-values is also good.

TABLE 2: New Atomic Mass Differences for Hg (μu)

	Doublet	This Work	Mass Evaluation[a] – This Work
F	^{198}Hg^{35}Cl$_2$–^{196}Hg^{35}Cl^{37}Cl	3 885.91±1.66	+12.2 ± 9.8
G	^{200}Hg^{35}Cl$_2$–^{198}Hg^{35}Cl^{37}Cl	4 508.80±0.48	−2.7 ± 5.4
H	^{201}Hg^{35}Cl$_2$–^{199}Hg^{35}Cl^{37}Cl	4 972.65±0.37	+1.46 ± 5.4
J	^{202}Hg^{35}Cl$_2$–^{200}Hg^{35}Cl^{37}Cl	5 266.76±0.43	−0.65 ± 5.4
K	^{204}Hg^{35}Cl$_2$–^{202}Hg^{35}Cl^{37}Cl	5 800.67±0.53	−1.57 ± 5.4
L	^{202}Hg^{35}Cl$_2$–^{198}Hg^{37}Cl$_2$	9 774.87±1.06	−2.66 ± 5.6
M	^{204}Hg^{35}Cl$_2$–^{200}Hg^{37}Cl$_2$	11 066.85±0.55	−1.64 ± 7.6
N	^{199}Hg–^{12}C$_2$^{35}Cl$_5$	124 023.43±0.53	−18±6
P	^{200}Hg–^{13}C^{12}C^{35}Cl$_5$	120 707.97±1.22	−10±6
Q	^{201}Hg–^{12}C$_2$^{35}Cl$_4$^{37}Cl	128 995.43±0.61	−16±6
R	^{202}Hg–^{13}C^{12}C^{35}Cl$_4$^{37}Cl	125 976.01±1.32	−12±6
S	^{204}Hg–^{13}C^{12}C^{35}Cl$_3$^{37}Cl$_2$	131 776.05±1.25	−13±7

[a]Reference 7.

For the five doublets which give the absolute mass, it
appears that there may be, in this region, a systematic bias in
the 1977 Mass Evaluation by $\sim$ 14 μu. A closer examination of the
values confirms that, while each of the sub groups of mass spectroscopic data is internally consistent, the new even – A masses
appear to be biased systematically low by $\sim$ 5 μu. relative to the
odd-A masses. In contrast to the first 7 doublets, all 5 of these
doublets involve members which are chemically very dissimilar.

Our experience has been that the results for such doublets
appear to indicate that the two members may be produced at different locations in the source and may emerge from it with
different distributions in angle and energy. These problems are
further aggravated in this case by the use of fragmants ^{13}C from
a natural sample, a situation which required rather large total
ion currents.

Despite the suggestion of an unknown and unknowable bias is
these values, they form as a group, an important constraint on the
absolute values at the upper end of the region of stable nuclides.

Progress in developing system for Unstables

Finally, we report our progress in developing a system by
which we might use "Manitoba II" to measure directly the masses
of short-lived nuclides. (Fig. 3).

The University of Manitoba cyclotron may be used to produce
routinely $\sim$ 1 μA of protons at an energy of up to $\sim$ 40 MeV. For
our development work we are using the ^{63}Cu(p 3n)^{61}Zn reaction.

A helium jet system, in what is now a fairly common form, is
used to transport the unstable nuclides to the mass spectrometer
laboratory, a distance of some 80 m from the cyclotron. Helium
gas flows through an aerosol generator containing NaCl, carries
the aerosol particles to the target chamber and maintains the
pressure in the chamber around the target as $\sim$ 2/3 atmosphere.

The unstable nuclides emerging from the target adhere to the
aerosol particles which are carried by the helium flow and move
off through the capillary. Because the momentum transfer in the
reaction is relatively small, the number of unstable nuclides
which emerge from the back of the target is small. We have found
that the number of unstables transported through the capillary
could be increased by a factor of $\sim$ 10 by heating the target foil
to a temperature nears its melting point.

The capillary itself is a 2.2 mm diameter polyethylene tube,
for which the calculated transit time is $\sim$ 16 seconds. In

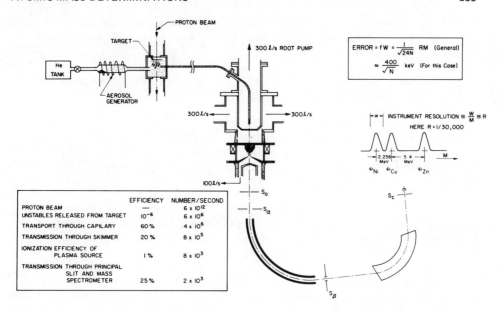

Fig. 3. System for Mass Measurement of Unstable Nuclides.

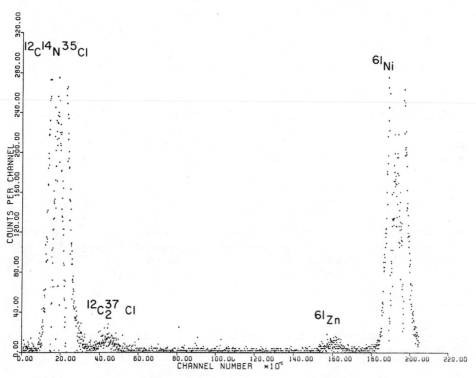

Fig. 4. Mass Spectrum at M = 61 showing 89 s ^{61}Zn.

preliminary experiments, the first part of the system was operated without the skimmer and the transport efficiency was found to be $\sim$ 60%.

We have followed the lead of the Chalk River group and have used a flat skimmer plate with a 1.5 mm diameter aperture. The region at the end of the capillary is evacuated by a 320 l/s Roots pump to a pressure of $\sim$ 0.1 torr, a value well below that required to achieve sonic flow. In the next region, two 300 l/s diffusion pumps reduce the pressure to between 10^{-3} with 10^{-4} torr so that 20 kV may be applied between the ion source and the skimmer.

The aerosol particles enter the ion source along its axis where the unstable nuclides are ionized in the plasma. To date we have used a modified Nielsen ion source for which efficiencies of $\sim$ 1% have been reported by others [13].

The mass spectrometer is operated at low resolution and high transmission for this work. With the principal slit wide open and the ion beam focused to a line within the slit, the resolving power is between 5 and 10 thousand and the transmission is $\sim$ 25%. The transmission is limited almost entirely by the relatively short length of the principal slit.

On the basis of these values for the efficiencies and transmission of the instrument one would expect, for a resolving power of $\sim$ 15,000, a count rate of $\sim$ 1000 c/s.

In Fig. 4 we show the spectrum accumulated over a period of $\sim$ 20 seconds at a resolving power of 18,000. All of the peaks have the nominal mass 61. The two on the left were caused by prior use of $C_2 Cl_6$ in the source to give a mass marker. The ^{61}Ni which has a natural abundance of 1.1% is also used as a marker. The ^{61}Zn peak was recorded with a count rate of $\sim$ 10^3 c/s, a value very near that which might be expected.

We have not as yet acheived sufficient stability or dependability in the entire system to determine a reliable value for the mass ^{61}Zn. In particular we hope to improve the yield of unstables which arrive at the ion source, the coupling of the ion source to the helium jet and the ion source itself.

ACKNOWLEDGEMENTS

The assistance of S.F. Howes, S.R. Loewen, C. Lander and R. Scharein in the experimental work is gratefully acknowledged. We are also grateful for the cooperation of J.S.C. McKee, the Director of the University of Manitoba cyclotron laboratory and

several members of the cyclotron staff, notably I. Gusdal,
A. McIlwain, and W. Mulholland.

REFERENCES

1. R.C. Barber, R.L. Bishop, L.A. Cambey, H.E. Duckworth,
    J.D. Macdougall, W. McLatchie, J.H. Ormrod, and P. van
    Rookhuyzen, Second Int. Conf. Nuclidic Masses, ed. by
    W.H. Johnson (Springer-Verlag, Wien, 1963), p. 393.
2. R.C. Barber, R.L. Bishop, H.E. Duckworth, J.O. Meredith,
    F.C.G. Southon, P. van Rookhuyzen and P. Williams,
    Rev. Sci. Instrum. $\underline{42}$, 1 (1971).
3. J.O. Meredith, F.C.G. Southon, R.C. Barber, P. Williams,
    H.E. Duckworth, Int. J. Mass Spec. Ion Phys. $\underline{10}$, 359 (1972).
4. K. Kozier, Ph.D. Thesis, University of Manitoba (1977),
    Winnipeg, Canada, unpublished.
5. F.C.G. Southon, J.O. Meredith, R.C. Barber, H.E. Duckworth,
    Can. J. Phys. $\underline{55}$, 383 (1977).
6. R.C. Barber, J.W. Barnard, D.A. Burrell, J.O. Meredith,
    F.C.G. Southon, P. Williams and H.E. Duckworth, Can. J. Phys.
    $\underline{52}$, 2386 (1974).
7. A.H. Wapstra and K. Bos, Atomic Data and Nuclear Data Tables
    $\underline{19}$, 177, 1977.
8. K.S. Sharma, J.O. Meredith, R.C. Barber, K.S. Kozier, S.S.
    Haque, J.W. Barnard, F.C.G. Southon, P. Williams and H.E.
    Duckworth, Can. J. Phys. $\underline{55}$, 1360 (1977).
9. K.S. Sharma, K.S. Kozier, J.W. Barnard, R.C. Barber, S.S. Haque,
    and H.E. Duckworth, Can. J. Phys. $\underline{55}$, 506 (1977).
10. J.N. Barkman, J.E. McFee, T.J. Kennett, W.V. Prestwich, (1977).
    private communication.
11. R.G. Lanier, J.T. Larsen, D.H. White, M.C. Gregory, (1972).
    Bulletin of the Am. Phys. Soc. $\underline{A17}$, 899.
12. A.H. Wapstra and K. Bos, Atomic Data and Nuclear Data Tables
    $\underline{20}$, 1, 1977.
13. K.O. Nielsen, Nucl. Instr. Meth. $\underline{1}$, 289 (1957).

RECENT MASS DOUBLET RESULTS FROM THE UNIVERSITY OF MINNESOTA[*]

J. Morris Blair, Justin Halverson[†], Walter H.
Johnson, Jr., and Ronald Smith[‡]

School of Physics and Astronomy
University of Minnesota
Minneapolis, Minnesota 55455

This paper will describe some of the recent mass doublet measurements made at the University of Minnesota during the period since AMCO 5. I will discuss two Ph.D. thesis experiments. The first, submitted by Dr. Justin Halverson in June 1977 involves doublet measurements of some isotopes of erbium, hafnium and osmium and also a remeasurement of the ^{13}C - ^{12}C mass difference. The second thesis, by Ronald Smith, was submitted in December 1977 and involved our first attempt at doublet measurements of short half-life unstable isotopes with conventional doublet techniques.

The major part of Dr. Halverson's thesis involved precise mass determinations of several isotopes in the region A = 160 to 190. These measurements were an extension of earlier work by Benson[1] and Kayser[2] in the light rare earths. The goal of this work was to determine a few precise masses to be used to determine the general slope of the mass surfaces as detailed in the Wapstra tables of masses. Although there are many mass difference results between neighboring isotopes, both from reaction data and β-decay results as well as the extensive chlorine doublet measurements by the Manitoba

---

[*]Supported by grants from the Graduate School and the Institute of Technology of the University of Minnesota and grant MPS 74-19408 from the National Science Foundation. Accelerator facilities were provided under the U.S. Energy Research and Development Administration Contract # EY-76-C-02-1265.
[†]Present address: Savannah River Laboratory, E. I. du Pont de Nemours and Company, Aiken, S.C. 29801.
[‡]Present address: Intel Corporation, 3065 Bowers Avenue, Santa Clara, CA 95051.

group, there have been very few precise measurements which yield isotopic masses in the region A = 140 to 190. Wapstra and Bos[3] have indicated this need once again in Part IV of the 1977 Atomic Mass Evaluation. With this need in mind, we chose to concentrate on the determination of a few atomic masses precisely. To do this we chose isotopes that were abundant and located near strong hydrocarbon reference masses. An example is the region of the mass spectrum used for the erbium measurements shown in Figure 1. By measuring the four doublets shown one overdetermines the resultant atomic masses of ^{167}Er and ^{168}Er. Included in this series of measurements is a scale calibration doublet $C_{13}H_{12}$ - $C_{13}H_{11}$ which should yield a mass difference equal to the mass of one hydrogen atom. This measurement can therefore be used to correct any proportional systematic errors such as those caused by electrostatic surface charging. For the erbium measurements a correction of 7 ± 3 ppm was subtracted from all doublets.

Doublet measurements were done with our 41 cm. magnetic radius double focusing mass spectrometer. The error signal method of peak matching was employed using Nicollet Instruments model 1062 signal averager. Details of our use of this technique have been given earlier.[1,4]

Table I gives the doublets measured, the raw results and the adjusted differences following a least squares adjustment of the raw results. Note that only small adjustments are made to the raw results indicating the consistency of the measurements.

Similar sets of doublets were chosen for isotopes of hafnium and osmium. Table I also shows these results. In the case of osmium, only one strong hydrocarbon could be found in this region. Although an overdetermined set of doublets could still be measured, a calibration hydrogen mass unit doublet could not be included. Instead, the calibration correction of 1 ± 2 ppm from the hafnium measurements was employed. The osmium measurements were the most difficult because of source shorting problems caused by the high

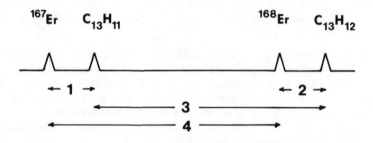

Fig. 1. Mass spectrum for the erbium doublets

Table I. Doublet Results

Doublet	Raw Data		Least Squares Adjusted	
	ΔM	$\sigma(\mu u)$	ΔM	$\sigma(\mu u)$
$C_{13}H_{12} - {}^{168}Er$	161543.3	5.1	161543.7	4.2
$C_{13}H_{11} - {}^{167}Er$	154040.4	6.2	154039.8	4.5
${}^{168}Er - {}^{167}Er$	1000320.9	4.3	1000321.2	3.8
$C_{14}H_{12} - {}^{180}Hf$	147356.6	4.8	147355.6	3.5
$C_{14}H_{11} - {}^{179}Hf$	140260.3	1.8	140260.4	1.7
${}^{180}Hf - {}^{179}Hf$	1000730.8	4.7	1000729.9	3.5
${}^{190}Os - C_{14}H_{21}$	794102.2	5.8	794100.4	4.7
$C_{14}H_{21} - {}^{189}Os$	206188.3	6.2	206186.2	4.9
${}^{190}Os - {}^{189}Os$	1000285.2	5.2	1000286.6	4.4

vapor pressure of the osmium chloride which served as the source of osmium ions. These problems led to focusing difficulties which caused a larger than normal spread in the individual osmium measurements and thus a somewhat larger final error.

The results obtained in this series of measurements are compared in Table II with values from the 1977 mass table[5] as well as a later adjustment which has been kindly provided recently by Professor Wapstra.[6] The comparison with the 1977 mass table results shows that our results are lower by about 20 μu. A similar difference was also found in the light rare earth region by Kayser[2] and by Benson.[1] Prof. Wapstra has provided me with an interim mass adjustment that was made earlier this year. These results are compared with the present results in the last column of Table II. The agreement between our present results and this new adjustment is considerably improved. A more detailed report of this work will appear in Physical Review C.[7]

As part of his thesis experiment Dr. Halverson remeasured the ${}^{13}C-{}^{12}C$ difference in order to obtain an accurate value for $S_n({}^{13}C)$ to be used to calculate the ground state γ-ray energy

Table II.   A Comparison of Present Mass Results with Values from
the Wapstra Mass Tables

Isotope	Present Mass u	Δ_{77} (μu)[a]	Δ_{79} (μu)[b]
^{167}Er	166.932 0356 ± 45[c]	-25 ± 8	-6 ± 7
^{168}Er	167.932 3568 ± 42	-26 ± 7	-8 ± 7
^{179}Hf	178.945 8150 ± 17	-12 ± 6	8 ± 5
^{180}Hf	179.946 5448 ± 35	-16 ± 7	4 ± 6
^{189}Os	188.958 1395 ± 49	-16 ± 9	10 ± 7
^{190}Os	189.958 4262 ± 47	-29 ± 8	-3 ± 7

[a]Δ_{77} = Present mass minus 1977 Mass Table value.

[b]Δ_{79} = Present mass minus 1979 Mass Table value.

[c]The error refers to the last significant figures of the
present result.

for the ^{12}C(n,γ)^{13}C reaction. Doublets used for this determination
are shown in Fig. 2. Ions are derived from ^{13}C benzene, deuterated
benzene and cyclohexane. In each case, the ion group employed is
a molecular ion. Raw doublet values again formed an overdetermined
set and were subjected to a least squares adjustment. The adjusted
values as well as the original results are shown in Table III. The
neutron separation energy for ^{13}C is found to be 4946.320 ± 0.050
kev as compared with 4946.392 ± 0.030 kev from the 1977 Wapstra
mass table.[5] Using our result, the ground state Q value for
^{12}C(n,γ)^{13}C reaction is 4945.310 ± 0.050 kev. This work has been
reported in Physical Review C Vol. 17.[8]

The thesis project of Dr. Ronald Smith under the direction of

$^{12}C_6H_{11}$ $^{13}C_6H_6$
                                                          $^{12}C_6D_6$
                                                          $^{12}C_6H_{12}$

Fig. 2.   Mass spectrum for the ^{13}C doublets

Table III. ^{13}C Doublet Results

Doublet	Raw Data		Least Squares Adjusted	
	ΔM	$\sigma(\mu u)$	ΔM	$\sigma(\mu u)$
$^{12}C_6H_{12}$ - $^{12}C_6D_6$	9289.7	0.3	9289.6	0.2
$^{12}C_6H_{12}$ - $^{13}C_6H_6$	26820.6	0.3	26820.7	0.3
$^{12}C_6D_6$ - $^{13}C_6H_6$	17531.2	0.3	17531.1	0.2

Prof. J. Morris Blair and myself was our first attempt to measure atomic masses of short half-life radioactive atoms. In this experiment, a 15 cm. magnetic radius double focusing mass spectrometer was installed on one of the beam lines of the University of Minnesota 20 MeV Tandem Van de Graaff Accelerator. A modified Bernas ion source[9] was employed to produce fission product ions using the U(p,fission) reaction. Because of the high ionization efficiency of this ion source for alkali ions, we chose to measure neutron rich rubidium and cesium isotopes. The instrument was operated in an ion counting mode using a Daly detector.[10] Because count rates were low, wider defining slits than normal for stable mass measurements were used. A full width at half maximum resolution of 1300 was employed for the measurements. A section of the mass spectrum containing one member of the doublet to be measured was stored in one 256 channel quadrant of a Nicollet Instrument Model 1062 signal averager operated in a multiscalar mode. The second member of the doublet was stored in another quadrant of the multiscalar memory. Mass differences were determined using a computer-aided centroid calculation technique similar to that described by Meredith et al.[11] An example of a stored ion peak is shown in Fig. 3. The peak illustrated is ^{91}Rb and consists of 256 accumulated spectral sweeps each of 1.28 second duration. This rubidium isotope has a half-life of 58.5 sec. At about the time this experiment had reached a data-collection stage, word was received that the funding for the operation of the accelerator was being phased out. Thus the beam time available for this experiment was limited and the experiment had a definite termination date of September 1977. Under different circumstances with more beam time available, we believe that the data accuracy could be improved substantially.

With the time available, we measured a variety of one to four unit doublets for the isotopes ^{89}Rb to ^{95}Rb. A similar set of cesium doublets was measured. Analysis of the cesium results is

Table IV.   Mass Difference Comparisons for Rubidium

Doublet	Value u	Source
^{90}Rb - ^{89}Rb	1.002 500 ± 220	Present
	1.002 300 ± 60	Wapstra[a]
	1.002 545 ± 29	Epherre et al.[b]
^{91}Rb - ^{90}Rb	1.001 710 ± 220	Present
	1.001 730 ± 70	Wapstra
	1.001 660 ± 46	Epherre et al.
^{92}Rb - ^{91}Rb	1.003 050 ± 250	Present
	1.003 050 ± 220	Wapstra
	1.003 178 ± 57	Epherre et al.
^{93}Rb - ^{92}Rb	1.003 440 ± 300	Present
	1.002 370 ± 280	Wapstra
	1.002 238 ± 82	Epherre et al.
^{94}Rb - ^{93}Rb	1.005 240 ± 560	Present
	1.003 710 Syst.	Wapstra
	1.004 316 ± 85	Epherre et al.
^{95}Rb - ^{94}Rb	1.003 080 ±2300	Present
	1.003 120 Syst.	Wapstra
	1.003 097 ± 117	Epherre et al.

[a]See reference 3.

[b]See reference 12.

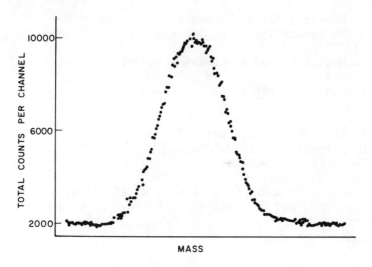

Fig. 3. ^{91}Rb Mass Spectrum

in progress and will be reported later. The 15 doublets that were measured among these seven isotopes of rubidium were subject to a least-squares analysis to produce the best set of mass differences. These differences are shown in Table IV and in that table are compared with similar results from Wapstra and Bos[5] and from the recent experimental paper by Epherre et al.[12] Although the errors for the present work are considerably larger than the results from Epherre et al., the agreement is satisfactory in four of the six comparisons. A similar result holds for the comparison with data from Wapstra and Bos. Only in the case of the ^{93}Rb-^{92}Rb difference is there a substantial disagreement. Although our results are not of comparable accuracy to that achieved in the work of Epherre et al., we believe that we have demonstrated that measurements of this sort can be accomplished using a medium energy accelerator and conventional mass doublet measurement techniques.

References

1. J. L. Benson and W. H. Johnson, Jr., Phys. Rev. 141, 1112 (1966).
2. D. C. Kayser and W. H. Johnson, Jr., Phys. Rev. C 12, 1054 (1975).
3. A. H. Wapstra and K. Bos, At. Data Nucl. Data Tables 20, 3 (1977).
4. R. R. Ries, R. A. Damerow, and W. H. Johnson, Jr., Phys. Rev. 132, 1662 (1963).
5. A. H. Wapstra and K. Bos, At. Data Nucl. Data Tables 19, 175 (1975).

6. A. H. Wapstra, Private communication, May 1979.
7. J. E. Halverson and W. H. Johnson, Jr., Phys. Rev. C 20, 345 (1978).
8. J. E. Halverson and W. H. Johnson, Jr., Phys. Rev. C 17, 1414 (1978).
9. R. Bernas in Recent Developments in Mass Spectrometry, K. Ogata and T. Hayakawa, eds. (University of Tokyo Press, Tokyo, 1970) p. 535.
10. N. R. Daly, Rev. Sci. Instr. 31, 264 (1960).
11. J. O. Meredith, F. C. G. Southon, R. C. Barber, R. Williams, and H. E. Duckworth, Int. J. Mass Spectrom. Ion Phys. 10, 359 (1972/73).
12. M. Epherre, G. Audi, C. Thibault, R. Klapisch, G. Huber, F. Touchard, and H. Wollnik, Phys. Rev. C 19, 1504 (1979).

CONSIDERING L.G. SMITH'S RF MASS SPECTROMETER[x]

E. Koets, J. Kramer, J. Nonhebel, J.B. Le Poole

Delft University of Technology, Applied Physics Department
Lorentzweg 1, 2628 CJ Delft, Netherlands

INTRODUCTION

The operating principles and the essential elements of this very high precision mass spectrometer, as well as its history up to mid 1975 have been described previously [1,2,3,4,5].

It is the purpose of this paper to cover the events between AMCO-5 and AMCO-6 and to add on to the accuracy discussions.

SINCE AMCO-5

Before the disassembly of the spectrometer in Princeton, an N_2-CO measurement had been made for reference purposes[5]. As soon as the machine came into operation in Delft, in April 1976, this measurement was repeated. Immediately a result was obtained, which was identical to the Princeton result. This indicates that the duplicatability of the spectrometer is intrinsic and confirms the importance of Smith's work.

Lack of peak-resolution and indications of systematic effects had initiated the development of the Telecentric Object Plane Selector with CEMA-image-transformer[5]. This unit has been attached to the machine since June 1976 to study the properties of the final image. This final image of the spectrometer is supposed to consist

[x] This work was also performed as part of the research program of the "Stichting Fundamenteel Onderzoek der Materie" (FOM), which is financially supported by the "Nederlandse Organisatie voor Zuiver Wetenschappelijk Onderzoek" (ZWO).

of two parallel line foci, which should coincide for particular
oscillator frequencies[1]. Actually the foci were not parallel, but
formed an 'x'. Furthermore they turned out to be frayed, cane-shaped
and of non-symmetrical intensity, while their bisector was not
aligned with the detector slit, nor with some other slits. Finally,
the foci were displaced lengthwise by about half their line length.

The peak-resolution [1,2], corresponding to the dimensions of
the 'x', was 3 to 4 times lower than what had been obtained before.
This higher resolution could only be reproduced for control settings
which made the beam reflect somewhere. The corresponding excessive
surface charges that were noticed, may have caused part of Smith's
time dependent effects. To indicate how crucial these may be:
Smith's results for ^{16}O before and after he made some modifications,
which reduced these surface charges, show a 180 nu discrepancy[4].

From these observations it became clear that the image properties of the machine had to be monitored permanently. So the Telecentric Object Plane Selector unit was modified to incorporate the
original detector, while a beam switch was added. It also seemed
dubious to rely on Smith's peak matching method to the extent he
did, since peak distortions, caused by the effects mentioned above,
are not sufficiently visible then.

Thus better peak-resolution became the primary objective. A
first step herein was the improvement of the geometry of the final
image. The foci forming an 'x' could mean that the magnetic field
had an axial gradient. Smith's notes reveal that he – already in
January 1968 – found, at 0.55 T, the magnetic field at the lower
pole cap to be 0.3 to 0.4 mT higher than the field at the upper
pole cap. This fits rather well with the observed dimensions of the
'x'. The magnet shim system was found to consist of five shim coils
on/in each pole cap, while most of their current ratios were fixed
by the factory. These ratios were made variable, after which the
foci could be made parallel and a bit narrower. While looking at
the channelplate image via a TV-system, the equivalent FWHM peak-
resolution of the 'x'-shape had been $0.15 \cdot 10^6$. The improvement of
the homogeneity of the magnetic field resulted in an image with an
equivalent FWHM peak-resolution of $2 \cdot 10^6$, apart from some distortions at the ends of the line foci.

Furthermore the peak-resolution is proportional to the
modulator voltage in first order. This voltage was raised from
some 130 V_{rms}, used by Smith, to about 650 V_{rms}, through the use
of new tubes and by better impedance matchings. The better matching
also takes away the tendency to oscillations of the power amplifiers.
This tendency was of course also dubious with respect to accuracy.

All together the FWHM peak-resolution was raised from a working
value for Smith of 0.3 to $0.4 \cdot 10^6$ to around $10 \cdot 10^6$ for a clean part

of the foci on the TV monitor. To be able to select the same part
for use with the original detector, its fixed slit was replaced by
an adjustable slitsystem.

This high peak-resolution made the existing instability of the
magnetic field totally unacceptable. These variations ranged from
a frequent 1 in 10^7 to an exceptional 1 in 10^5. This resulted in a
relative change in the magnetic induction of up to $3 \cdot 10^{-8}$ over the
33 ms sample time for each of the doublet species. Partial improvement of the power supply has resulted in relative variations in the
magnetic induction of less than 10^{-7} over a 100 s time interval and
not more than 10^{-9} over a sample period. This did result in an
improvement of the peak-stability, though not as much as that of
the magnetic induction. The effects of changing surface charges
seem to be dominant now.

Apart from vignetting caused by rotated (images of) slits, the
dominant alignment problem is for the moment located between the
'90°-displacer' (one quarter turn into the magnet) and the first
modulator transition, where a deflector and an auxiliary lens are
located. The beamcurrent at the 90°-displacer location is typically
1 to 2 nA. At the modulator it is some 100 times lower. In addition
Smith had many problems with the alignment of the electron beam in
the ion source. But allowing the filament freedom of expansion,
mounting all parts of the electron gun on the same flange, proper
pre-alignment of the electron gun and the introduction of a set of
deflection coils eliminated these problems.

Smith designed the magnification of the spectrometer to be
unity from source to detector. Actually it is somewhat smaller,
which results in a tendency of the plane, which is achromatically
conjugated with the source object plane, to lie behind the detector.
On top of that appreciable astigmatism exists in the final image.

ACCURACY CONSIDERATIONS

Let T_1 and T_2 be the respective time intervals between two
modulator transitions of the two species of a doublet. Smith then
proved that T_1/T_2 can be measured with adequate accuracy [1,2,3,4,5].
This ratio is however not necessarily equal to the required ratio
of cyclotron times. These ratios may differ if radial or tangential
electrostatic fields exist and these do exist due to the surface
charges. To be able to compensate for these effects Smith introduced
an adjustable radial deflection field, which was set during separate
calibration runs with relatively well known, wide doublets[2]. However
an hour elapsed between his calibration runs and the corresponding
measurement runs, with a shut-down in between to replace sample
bottles. This long interval should of course be avoided if at all
possible. Up to now a third leak valve has been mounted, so both

the calibration and the measurement doublet are in place continuously
and already the switch-over can be done much more rapidly. Furthermore a modification of the electronics is planned which will turn
the machine's present 2 step sequence into a 3 step sequence. The
machine may then be calibrated every 100 ms, while the circumstances
during calibration and measurement runs will then be much more
similar too. It is also planned to lower the beam current considerably
and to present much cleaner surfaces to the beam to reduce surface
charging. In the meantime the effect of beam current on each measurement will be determined and taken into account.

The ratio of a particle's mass and its cyclotron time is proportional to the magnetic induction. By measuring the ratio of the
cyclotron times of two masses the magnetic induction is eliminated.
It is however only allowed to do so, if the two species of a doublet
experience the same magnetic induction. This will only be so if the
magnetic induction is constant in time and space. In general,
variations in time will however not result in systematic effects.
Variations in space do give systematic effects, if the trajectories
of the two doublet species don't coincide exactly. They do not, due
to the surface charges.

The 'x' formed by the foci was due to an axial gradient of the
magnetic field. This gradient resulted in a systematic effect of
1 in 10^9 if the two doublet beams were axially misaligned by 0.1 μm,
which corresponds to a typical surface potential of 1 V. The increased
homogeneity of the magnetic field thus did not only raise the peak-
resolution, but also decreased the sensitivity for axial misalign-
ment. Hence typical surface potentials of 100 V can now be tolerated.

During his very last series of measurements Smith was aware of
most of these problems and also had the means to do at least something about them. He was however not aware of this axial alignment
problem, nor was there a way to check this alignment.

To start with no measurements have been made in Delft, because
first of all the properties and limitations of the machine had to
be studied. Then some of the above mentioned aspects were given
priority, since otherwise the attainable accuracy level would not be
acceptable. Recently a variety of break-downs in the 15 year old,
extensively modified equipment prevented some planned measurements.

AFTER AMCO-6

For the moment the activities are focussed on the following items:
- Work on a mini probe NMR system will be continued. It will be used
  to find optimum shim conditions for higher resolution and smaller
  systematic effects / less cricital settings / congruence demands.
- The analysis of RF and signal reduction systems will go on.

Preliminary results reveal that, with state of the art electronics, possibilities exist for both a considerable easing on some RF specifications and an increased performance.

- The scattering action of the Phase Defining Slit[1] is one of the causes for surface charges. The possibility will be investigated to replace its bunching action by that of an energy modulator mounted immediately after the ion source. Bunching will then take place in a klystron-like way. The machine's transmission might also be increased by this modification.
- The present single stage multiplier-electrometer amplifier detector system will be replaced by a dual channel counting detector, which will detect for each individual bunch how many particles land on either side of the reference position. This will create the possibility to handle much smaller ion currents, which not only will increase the accuracy since the surface charge effects will become negligible, but also will extend the field of use to low output ion sources. Apart from this it will make the need to sweep the oscillators obsolete, thereby eliminating the systematic phase slip errors. Another feature will be the independent information that will become available on the distortion of the final images corresponding to either type of bunch.
- The opening angles of the beam will be reduced to cut down the aberrations and part of the surface charges. To be able to do this in an extreme way, work on an ion source with high chromatic brightness $[A \cdot m^{-2} \cdot sr^{-1} \cdot V^{-1}]$ will be continued.
- Two extra lenses – one ahead of the injector, the other after the ejector – will result in higher dispersion for each half of the machine, since then all contributions can be made to add, instead of partially subtract as in the present situation. They will also introduce an independent means to adjust the magnification.

REFERENCES

1. L.G. Smith, First results with the Princeton RF Spectrometer, in: "Proceedings of the Third International Conference on Nuclidic Masses", R.C. Barber, ed., University of Manitoba Press, Winnipeg, Manitoba, 1967; p. 811
2. L.G. Smith, Measurements of Six Light Masses, Phys. Rev. C 4; 22 (1971)
3. L.G. Smith, Recent Precision Mass Measurements at Princeton, in: "Proceedings of the Fourth International Conference on Atomic Masses and Fundamental Constants", J.H. Sanders and A.H. Wapstra, eds., Plenum, New York, 1972; p. 164
4. L.G. Smith and A.H. Wapstra, Masses of isotopes of H, He, C, N, O and F, Phys. Rev. C 11, 1392 (1975)
5. E. Koets, L.G. Smith's Precision RF-Mass-Spectrometer transferred from Princeton to Delft, in: "Proceedings of the Fifth International Conference on Atomic Masses and Fundamental Constants", J.H. Sanders and A.H. Wapstra, eds., Plenum, New York, 1976.

MAPPING OF NUCLEAR MASSES IN THE REGION ($N \leq 126$, $Z \geq 82$) FROM DIRECT MEASUREMENTS OF FRANCIUM ISOTOPES

G. Audi, M. Epherre, C. Thibault, R. Klapisch, G. Huber[*],
F. Touchard, H. Wollnik[†].

Laboratoire René Bernas, C.S.N.S.M. - 91406 Orsay (France)
and the ISOLDE Collaboration, CERN - 1211 - Genève (Suisse)

INTRODUCTION

In the preceding joint report to this conference[1], the experiment installed at CERN on-line with ISOLDE for measuring masses of exotic nuclei, with a two stage double-focusing mass-spectrometer, has been described.

We shall report here more particularly on the measurements and the results obtained for the francium isotopes.

ISOTOPE PRODUCTION

The francium isotopes were produced by spallations induced by 600 MeV protons from SC at CERN in a $12-14 \text{g/cm}^2$ uranium impregnated graphite cloth target at about 2000°C.

Figure 1 shows the yield curve for $1 \mu A$ proton beam. The mean intensity during the run was about .8 μA. The lightest and heaviest isotopes for which masses were measured were ^{204}Fr and ^{228}Fr. In the very near future, we should be able to extend the measurements to ^{202}Fr and ^{230}Fr. With these measurements a link can be established between the α-and $\beta-$ decay chains from ^{202}Fr and ^{203}Fr to ^{186}Os and ^{191}Ir respectively, and the directly measured mass chain.

----

[*] Present address : Gesellschaft für Schwerionenforschung, 6100, Darmstadt, Germany.
[†] Permanent address : II. Physikalisches Institut, 6300, Giessen, Germany.

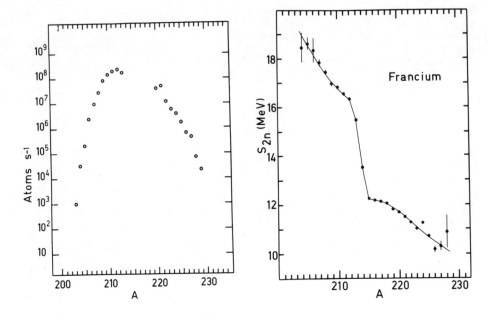

Fig. 1 - Experimental yields of
francium isotopes from a $14g/cm^2$
uranium impregnated graphite cloth
target at 2050°C. The intensity of
the 600 MeV proton beam was 1μA,
from Ref. 16.

Fig. 2 - Experimental two-
neutron separation energies
S_{2n} versus atomic mass number
A. The line here is only a
guide for the eye.

The hole in the curve in fig. 1 is due to the very short-lived
francium isotopes (a few μsec.) between mass number 214 and 219,
compared to the release time of a few seconds. This restricted the
number of reference masses that could be used for the mass measurements to $^{211,213,220-223}Fr$.

The masses of the missing isotopes are however known through
α-and β-chains which connect them to bismuth and lead.

EXPERIMENTAL SET-UP

The experimental set-up and procedure are described else-
where^{1,2}, and details on the ion detection by a movable dynode
electron-multiplier, can be found in Ref. 3. This device plays an
essential part in the mass measurements of the francium isotopes :
due to long α-chains, the high activity on the first dynode of the
electron-multiplier gives a high background signal that cannot be
discriminated from the real signal, and the mass measurements would
thus have been impossible without this moving collector. For example
for ^{220}Fr after 2 minutes the background is 2 times larger than the
signal itself.

For this experiment, the resolving power of the mass spectrometer was 5000 at 20% of the peaks height, and the transmission
was about 10^{-3}. Depending on statistics, the relative precision
of these mass measurements was between 2×10^{-7} (the limit of the
apparatus) and 3×10^{-6} (which means at mass 200 : between 40 and
600 keV).

RESULTS

Francium isotopes

The results obtained for the masses of the francium isotopes
are reported on table 1. They are presented as mass excesses and
compared with the values published in the last Atomic Mass Table[4].
There is a good agreement with the two previously known masses
226 and 227. The particular case of ^{224}Fr is discussed below.

From these values the two neutron separation energy curve can
be deduced (figure 2). The values for ^{204}Fr and ^{228}Fr appear to
follow the smooth trends exhibited in the two regions (within
experimental errors). On the other hand the two points corresponding
to ^{224}Fr and ^{226}Fr deviate from the regular behaviour displayed by
the neighbouring points, indicating that our mass value for ^{224}Fr
could be too low, as an increase of this mass would lower the S_{2n}
value of ^{224}Fr and increase that of ^{226}Fr. No explanation has been

Table 1 : Mass Excesses of the Francium Isotopes

	(M – A) keV present Measurements	(M – A) keV 1977 Atomic Mass Table		(M – A) keV present Measurements	(M – A) keV 1977 Atomic Mass Table
202		3160 ± 310	216		2975 ± 14
203		1230 ± 210	217		4307 ± 15
204	840 ± 510	870 SYST	218		7050 ± 6
205	-1270 ± 130	-1040 SYST	219		8617 ± 9
206	-1380 ± 100	-1180 SYST	220		11470 ± 8 [a]
207	-2970 ± 100	-2650 SYST	221		13265 ± 12 [a]
208	-2700 ± 60	-2770 SYST	222		16338 ± 21 [a]
209	-3810 ± 50	-3760 SYST	223		18382 ± 4 [a]
210	-3380 ± 50	-3640 SYST	224	21215 ± 60 [b]	21710 SYST
211		-4220 ± 50 [a]	225	23820 ± 80	23790 SYST
212	-3575 ± 40	-3690 SYST	226	27140 ± 110	27460 ± 330
213		$-3556 + 11$ [a]	227	29650 ± 170	29580 ± 100
214		-965 ± 13	228	32420 ± 660	
215		309 ± 13			

a   Adopted values for the primary reference masses.
b   As explained in the text this value is considered uncertain

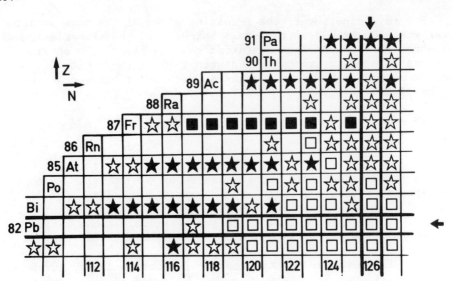

Fig. 3 : Known masses in the region (N $<$ 126   Z $\geqslant$ 82)
  □ primary data [4] known previously
  ☆ secondary data [4] known previously
  ■ this experiment directly measured masses
  ★ deduced from the preceding ones and from Q_α - values[4,7].

found for such a result : this mass has been measured through 3 sequences, but during the same run and will have to be remeasured.

Nevertheless, some of the measurements carried out in this work, using the mass of [224]Fr as a reference mass, give consistent results with those which did not use it, showing that whatever contaminations occured at mass 224, they remained unchanged during the run and did not disturb the other results.

Furthermore, no isomer is known for any of the isotopes considered here,[5] so that the measured masses are very likely not isomer-mixed masses.

Neighbouring Odd-Z Elements

In the region (N$\leqslant$126, Z$\geqslant$82) of the chart of the nuclides, very few masses were known prior to this work, but long chains of α-decay energies [6,7] were determined without connection to known masses.

The 8 isotopes of francium : [204-210,212]Fr crossing these α-chains set up the connection for the odd-Z elements, and lead to the first determination of the masses of 28 isotopes of Pa, Ac, At, Bi and Tl (figure 3).

Table 2 : Mass excesses of isotopes of astatin, bismuth, thallium, actinium and protactinium.

Isotope	MASS EXCESS (keV)		Isotope	MASS EXCESS (keV)	
	deduced from Fr mass measurement	predicted by systematics [4]		deduced from Fr mass measurement	predicted by systematics [4]
200_{AT}	- 8750 ± 510	- 8670	209_{AC}	8890 ± 140	9120
201_{AT}	-10740 ± 140	-10520	210_{AC}	8650 ± 110	8860
202_{AT}	-10730 ± 110	-10520	211_{AC}	7080 ± 110	7400
203_{AT}	-12300 ± 110	-11970	212_{AC}	7250 ± 80	7180
204_{AT}	-11890 ± 80	-11970	213_{AC}	6120 ± 70	6170
205_{AT}	-13010 ± 50	-12960	214_{AC}	6400 ± 70	6140
206_{AT}	-12470 ± 70	-12730	216_{AC}	8090 ± 60	7980
208_{AT}	-12530 ± 40	-12640			
			215_{PA}	17740 ± 120	
196_{BI}	-17780 ± 510	-17760	216_{PA}	17680 ± 90	
197_{BI}	-19640 ± 150	-19410	217_{PA}	17040 ± 90	
198_{BI}	-19510 ± 110	-19300	218_{PA}	18620 ± 70	
199_{BI}	-20930 ± 120	-20610			
200_{BI}	-20390 ± 90	-20460			
201_{BI}	-21455 ± 50	-21410			
202_{BI}	-20780 ± 70	-21040			
204_{BI}	-20710 ± 40	-20820			
197_{TL}	-28380 ± 60	-28330			

Table 2 compares the masses thus obtained to the values predicted by the systematics of Wapstra and Bos[4]. The agreement is quite good and the differences do not exceed 330 keV.

The extension brought by this mapping to the two neutron separation energie curves can be seen on figure 4. The behaviour of these curves is quite regular.

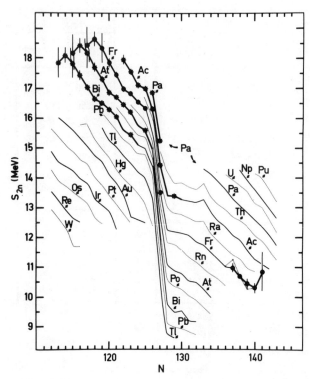

Fig. 4 - Experimental two-neutron separation energies S_{2n} versus
the neutron N in the francium region, from the reported
mass measurements (dots), the deduced masses (stars) and
from 1977 Atomic Mass Table (Ref. 4) (excluding systematic
extrapolations). The dotted line on the francium curve
corresponds to the value of ^{224}Fr as measured here, the
full line correspond to an interpolated value to ^{224}Fr.

COMPARISON WITH MASS PREDICTIONS

Francium Isotopes

The experimental masses are compared with the predictions of the current mass formulae on figures 5 and 6. One can notice the increasing deviation of the values predicted by the Garvey-Kelson type mass relations [8,9,10] towards the neutron-deficient isotopes. Among the macroscopic-microscopic calculations [11-14] (Fig. 6) the values obtained by Groote, Hilf and Takahashi [13] are within 600 keV of the measured masses throughout the chain of isotopes. The best predictions are obtained with the semi empirical Shell Model formula of Liran and Zeldes [15].

Neighbouring Odd-Z Elements

The masses predicted for the neighbouring odd-Z elements are compared on figure 7 with the experimental values for the two best predictions we noted above : the Groote, Hilf and Takahashi calculations [13] and the Liran and Zeldes formula [15], for which the agreement is remarkable.

We point out that this formula gave the best agreement also for the cesium masses (Z=55), but was not one of the best for the lighter (Z=37) rubidium[2].

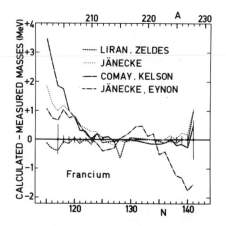
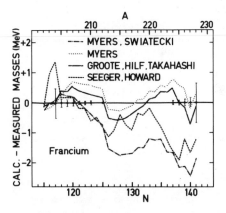

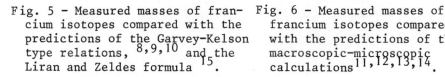

Fig. 5 – Measured masses of francium isotopes compared with the predictions of the Garvey-Kelson type relations, [8,9,10] and the Liran and Zeldes formula [15].

Fig. 6 – Measured masses of francium isotopes compared with the predictions of the macroscopic-microscopic calculations [11,12,13,14].

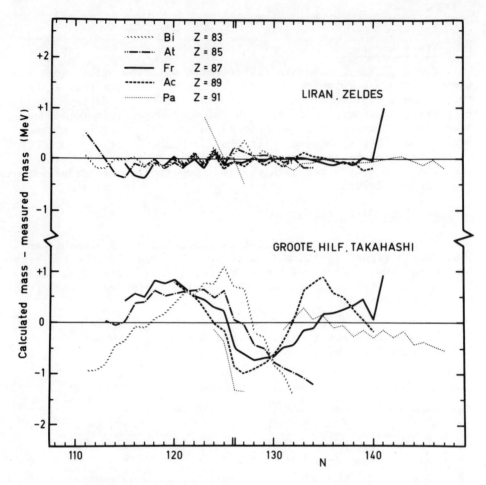

Fig. 7 – Experimental masses of Bi, At, Fr, Ac and Pa isotopes
        compared to the predicted masses of Liran and Zeldes [15]
        and Groote, Hilf and Takahashi [13].

CONCLUSION

     Using a double focussing mass spectrometer,on-line with the
ISOLDE isotope separator, we have determined the masses of 12
isotopes of francium, and using previously known Q_α-values, 28
masses of neighbouring elements were deduced, thereby mapping fairly
well the nuclear masses in the region (N≤126, Z≥82).

     From the results obtained it seems that the semi-empirical
Shell Model formula gives the best overall fit to our results.

In the very near future, when the mass spectrometer will be reinstalled at ISOLDE, we plan to remeasure the mass of ^{224}Fr and to measure the masses of at least two more exotic nuclei at each end of the francium series.

REFERENCES

1 - M. Epherre, G. Audi, C. Thibault, R. Klapisch, this conference, "Direct Measurements of Masses of Short Lived Nuclei : Rb and Cs".

2 - M. Epherre, G. Audi, C. Thibault, R. Klapisch, G. Huber, F. Touchard, and H. Wollnik, Phys. Rev. C 19, 1504 (1979).

3 - F. Touchard, G. Huber, R. Fergeau, C. Thibault and R. Klapisch, Nucl. Instrum. Methods 155, 449 (1978).

4 - A.H. Wapstra and K. Bos, At. Data Nucl. Data Tables 19, 175 (1977).

5 - C. Ekström, S. Ingelman, G. Wannberg and M. Skarestad, Phys. Scripta 18, 51 (1978).

6 - Table of Isotopes edited by C. Michael Lederer and Virginia S. Shirley, 1978, Wiley Interscience Pub.

7 - K.H. Schmidt, W. Faust, G. Münzenberg, H.G. Clerc, W. Lang, K. Pielenz, D. Vermeulen, H. Wohlfarth, H. Ewald and K. Güttner, Nucl. Phys. A 318, 253 (1979).

8 - J. Jänecke, At. Data and Nucl. Data Tables 17, 455 (1976).

9 - E. Comay and I. Kelson, At. Data and Nucl. Data Tables 17, 463 (1976).

10 - J. Jänecke and B.P. Eynon, At. Data and Nucl. Data Tables 17, 467 (1976).

11 - W.D. Myers and W.J. Swiatecki, Nucl. Phys. 81, 1 (1966).

12 - W.D. Myers, At. Data and Nucl. Data Tables 17, 411 (1976).

13 - H.V. Groote, E.R. Hilf and K. Takahashi, At. Data and Nucl. Data Tables 17, 418 (1976).

14 - P.A. Seeger and W.M. Howard, At. Data and Nucl. Data Tables 17, 428 (1976).

15 - S. Liran and N. Zeldes, At. Data and Nucl. Data Tables 17, 431 (1976).

16 - L.C. Carraz, S. Sundell, H.L. Ravn, M. Skarestad and L. Westgaard, Nucl. Instr. and Methods 158, 69 (1979).

NEW MASS SPECTROMETRIC MEASUREMENTS ON SODIUM ISOTOPES
MASSES OF $^{31-34}$Na

C. Thibault, M. Epherre, G. Audi, R. Klapisch,
G. Huber, F. Touchard, D. Guillemaud*, F. Naulin*

Laboratoire René Bernas
du C.S.N.S.M.
BP1, 91406 Orsay (France)

INTRODUCTION

In this paper some new unpublished mass-spectrometric measurements on sodium isotopes will be presented. In AMCO V[1] mass measurements for A = 26 to 32 were already presented and had shown the
onset of a deformation for A = 31 and 32. This had encouraged us
to pursue these experiments in order to investigate what happens
for isotopes further from stability : ^{33}Na which was already observed
to be bound but for which the mass could not be measured, and ^{34}Na,
the existence of which was questionable according to predictions.
In addition, it seemed also important to improve the precision on
^{31,32}Na for which an error bar of 1 MeV had been obtained while the
observed anomaly was 2-4 MeV only.

EXPERIMENT

The principle of the experiment has already been described[2]
(Fig. 1). The required increase of the yields by a factor 100 has
been obtained through the combination of several improvements :
(i) the ionization efficiency and transmission of the mass spectrometer have been improved resulting in an overall efficiency of 10%
for ^{30}Na (instead of 1% previously). (ii) The thichness of the
target has been multiplied by 3 (3g/cm^2). (iii) The proton beam
intensity from the P.S. has been multiplied by 4 (6x10^{12}p/pulse).

----

*Institut de Physique Nucléaire - 91406, Orsay (France).

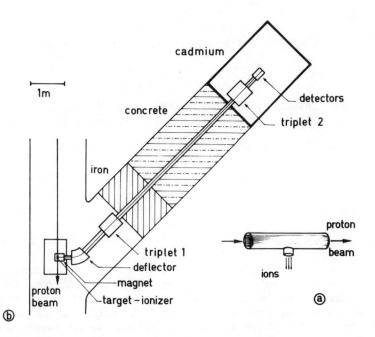

Fig. 1 - Schematic view of the set up.
   (a) The target is a stack of graphite foils covered with
       uranium oxyde (20mg/cm^2) and wrapped in a rhenium
       oven heated at 1600°C. This target is bombarded by the
       20 GeV proton beam from the P.S. at CERN. The products of
       the reaction U+p are stopped in the graphite. The alkalis
       diffuse out fast ($\sim$ 100ms) and are selectively ionized
       by thermionic effect on the rhenium surfaces.
   (b) The target serves directly as ion source for the single
       stage mass spectrometer. The mono-isotopic ion-beam is
       transported through the shielding into the counting room
       where is the electron multiplier.

    The cross sections that we measured for the production of the
sodium isotopes in the fragmentation of uranium by 20GeV protons is
shown on Fig. 2. One important feature to be pointed out is that the
distribution is spread over 8 orders of magnitude. As a consequence,
while ^{34}Na may be observed with less than 1cnt/pulse, 100 cnts/pulse
are already obtained for ^{32}Na and more than 10^7 cnts have to be
counted in less than 100ms for each pulse, for all the reference
sodium isotopes from A = 22 to 26. This makes very difficult to avoid
pile-up on them. One has also to keep in mind that for A≥28, all
half-lives are shorter than 50ms[3] - down to 5ms for ^{34}Na - thus
requiring very short delay times in the target.

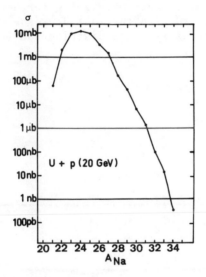

Fig. 2 : Isotopic distribution of the sodium produced in the
          fragmentation of a uranium target by 20GeV protons.
          The cross sections are normalized on $\sigma(^{24}$Na$)$ = 11.4mb.

The method of measurement is similar to what explained in precedent papers[2,4], and uses the same basic equations between the masses M_A, M_B, M_C, of the isotopes A, B, C :

$$M_A(V_A+\delta) = M_B (V_B+\delta) = M_C (V_C+\delta)$$

where δ is a small systematic correction. V_A, V_B, V_C, are the accelerating potentials. They are measured with an accuracy of 10^{-6} and have to be kept constant to 10^{-5} while the pulsed proton beam passes through the target in $2\mu s$ every 5 seconds.

RESULTS

Fig. 3 shows the accuracy obtained for the different masses. The increase of the error bars when going further from stability is here mainly connected with the fast decrease of statistics as pointed out in the cross sections distribution (Fig. 2). For ^{33}Na, the accuracy is 500 keV.

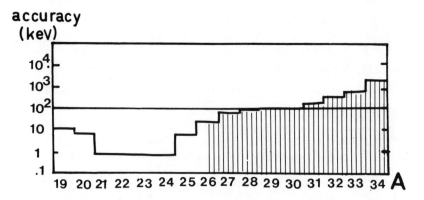

Fig. 3 : Precision of the mass measurements on sodium isotopes. The on-line mass spectrometric measurements are presented with hatched area.

Mass 34

For ^{34}Na, one of the best spectra is shown on Fig. 4, corresponding to 3 hours of recording. The decay of ^{30}Na is due partly to
the diffusion time in the target and partly to its half life of 50ms.
For ^{34}Na, two pairs of peaks only are clearly seen and its very
fast decay is mainly due to its very short half life of 5ms. From
this spectrum and some others, the mass could be determined for the
two first pairs of peaks with an accuracy of some MeV. Unfortunately,
the results clearly indicate the presence of a residual more bound
contaminant.

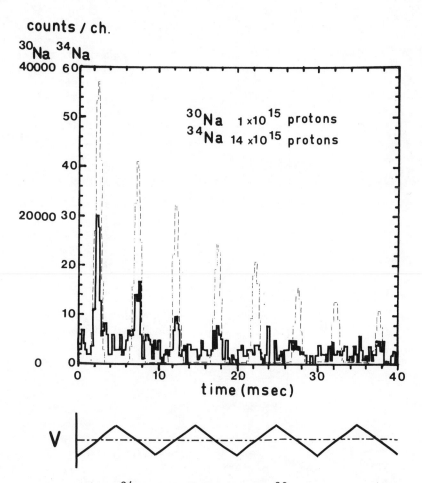

Fig. 4 : Spectra of ^{34}Na (solid line) and ^{30}Na (dashed line).
The successive peaks are produced by the 100Hz modulation
of the accelerating voltage which sweeps the ion-beam back
and forth across the slit. The proton beam passes through
the target at time zero. The time per channel is 240 µs.

This finally restricted our measurement to the determination of a lower limit of 24 MeV for the mass excess. However, the fact that ^{34}Na is bound towards n-emission provides an upper limit :

M(^{34}Na) $\leqslant$ M(^{33}Na) + Mn

These two limits (24, 29) MeV constitute our present result for the mass excess of ^{34}Na.

Mass 35

A search for mass 35 has been done and one pair of peaks could finally be observed. The measured yield is represented on Fig. 5 and appears to be in agreement with a systematic extrapolation based on the $^{28-34}$Na measured yields. However measurement of the yield of mass 36 (which is overwhelmingly ^{36}K) in the same conditions shows that the mass-35 yield is also in agreement with the potassium yields. Due to the very low yields of mass 35 the half life could not be determined but an approximate value of the mass excess could be obtained : -10 ± 5MeV. This is in good agreement with

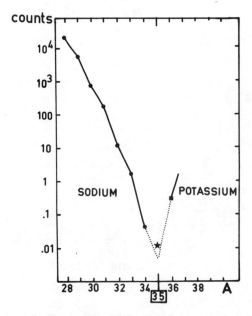

Fig. 5 : Normalized yields recorded for masses 28-36. $^{28-34}$Na yields (points) and ^{36}K yield (square) have been extrapolated towards mass 35 (dashed lines) with the assumptions y(^{35}Na)/y(^{33}Na)=y(^{33}Na)/y(^{31}Na) and y(^{35}K)/y(^{36}K)= y(^{37}K)/y(^{38}K).

The measured mass -35 yield (star) appears in relative agreement with both predictions.

the known mass of ^{35}K (-11MeV) and very far from the predicted mass of ^{35}Na (30 to 37MeV). In future experiments, we hope to increase again the yields in order to improve the accuracy of the masses of sodium up to 34. But even if the intensity increases at mass 35, we shall probably not be able to measure the mass of ^{35}Na, since we cannot resolve it completely from ^{35}K. A clear distinction between both would be to observe ^{35}Na through delayed neutrons.

DISCUSSION

Fig. 6 shows the two-neutron separation energy of the isotopes of sodium and of neighbouring elements as a function of neutron number. The N=20 known shell effect is observed for calcium and

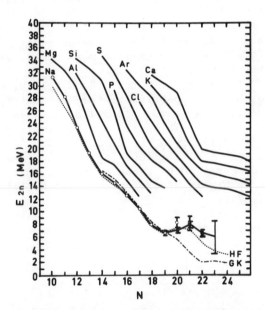

Fig. 6 : Two-neutron separation energy of sodium to calcium elements. The present mass spectrometric measurements (solid circles) appear to be in good agreement with the previous ones[1,2] (open circles) and show that the hump already observed for N=20,21 extends up to N=23. This was not predicted by systematic extrapolations based on Garvey Kelson relations (G.K)[5] but is well reproduced by Hartree-Fock calculations (H.F)[6] as due to the onset of a deformation at N=20 while calcium and potassium isotopes exhibit shell effects.

potassium isotopes only, while for sodium a hump very similar to the one observed in the rare earth deformed region is observed. This had been interpreted as the sudden onset of a deformation for ^{31,32}Na[1,2]. The new more accurate results confirm this anomalous behaviour for A=31,32 and show that it extends up to ^{34}Na, as calculated by the Hartree Fock method[6] in which a shape transition had been shown to occur for ^{31}Na. In conclusion, these new mass spectrometric measurements on $^{31-34}$Na indicate that $^{31-34}$Na are certainly strongly deformed nuclei.

REFERENCES

1 - C. Thibault, R. Klapisch, C. Rigaud, A.M. Poskanzer, R. Prieels, L. Lessard, and W. Reisdorf, Mass Spectrometry of Unstable Nuclei, in "Atomic Masses and Fundamental Constants", Vol. 5, p.205-J.H. Sanders, and A.H. Wapstra ed., Plenum Publishing Corporation, New York (1970).

2 - C. Thibault, R. Klapisch, C. Rigaud, A.M. Poskanzer, R. Prieels, L. Lessard, and W. Reisdorf, Direct measurement of the masses of ^{11}Li and $^{26-32}$Na with an on-line mass spectrometer - Phys. Rev. C 12 : 644 (1975).

3 - E. Roeckl, P.F. Dittner, C. Detraz, R. Klapisch, C. Thibault, and C. Rigaud, Decay properties of the neutron-rich isotopes, ^{11}Li and $^{27-31}$Na, Phys. Rev. C 10 : 1181 (1974).

4 - M. Epherre, G. Audi, C. Thibault, and R. Klapisch, Direct mass measurements of short-lived nuclei : $^{74-79}$Rb , $^{90-99}$Rb, $^{117-124,126}$Cs and $^{138,140-147}$Cs, in these proceedings.

5 - N.A. Jelley, J. Cerny, D.P. Stahel, and K.H. Wilcox, Predictions of the masses of highly neutron-rich light nuclei, Phys. Rev. C 11 : 2049 (1975).

6 - X. Campi, H. Flocard, A.K. Kerman, and S. Koonin, Shape-transition in the neutron rich sodium isotopes, Nucl. Phys. A 251 : 193 (1975).

DIRECT MEASUREMENTS OF MASSES OF SHORT LIVED NUCLEI :
$^{74-79}$Rb, $^{90-99}$Rb, $^{117-124,126}$Cs and $^{138,140-147}$Cs

M. Epherre, G. Audi, C. Thibault, R. Klapisch

Laboratoire René Bernas

du CSNSM, 91406 Orsay (France)

INTRODUCTION

The possibility of using mass spectrometric techniques to determine the masses of short lived radioactive isotopes was investigated in 1970 after the successfull measurements of short half lives using on line mass spectrometry[1]. At the AMCO IV conference -1971- two projects were presented ; one from the Minnesota group[2] using a double focusing mass spectrometer and the other from the Orsay group[3] using a single stage instrument. At the AMCO V conference - 1975- the results of the direct measurements of the masses of ^{11}Li and $^{26-32}$Na were presented by Thibault et al.[4]. They showed that an accuracy of 100 keV could be reached for light elements even with a single stage mass spectrometer and that the masses of a chain of isotopes could be determined in a single experiment, both facts would allow us to gain information on the global properties of the nuclear mass surface as well as on the local peculiarities of nuclear structure.

To pursue these measurements on heavier nuclei with at least the same accuracy of 100 keV, it requires an increase of the precision on the mass determination. An accuracy of 100 keV corresponds to a relative precision of 10^{-5} for A=10 and of 10^{-6} for A=100. It was not possible to reach such a precision with the former single stage mass spectrometer but well within the possibilities of a double focusing instrument[5]. Consequently a double focusing mass spectrometer has been designed, and installed on line with the ISOLDE isotope separator at CERN in order to initiate a systematic program of mass determinations of the short lived nuclei. This new experimental set up will be presented here with the results obtained on the first extensive series of mass measurements on rubidium and cesium. The masses of the francium isotopes that we measured are presented in another communication to this conference[6].

EXPERIMENTAL

The masses of isotopes as far from stability as, for example, ^{74}Rb and ^{99}Rb could be measured owing to their high yields from ISOLDE. As an example Fig. 1 presents the production yields for the isotopes of cesium. With a rate of 10^4 atoms/s it takes about one hour data accumulation time to reach an experimental accuracy of 100 keV. Details on the whole facility can be found in Ref. 7.

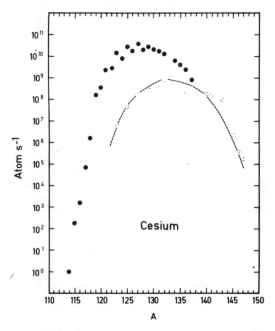

Fig. 1 - Yields of the isotopes of cesium produced at ISOLDE in the spallation of a La target (140g/cm^2) -solid points- and in the fission of an U target ($\sim$14g/cm^2)- open points- normalized to 1μA proton (600 MeV) beam.

The experiment is schematically presented on Fig. 2. The mass separated ions coming from ISOLDE are focused and stopped on the inner face of a conical tube heated to $\sim$1200°C. The implanted atoms diffuse back quickly to the foil surface where they are ionized. The 3 volts dc heating potential directs them to the entry

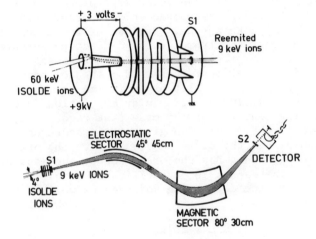

Fig. 2 - Schematic view of the experiment. The upper part shows
         the 60 keV ISOLDE ions stopped in the tantalum tube and
         the 9 keV ion beam formed subsequently in the ion source
         of the mass spectrometer. The lower part show the trajec-
         tory of this beam through the mass spectrometer.

of a thick lens optical system. This forms an ion beam of 9 keV which
is focused on the entrance slit S_1 of the mass spectrometer. The
apparatus has a modified Mattauch Herzog geometry with a spherical
electrostatic analyzer and an homogeneous magnetic sector. After
passing slit S_2 the ions are detected by a high gain e- multiplier.
However the radioactive ions which normally accumulate on the first
dynode create via α and β decays a high background making impossible
the measurements on rare isotopes. For that reason a new detection

system has been studied in which the 1st dynode has been replaced by a moving tape. Details on this device can be found in Ref. 8

The principle of our measurements is still derived from the well known Swann theorem which can be stated as followed : in a constant magnetic field ions of masses M_A, M_B, M_C will follow the same trajectory if all the electric potentials obey the relation

$$M_A V_A = M_B V_B = M_C V_C$$

where $V_{A,B,C}$ stand for accelerating and deflecting voltages. Therefore both potentials have to be kept proportional and one of them has to be measured very precisely in order to determine one unknown mass from a known one. Because of the energy focusing properties of the mass spectrometer the deflecting potential V is the one we measure very precisely. For its measurement the same method as in the sodium experiment[4] was used : the ion beam was swept back and forth across the exit slit by adding to the main dc potential $V°$ a small calibrated -100Hz- triangular modulation (see Fig. 3). The position of the centroïd of the mass peak recorded by a multiscaler synchronized to the modulation gives the measure of v at the time the ion beam passes the slit and $V = V° + v$. The precision on the determination of V depends the precision on the mass of the ion and in particular to reach a 10^{-6} precision it requires a corresponding stabilization of the deflecting voltages as well as of the constant magnetic field.

$$\frac{dM}{M} = \frac{dV}{V} = \frac{dV°}{V} + \frac{dv}{V}$$

A precision of 10^{-7} can be obtained on the measurement of $V°$[9] and the accuracy with which v is determined depends on statistics. For good statistics (10^5 counts on a single peak) a precision of 10^{-7} could thus be reached. In fact, owing to the stability of the magnetic field, $2\ 10^{-7}$ was the best precision obtained.

The experimental procedure is illustrated on Fig. 3. Three masses are always compared, 2 of which being known, in order to calibrate possible systematic errors. In this particular example the mass of ^{98}Rb was being determined. The ISOLDE magnetic field was programed to jump to values corresponding to the 3 masses 95,97 and 98 and the accelerating and deflecting dc potentials of the mass spectrometer were set to the appropriate values. In the timing the 4.5s was chosen to average the fluctuations with time and the number in front was chosen according to the relative abundances of the 3 isotopes. Each pair of peaks was registered in a given section of the multiscaler memory. Simultaneously the measurement of the electrostatic potential $V°$ occurred repeatedly every 2 seconds and was stored on a PdP 9 computer. After each jump before starting again accumulation and voltage measurements we waited 4.5s for all power supplies to reach stable values.

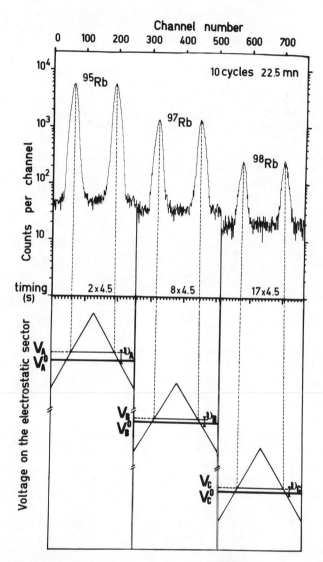

Fig. 3 - Spectrum from the multiscaler showing the recording of
3 rubidium isotopes with the corresponding modulated
voltage.

The sequence of jump 95, 97, 98 was repeated until adequate statistics were reached. The data were then transferred to the computer. The effect of the moving tape could be observed on the spectrum : the background on each peak was due to the radioactivity of the corresponding isotope. On the ^{95}Rb peak the background was 50 counts. Without the moving tape the radioactivity of all the isotopes would have contributed to the background. For ^{98}Rb it would have been at least 8x50 = 400 counts hiding completely the peak itself.

To evaluate the data the following procedure is employed :
- Each measurement is self calibrated by introducing a parameter when comparing 3 masses :

$$M_A(V_A + \delta) = M_B(V_B + \delta) = M_C(V_C + \delta)$$

δ is determined from two known masses and used subsequently to determine a third one. The consistency of the method has been checked on sequences of 3 known masses.
- The masses of two isotopes have to be known to start the measurements for example M_A and M_B. The mass M_x of the isotope X is determined from M_A and M_B and from the measurements of the corresponding voltages

$$M_X = \frac{M_A M_B (V_A - V_B)}{V_x(M_A - M_B) + M_A V_A - M_B V_B}$$

The error on M_x is evaluated taking into account the experimental error and the errors on the reference masses M_A and M_B.
- Then M_x being determined, can be used as a reference to determine the next unknown mass M_y. M_y can be measured from M_A and M_x. M_A and M_B, or M_B and M_X; as often as possible several sequences like (A,X,Y) ; (A,B,Y) ; (B,X,Y) are used to determine each mass. All the masses are thus determined in a step by step procedure in which the propagation of errors through the successive measurements and the correlations due to the reference masses have to be taken into account. A detailed explanation of the procedure is given in Ref. 9. The accuracy finally obtained for all the masses measured is shown on Fig. 4. In many cases it is better than 100 keV which corresponds to a precision of a few 10^{-7}. One can notice the increase of the error as going away from stability which is essentially due to the propagation of the errors through the reference masses in the step by step procedure. Also note the importance of having precisely known primary reference masses to increase the overall precision of the masses we determined.

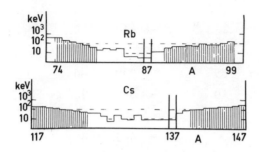

Fig. 4 - Accuracy on the determined masses. The hatched sections
        correspond to the masses we have measured, and the unhat-
        ched sections in between represent the previously known
        masses with the accuracy adopted in the 1977 mass adjuste-
        ment of Wapstra and Bos[10].

RESULTS

     The results finally obtained are presented on Tables 1 and 2
and compared to the adopted mass excesses in the 1977 Atomic Mass
Table[10].

     Several points have to be emphasized. Measurements were started
on either side of the stability at mass number 90 and 79 for the
rubidium isotopes and at mass number 138 and 126 for the cesium
ones because of the rather good precision of the masses known in
between among which we chose the primary reference masses. Now it
appears that some of them had just been determined through a Q_β
measurement and would have to be confirmed in a future experiment.
In particular, precise Q_β measurements[11] have recently been performed
on n-rich rubidium and cesium isotopes and it appeared that the
masses of ^{88,89}Rb and ^{139}Cs as given in the 1977 mass adjustment[10]
would have to be slightly changed. These masses have been used as
primary references in our measurements. If these new values are

     M. EPHERRE, G. AUDI, C. THIBAULT AND R. KLAPISCH

Table 1 : MASS EXCESSES OF THE RUBIDIUM ISOTOPES

A	(M−A)keV our results	(M−A)keV 1977 Atomic Mass Table	Deviation	
74	-52000 ± 380			
75	-57460 ± 190	-57510 ± 600	(.08σ)	
76	-60670 ± 150 [n]	-60610 ± 270	(.2σ)	
77	-65040 ± 110 [n]	-65110 ± 120	(.5 σ)	
78	-67145 ± 65 [a]	-67090 ± 180	(.3 σ)	
79	-70892 ± 40 [n]	-70860 ± 110	.3σ	
80		$-72190 \pm 23.$		
81		$-75445 \pm 35.$		
82		$-76213 \pm 20.$		
83		$-78987 \pm 32.$		
84		-79752 ± 4		
85		$-82159 \pm 4.$		
86		-81738 ± 3		
87		-84596 ± 3		
88		$-82602 \pm 12.$		
89		$-81717 \pm 13.$		
90	-79346 ± 24 [b]	-79570 ± 60	3.5σ	
91	-77800 ± 35 [n]	-77970 ± 40	3.5σ	
92	-74840 ± 40 [n]	-75120 ± 200	1.4σ	
93	-72755 ± 65	-72920 ± 170	.9σ	
94	-68735 ± 45	-69460 SYST		+725keV
95	-65850 ± 100	-66550 ± 310	2.1σ	
96	-61220 ± 70	-62770 SYST		+1550keV
97	-58370 ± 80			
98	-54180 ± 100			
99	-50850 ± 160			

a,b − Known isomer − a correction has been applied

n − no isomer observed in the spin determination experiments at ISOLDE.

Table 2 : MASS EXCESSES OF THE CESIUM ISOTOPES

A	(M-A)keV our results	(M-A)keV 1977 Atomic Mass Table		Deviation
117	-65320 ± 240	-66850	SYST	+1530keV
118	-67510 ± 200 j	-68670	SYST	+1160 "
119	-71600 ± 170 i	-72530	SYST	+ 930 "
120	-73280 ± 140 j	-73640 ±	320	1.0σ
121	-76660 ± 110 i	-77150	SYST	+ 490keV
122	-77700 ± 85 i	-78010	SYST	+ 310 "
123	-80760 ± 70	-80890	SYST	+ 130 "
124	-81500 ± 55	-81530 ±	480	.1σ
125		-84040 ±	40	
126	-84244 ± 40	-84330 ±	140	.6σ
127		-86206 ±	21 .	
128		-85935 ±	6 .	
129		-87563 ±	24 .	
130		-86863 ±	12	
131		-88066 ±	8 .	
132		-87175 ±	23 .	
133		-88089 ±	8	
134		-86909 ±	8	
135		-87665 ±	9	
136		-86358 ±	8	
137		-86560 ±	7 .	
138	-82870 ± 40 i	-82770	SYST	- 100keV
139		-80630 ±	70 .	
140	-76980 ± 90 n	-77240 ±	250	1.0 σ
141	-74380 ± 90 n	-75000 ±	100	4.6 σ
142	-70420 ± 110 n	-70950 ±	130	3.1 σ
143	-67590 ± 130 n	-68360	SYST	+ 770keV
144	-63160 ± 160 n	-63930	SYST	+ 770 "
145	-60010 ± 170	-61720	SYST	+1710 "
146	-55630 ± 190			
147	-52370 ± 310			

i - Isomer observed, level not known, no correction could be applied.
j - Evidence found for an isomeric state - no correction applied.
n - No isomer observed in the spin determination experiments.

confirmed at this meeting, the masses of the n-rich rubidium and cesium isotopes which we have determined would have to be modified. Nevertheless the correction will not exceed the order of 100 keV. These new values will be published in a forth coming paper. One can already notice from two other communications[12,13] at this conference the fairly good agreement between the masses deduced from these recent Q_β measurements and the masses we directly measured (a correction due to the change in the reference has been applied). On the neutron deficient side no new experiment has been performed and we plan to measure directly the masses of isotopes very near stability as $^{80-83}$Rb ^{125,127,129}Cs. We also plan for a future experiment to increase the precision of the very last masses measured, which for ^{74}Rb will be particularly interesting. Indeed this isotope is the heaviest odd N=Z known nucleus and a 100 keV accuracy on its mass would lead to a meaningful determination of the $\log \mathcal{F} t$ value of this superallowed β decay towards ^{74}Kr.

An other problem raised concerns the isomers : some isotopes can be produced not only in their ground state but also in an isomeric state and our present technique does not allow us to separate the contributions of the two if the isomer has a half life greater than several tens of ms. In some cases the excitation energy of the isomer was known as well as its production relatively to the ground state and a correction has been applied (cases a and b on table 1). In other cases no isomer was observed in spin determination experiments at ISOLDE either with magnetic resonance[14] or laser spectroscopy[15] (cases n on tables 1 and 2) and finally in some cases isomers were observed and no correction could be applied because their level was not known (cases i). It is obviously not possible to evaluate the error thus introduced and we consider to study particularly this problem in the future.

DISCUSSION

The masses of these two long chains of isotopes are compared to the prediction of the standard mass formulae[16] on Fig. 5 and 6. On each figure the zero horizontal line represents the experiment and points in the lower plane correspond to isotopes predicted too strongly bound. We will restrict to several general remarks :

- the importance of the Wigner term (in the macroscopic calculations) which takes into account the increase of the symmetry energy when approaching N=Z. The old calculation of Myers and Swiatecki presented on Fig. 5a did not take it into account.
- The general disagreement of all the calculated masses for the n-rich rubidium isotopes (Fig. 5) correlated as will be seen later to the deformation appearing at N=60 : the droplet type calculations predict too early a deformation ($55 < N < 57$) whereas the local formula cannot predict it at all.

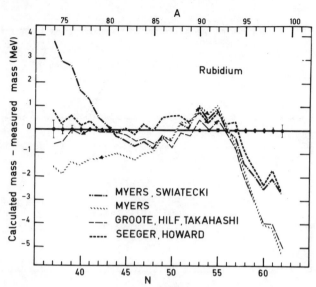

Fig. 5a : Comparison between experimental calculated masses for
the rubidium isotopes. Solid points are the reported measurements
with their errors. The arrows indicate the limits of the known
masses which were used to fit the predictions.

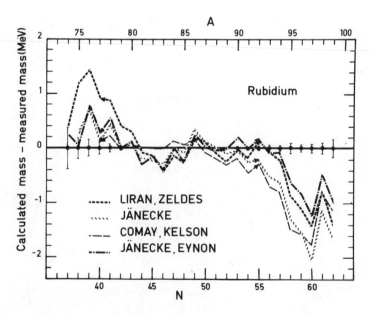

Fig. 5b : Comparison between measured and calculated masses for
the rubidium isotopes (continuation of Fig. a).

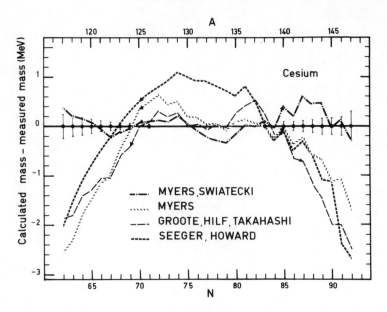

Fig. 6a : Comparison between measured and calculated masses
          for the cesium isotopes.

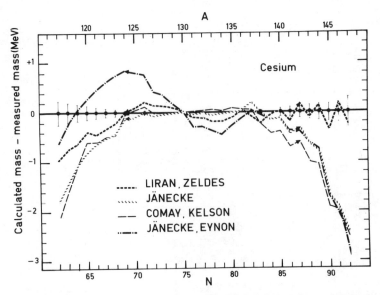

Fig. 6b : Comparison between experimental and calculated
          masses for the cesium isotopes (continuation of
          Fig. 6a).

- The same general behaviour of predicting too strongly bound nuclei
for both n-rich and n-deficient cesium isotopes (Fig. 6) for all the
calculations except the old liquid drop model of Myers and Swiatecki
and the semi empirical shell model of Liran and Zeldes. It is striking to observe more similarity between the different calculations
than between the experimental and calculated masses.

    But masses depend on nuclear structure and interesting features
show up on the differential representation of the mass surface as
presented on Fig. 7 and 8. One can recognize, for example, on these
2-neutron separation energy plots the rapid drops corresponding to
the major neutron shells (N=50 and N=82) and the bumps corresponding
to a region of deformed nuclei (i.e. the rare earth region on Fig. 7).

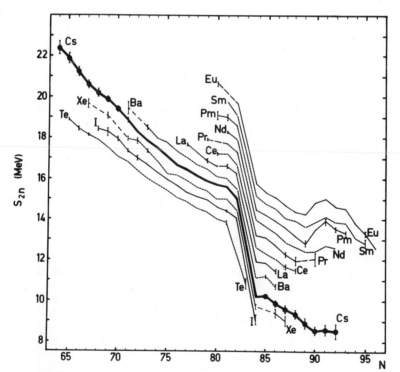

Fig. 7 - Experimental 2-neutron separation energies S_{2n} for the
       elements around cesium.
       (from our measurements and from the 1977 Atomic Mass Table[10]).

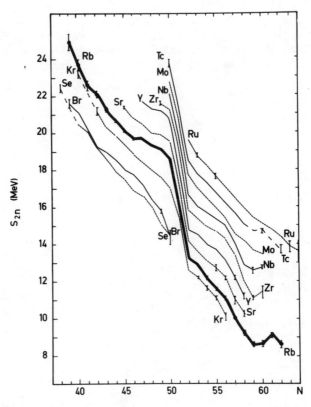

Fig. 8 - Experimental 2-neutron separation energies S_{2n} for the
elements around rubidium (from our measurements and from
the 1977 Atomic Mass Tables[10]).

Our results on the neutron rich rubidium isotopes show a smaller drop at N=56 which could correspond to the closure of the $d_{5/2}$ subshell and a bump at $N \geqslant 60$, signifying a sudden change in nuclear shape. Earlier investigations of systematics of the $E_2{}^+$energy in the even nuclei indicated the existence of a deformation region centered around ^{102}Zr. Our results indicate that the deformation sets in suddenly at N=60 in the rubidium isotopes. Two other experimental facts have since confirmed that these nuclei are strongly deformed ; the low values of the energies of the first 2^+ levels in ^{98}Sr and ^{100}Sr[17,18] and the anomalously large isotope shift for ^{97}Rb and ^{98}Rb.

CONCLUSION

We have presented here the new possibilities that arise when connecting a double focusing mass spectrometer with the ISOLDE facility. The first results obtained on two long chains of isotopes have shown that interesting features can be revealed when studying unexplored regions of nuclei. We plan to bring some improvements to the experiment and to extend these mass measurements on other neighbouring elements.

REFERENCES

1 - R. Klapisch, C. Thibault-Philippe, C. Detraz, J. Chaumont, R. Bernas and E. Beck, Half-life of ^{11}Li, ^{27}Na and of the new isotopes ^{28}Na, ^{29}Na, ^{30}Na and ^{31}Na produced in high energy nuclear reactions, Phys. Rev. Letters 23 : 652 (1969).
2 - D.C. Kayser, R.A. Britten and W.H. Johnson, Recent Minnesota mass results, in Atomic Masses and Fundamental Constants 4, J.H. Sanders and A.H. Wapstra ed., Plenum Press, London (1972).
3 - R. Klapisch and C. Thibault, On line mass spectrometric analysis of far unstable light nuclei produced in high energy nuclear reactions. Present status and the prospect of direct mass measurements, in Atomic Masses and Fundamental Constants 4, J.H. Sanders and A.H. Wapstra ed., Plenum Press, London (1972).
4 - C. Thibault, R. Klapisch, C. Rigaud, A.M. Poskanzer, R. Prieels, L. Lessard and W. Reisdorf, Mass Spectrometry of unstable nuclei, in Atomic Masses and Fundamental Constants 5, J.H. Sanders and A.H. Wapstra ed., Plenum Press, New York (1976).
5 - W.H. Johnson, The feasibility of direct mass determinations of short half life isotopes in Proceedings of the International Conference on The Properties of Nuclei far from the region of β Stability, Leysin, 1970 CERN Report N°70-30 p.967 (unpublished)/
6 - G. Audi, M. Epherre, C. Thibault, R. Klapisch, G. Huber, F. Touchard, H. Wollnik, Mapping of Nuclear masses in the region (N<126, Z>82) from direct measurements of francium isotopes (this volume).

7 - H.L. Ravn, to be published in Physics Report - 1979.

8 - F. Touchard, G. Huber, R. Fergeau, C. Thibault and R. Klapisch, A movable dynode electron multiplier for the detection of radioactive ions in an on-line mass spectrometer, Nucl. Instrum. Methods 155 : 449 (1978).

9 - M. Epherre, G. Audi, C. Thibault, R. Klapisch, G. Huber, F. Touchard and H. Wollnik, Direct Measurements of the masses of rubidium and cesium isotopes far from stability, Phys. Rev. C 19 : 1504 (1979).

10 - A.H. Wapstra and K. Bos, the 1977 Atomic Mass Evaluation, Atomic Data and Nuclear Data Tables, 19 : 177 (1977).

11 - R. Decker, K.D. Wünsch, H. Wollnik, E. Koglin, G. Siegert and G. Jung, preprint submitted to Z. Physik (1979).

12 - U. Keyser, H. Berg, F. Münnich, B. Pahlmann, R. Decker, G. Jung and B. Pfeiffer, Beta-Decay energies of neutron rich fission products in the vicinity of mass number 100, this Volume.

13 - H. Wollnik, Comparison between precise Q -Values and direct mass measurements on neutron rich Rb and Cs isotopes, this volume.

14 - C. Ekström, S. Ingelman, G. Wannberg and S. Skarestad, Nucl. Phys. A 292 : 144 (1977) ; C. Ekström, G. Wannberg, and J. Heinemeier, Phys. Lett. 76B : 565 (1978).

15 - G. Huber, F. Touchard, S. Buttgenbach, C. Thibault, R. Klapisch, S. Liberman, J. Pinard, H.T. Duong, P. Juncar, J.L. Vialle, P. Jacquinot and A. Pesnelle - to be published.

16 - 1975 Mass Predictions in Atomic Data and Nuclear Data Tables 17 : 411 (1976) S. Maripun Ed.

17 - H. Wollnik, F.K. Wohn, K.D. Wünsch and G. Jung, Nucl. Phys. Experimental indication of the onset of nuclear deformation in neutron-rich Sr isotopes at mass 98. Nucl. Phys. A 291 : 355 (1977).

18 - R.E. Azuma, G.L. Borchert, L.C. Carraz, P.G. Hansen, B. Jonson, S. Mattson, O.B. Nielsen, G. Nyman, I. Ragnarsson and H.L. Ravn, The strongly deformed nucleus ^{100}Sr, to be published in Physics Letters B.

NUCLEAR MASSES FROM FIRST PRINCIPLES

Hermann G. Kümmel

Institut für theoretische Physik
Ruhr-Universität Bochum and
Max Planck Institut für Chemie, Mainz

I. INTRODUCTION

In this talk I want to review the present status of the calculation of nucleon masses from "first principles". How "first" the
principles are, depends on one's viewpoint. I shall take the position that we want to compute nuclear properties from two body
nucleon-nucleon (NN) potentials which in turn have been fixed by
optimizing the description of two nucleon data (scattering and
bound state of the deuteron). What the theorist then has to do is
to solve the $A = N + Z$ body problem as good as he can and sell the
results to the experimentalists. Such a procedure is legitimate as
long as the following two points are observed:

  i) it is made sure that the (necessarily approximate) proce-
     dure used in many body theory converges
 ii) the basic concept of using two body potentials always is
     kept in mind: knowing that in fact there is meson exchange
     instead of potentials and baryonic resonances in addition
     to mere nucleons, a complete agreement between such a simp-
     le description and nature cannot be expected. Yet it is of
     course interesting and important to know to what extent
     "pure many nucleon theory" is valid.

We know more than one thousand binding energies, most with extreme
accuracy - an accuracy no theorist can compete with. Yet, if we
would be able to compute all these many ground state energies we
would have a very stringent test of NN-potentials, even if the
accuracy of numerical results would be well below the experimental
one. At present there is no hope that we can do that. All we can
do is to compute deuteron, triton, light closed shell nuclei up to
^{48}Ca, nuclear matter, and - more recently - some nuclei around ^{16}O.

315

This now constitutes a far less stringent test of NN interactions
than knowing all binding energies. Thus, to have any valuable test
at all we also have to look at some other features of these few
nuclei in addition to the binding energy. The most important of
these is the density distribution. To make things simple, we rely
mainly on the radius as the most relevant parameter of the density
distribution.

The inaccuracy of the results obtained by the theorists is due
to the enormous difficulties in reliably solving the many body
problem. It is in the frame of the now fashionable expS-method
(or coupled cluster method) that these difficulties have been overcome, such that we now can, at least numerically, control the convergence of approximations. For three body and infinite systems
other useful methods have also been devised.

Thus, I shall first give an outline of the ideas underlying
this method and its generalizations. As far as needed, the other
methods will be described als well. Then I shall give the results
obtained and compare them with experiment and indicate, how people
are trying to incorporate mesonic degrees of freedom, etc.

II. EXPS-METHOD - CONCEPTS AND THEIR REALIZATIONS

We start with ground states of closed shell systems. Assuming
a Hamiltonian (with two body potentials for simplicity)

$$H = \sum T_i + \sum V_{ij} ,$$ (1)

we need to solve the Schrödinger equation

$$H \Psi_0 = E_0 \Psi_0$$ (2)

for the ground state. Assume that a determinant ϕ_o of $A = N + Z$
"occupied" single particle states can be used as a starting point
in the sense that this ϕ_o has a nonvanishing overlap with Ψo:

$$\langle \phi_0 | \Psi_0 \rangle \neq 0 ,$$ (3)

and that

$$\Psi_0 = (1+F) \phi_o ,$$ (4)

where $F = \sum_{n=1}^{A} F_n$ and F_n is an n particle-hole operator (exciting n
nucleons out of the "Fermi sea" of occupied levels). Eq. (4) is a
quite general and exact decomposition of the wave function we are
looking for. One could try to use the Schrödinger equation to determine all F_n. However, such a procedure is known to converge very
slowly as there occur "macroscopic" terms (proportional to the
particle number A) which finally must cancel in addition to

"microscopic terms" which one needs to compute. It is rather evident
- and well known in quantum chemistry - that such trouble must show
up: on the right hand side of (2) there occurs the macroscopic
energy which must be cancelled by some terms on the l.h.s. There
is a very simple way out: write

$$\Psi_0 = e^S \phi_0, \qquad (5)$$

with

$$S = \sum_{n=1}^{A} S_n , \qquad (6)$$

where S_n (like F_n) is a n particle hole operator best to be described by the diagram of Fig. 1

Fig. 1   Graphical representation of S_2

(downgoing lines correspond to emptied levels originally occupied,
upgoing lines to filled levels originally empty). A general wave
function will consist of ϕ_0 plus a part where two particles excite
each other $S_2\phi_0$, where two pairs of particles independently excite
each other $\frac{1}{2}S_2 S_2\phi_0$ (the factor $\frac{1}{2}$, being necessary to avoid
counting pairs twice), and so forth, such that the contri-
bution of any independent pair excitation is

$$\left(1 + S_2 + \frac{1}{2}S_2^2 + \frac{1}{3!}S_2^3 + \cdots\right)\phi_0 = e^{S_2}\phi_0.$$

The same consideration leads to $\exp S_3\phi_0$ for each "cluster" of three
independent excitations. To be convinced that (5) is a natural decomposition into all possible independent n particle cluster excitations, one furthermore has to know that $\exp S_1\phi_0$ is the most general
determinant not orthogonal to ϕ_0 (Thouless theorem[2]). The wave
function is represented in Fig. 2.

$$\psi_0 = \qquad \qquad + \qquad \qquad + \qquad \qquad$$

$$\phi_0 \qquad + \qquad S_1\ \phi_0 \qquad + \ \tfrac{1}{2}S_1^2\ \phi_0 + \cdots$$

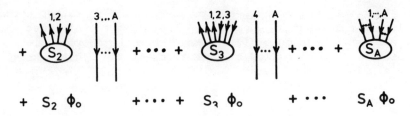

$$+ S_2 \phi_0 \qquad + \cdots + \quad S_3 \phi_0 \qquad + \cdots \qquad S_A \phi_0$$

Fig. 2   Graphical representation of ground state wave function

By a simple trick, writing (2) as

$$e^{-S} H e^{S} \phi_0 = E_0 \phi_0$$

we get rid of the problem mentioned before: we now project on the states ϕ_0, ϕ_{1o} (1 particle hole in ϕ_0), ϕ_{2o} etc. Thus, we have an explicit representation of the Schrödinger equation

$$\langle \phi_0 | e^{-S} H e^{S} | \phi_0 \rangle = E \qquad \text{energy}$$

$$\langle \phi_{1o} | e^{-S} H e^{S} | \phi_0 \rangle = 0 \qquad \begin{array}{l}\text{generalized HF equation} \\ \text{(one body equation for } S_1)\end{array}$$

$$\langle \phi_{2o} | e^{-S} H e^{S} | \phi_0 \rangle = 0 \qquad \begin{array}{l}\text{generalized Bethe-Goldstone} \\ \text{equation (two body equation} \\ \text{for } S_2)\end{array}$$

$$\langle \phi_{3o} | e^{-S} H e^{S} | \phi_0 \rangle = 0 \qquad \begin{array}{l}\text{generalized Bethe-Faddeev} \\ \text{equation (3 body equation} \\ \text{for } S_3)\end{array}$$

etc.

Except in the expression for the energy itself, E_0 is eliminated and it comes as no surprise that there occur no macroscopic terms together with microscopic ones.

These equations can be written down to any order since $\exp(S) H \exp S$ is a <u>finite</u> number of terms. Also, they can be arranged in such a way that a hard core potential poses no problem. Furthermore the n-body equation (for S_n) is coupled to the equations for S_{n+1} and S_{n+2} (for two body potentials). This allows a straightforward truncation; for instance, the "three body approximation" implies solving equations for S_1, S_2 and S_3, omitting coupling to S_4 and S_5. This is a self-contained set of nonlinear coupled equations solvable by some kind of self-consistent iteration procedure. The techniques are due to Zabolitzky[3] and shall not be described here.

In a more physical language the three body approximation
referred to above involves all two and three body correlations and
corrections to the starting s.p. (single particle) wave functions
(due to S_1). All those quantities then have been computed taking
into account all mutual couplings of up to three body clusters.
The strength of the method is its conceptual simplicity, leading
to a superiority in bookkeeping of the many terms occurring in
higher orders and to clean decisions how to truncate. Indeed, the
resulting equations clearly tell us which terms should be kept
together. The numerical tests made in some cases, comparing with
diagrammatic methods which sometimes depend on the often misleading
intuition of the individual using them, always have favoured the
truncation scheme suggested by the expS-method.

Having spent so much time on the ground state properties of
closed shells, there is very little left for describing the generalization to non closed shell systems [4,5]. Assume we want to
calculate a nucleus with one additional "valence" nucleon outside
the closed shell, for example ^{17}O. The idea is to feed the "known"
wave function of ^{16}O, expS ϕ_o, into the theory, writing the wave
function now as

$$\psi_j = e^S (1 + \sum_{i=1}^{A} F_i) a_j^+ \phi_o .$$

Here $a_j^+ \phi_o$ is the determinant of the 17 nucleons, one of them in
the "valence state" with quantum numbers labelled by j.
F_1 describes the s.p. excitation of the valence nucleon, i.e. the
change of the valence state due to the interaction with the core;
F_2 includes the correlations of the valence nucleon with any one
of the core nucleons, and so forth. By a technique similar to the
closed shell case, equations for the F_i and the energy difference
between ^{17}O and ^{16}O are obtained. This procedure can be generalized to two, three, ... valence nucleons (or holes). Again, one
proceeds by feeding in the results of the foregoing system with
one nucleon less. The advantage of this method compared to diagrammatic ones is even more striking. At present this method is
the only one which works for hard core interactions.

Concluding this section, it should be noted that in the three
body case the direct approach is feasible with present day computers [6]. No expS method is required. People working on this problem
usually use the Faddeev representation of the three body Schrödinger
equation. Highly technical problems arise and have plagued all
calculations so far. I shall come back to this problem later dis-
cussing the triton. Although this theory has been formulated also
for the four body problem, no calculations of comparable quality
exist.

However, there also is the completely different variational
method for infinite homogeneous systems such as nuclear matter, mainly promoted by Pandharipande and collaborators [7]. It uses a wave

function of a prescribed form $\Psi_0 = \Pi f(r_{ij})\phi_0$, where the $f(r_{ij})$
describe correlations as a function of the distance r_{ij} between
two nucleons; for instance, $f(r) = 0$ inside the hard core region,
$f(r) \to 1$ for r greater than the range of the potential. Very
sophisticated methods have been invented to carry out the many
dimensional integrations for the expectation values $\langle H \rangle = \langle \Psi_0 | H | \Psi_0 \rangle$
of the energy. The energy then is determined as the minimum of $\langle H \rangle$
by a suitable variational procedure. Very reliable calculations
did emerge from this method, recently even for potentials as
complex as the NN interaction (including tensor and L.S.-coupling).

I will not dwell upon these recent developments as we here are concerned with finite systems and many of you may even doubt that the
consideration of nuclear matter makes any sense at all. Let me note
only that due to recent improvements in both the variational and
expS-method the long standing disagreements between the results of
both methods did disappear to the extent that most (not all) many
body theorists believe that their theories are well understood.

III. RESULTS

One of the problems we have to live with is that there are
many different phenomenological potentials reproducing the two nucleon data equally well. The occurrence of such "phase shift equivalent potentials" has two sources: first, the sometimes quite
large error bars of the two body data; second, even with no errors
it is known that potentials need be equivalent only "on the energy
shell". Off the energy shell they are quite arbitrary. Thus I am
forced to present to you the results for a series of fashionable
potentials. Although in this sense all phase shift equivalent potentials are "equal", some are "more equal" than others: they have
been derived from more basic concepts, namely from one boson-
exchange. As we shall see soon, the results obtained with these
potentials do not justify their use more than others, in spite of
having a lot more snob appeal.

Let me now present the results.

A. TRITON AND ^{3}He

The results obtained by different authors (based on Faddeev and
variational methods) are presented in table 1. For simplicity only
the RSC (Reid soft core) potential is considered. It is evident that
the results differ by an intolerable amount with each other and with
experiment. By now it is not clear whether the former is due to a
trivial error or to the not yet sufficiently well established convergence of the expansion into partial waves. I should mention that
until very recently there was a large discrepancy between experiment
and theory concerning the density distribution. The reinterpreta-
tion of the electron scattering data by inclusion of baryonic

Table 1   Triton binding energies

Ref.	BE (MeV)
8	6.4 ± 0.5
9	6.98 ± 0.2
10	7.0
11	7.58 ± 0.1
Exp.	$8,481.92 \pm 0.20$

resonances has removed this discrepancy [12].

Coming to ^{3}He, we would be happy if we could understand the ^{3}H $-$ ^{3}H energy difference of experimentally 0.764 MeV. Indeed, due to an almost model-independent analysis [12], one obtains 0.683 $\pm$ 0.029 MeV. The remaining difference could be accounted for by symmetry breaking of NN potentials and three body forces due to mesonic effects.

2. LIGHT CLOSED SHELL NUCLEI

In looking at the results obtained by us [1] for ^{4}He, ^{16}O and ^{40}Ca in Figs. 3-5 we observe the good convergence of the expS-method: In these binding energy per nucleon vs. charge radius plots, lines connect equal approximations (labelled SUB(2), SUB(3) and SUB(4) for two, three and four body approximations, respectively) for different potentials (RSC-Reid soft core [14], HJ-Hamada Johnston, SSC(B)-super soft core B [15]). It is seen that the convergence is very good and that we do not reach the experimental region by any of the phase shift equivalent potentials. Furthermore the "softer" the potential, the larger is the binding energy and the smaller the radius.

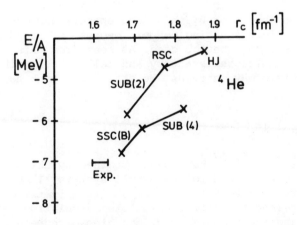

Fig. 3   Binding Energy vs. Charge Radius for ^{4}He

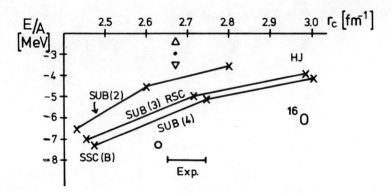

Fig. 4   Same as Fig. 3, for ^{16}O.

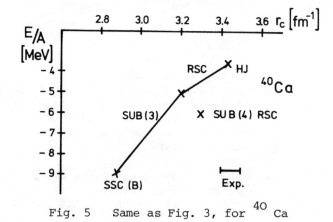

Fig. 5   Same as Fig. 3, for ^{40}Ca

We are not surprised by this discrepancy: the description of
nuclei by potentials ignoring mesonic degrees of freedom and baryonic resonances cannot be too good. As seen from this viewpoint
the results are quite satisfactory and tell us something about how
far the potentials may carry us. We shall come back to this point
later.

3. NUCLEI AROUND ^{16}O

These results [16, 17] have been obtained by using the generalization of the expS-method to open shell systems. They are given in
table 2. (No radii have yet been computed). Only the one valence
values are obtained from first principle calculations since in the
two valence case experimental s.p. energies have been used. In ^{15}N
some additional approximations beyond those described in the last
chapter have been made. Thus the only "clean" calculation is done

Table 2   Ground State Energy Differences

Nuclei	Theory (MeV)						Exp. (MeV)
	SUB (2)			SUB (3)			
	HJ	RSC	SSC (B)	HJ	RSC	SSC (B)	
$^{15}_{N}$-$^{16}_{O}$	-13	-16	-20	-6.3	-9.5	-11.5	-12.48
$^{17}_{O}$-$^{16}_{O}$	2.3	4.2	5.6		2.3		4.10
$^{18}_{F}$-$^{16}_{O}$	9.3	9.15	9.05		10.85		9.75
$^{18}_{O}$-$^{16}_{O}$	10.7	10.9	10.4		11.3		12.19

with ^{17}O. Generally it is seen that the results again depend
strongly on the phase shift equivalent potentials. This indicates
again the necessity of introducing mesons. Furthermore, the trends
to more binding in going from the hard to soft core potentials is
recovered again. Finally, it is gratifying that there is satisfactory convergence in going from the two body SUB(2) to the three
body approximation SUB (3). No SUB (4) calculations have yet
been attempted.

4. NUCLEAR MATTER

     As mentioned before, this subject has been a matter of great
controversy which only recently has been removed due to the careful third order calculations by Day [19] using a version of many body
theory equivalent to the expS-method. In nuclear matter the technical difficulties are larger than in finite system. In the latter
the density typically is much lower and only a few oscillator orbitals usually suffice. This is why here the variational calculations [18] are competitive. For the latter the RSC potential has not
yet been taken fully into account so that a complete comparison is
not possible. In Fig. 6 we show the present situation.

     There is no disagreement between both calculations considering
the large error bars (which also are present for variational results,
although they have not been plotted). The convergence of the many
body methods seems not to be as good as for finite systems. There
are some disturbing problems, however. For instance, the results
depend rather strongly on the "starting s.p. potential". Of course,
this should not be the case in an exact calculation and the results
should still depend only weakly on this potential in an approximate
calculation. (For finite nuclei the situation is much better.)
I think you will agree with me that here some more work has to be
done to reach final conclusions. No three body calculations exist
for potentials other than RSC. Since for nuclear matter their contributions are pretty large, it does not make sense to look too

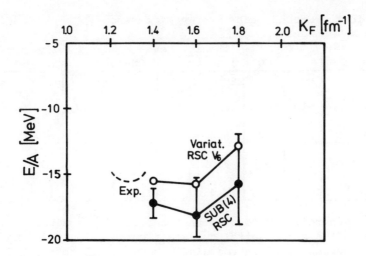

Fig. 6   Binding Energy vs. Fermi Momentum for Nuclear Matter

close at the many existing second order calculations with other
potentials. I would thus like to finish the discussion of nuclear
matter by just stating that again the experimental region has not
been reached.

IV. OUTLOOK

     Naturally, one would still like to derive the experimental
binding energies by trying harder. This implies that now mesons
have to be taken into account. This in turn implies that baryonic
resonances Δ , too, are needed since they somehow are bound states
of mesons plus nucleons - at least they have corresponding
symmetries.

     This, now, is an extremely complex problem, as some particles
are created and annihilated whereas others are changed from nucleons to Δ 's and vice versa: the standard many body problem, by
itself a rather complex one, is just child's play compared to this.
Nevertheless, attempts have been made to feed in these new concepts.
The first step is to derive potentials from meson exchange: this is
not trivial since the exchanged meson "feels" the presence of the
medium of all other nucleons: thus one has to incorporate this
effect into the NN potential (defined by summing all ladders of one
boson exchanges (see Fig. 7a). This has been done for instance by
the Bonn-Jülich-group [20]. I have plotted (as ∇) into Fig. 4 the

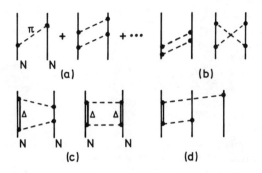

Fig. 7   Diagrams Occurring in Realistic Calculations

results obtained by including ladders and "stretched box" and
"crossed box" diagrams of Fig. 7b. It should be noted that these
calculations use only the two body approximation. Evidently this
theory carries us even further away from experiment. Even the inclusion of Δ 's according to Fig. 7c does not help at all (Δ in
Fig. 4). A small consolation comes from the fact that there also
occur three body forces due to the effects indicated in Fig. 7d:
indeed their effect is an increase of the binding energy by
0,5 MeV per particle in ^{16}O (plotted too, as a dot, in Fig. 4).
There are other three body plus ordinary mesonic effects, however;
by using an old version of an effective three body force [22] we
obtained an essential improvement in the right direction
(circle in Fig. 4).

So, theorists are not able to compete with experimentalists
in accuray - they never will in this field. The enormous complexity
of the problem even without mesons, will not allow calculations
with small error bars. Still, to quite a good extent we may consider
the nucleon as a system of nucleons interacting via potentials and
we may hope that by including mesons etc. we finally will reach a
tolerable agreement between theory and experiment - and a better
understanding of nature.

References

1.   H. Kümmel, K.H. Lührmann, J.G. Zabolitzky
     Phys. Reports 36C 1 (1978)
2.   D.J. Thouless
     The Quantum Mechanics of Many Body Systems
     2nd Ed., Academic Press, New York and London, 1972
3.   J.G. Zabolitzky
     Nucl. Phys. A228 272, 284 (1974)
4.   R. Offermann, W. Ey, H. Kümmel
     Nucl. Phys. A273 349 (1976)
5.   W. Ey
     Nucl. Phys. A296 189 (1978)
6.   Y.E. Kim
     in Few Body Nucleon Physics, Internat. Centre for
     Theoretical Physics, Trieste, 1978
7.   V.R. Pandharipande, R.B. Wiringa
     Rev. Mod. Phys. (in press)
8.   R.A. Malfliet, J.A. Tjon
     Ann. Phys. (NY) 61 425 (1970)
9.   R.A. Brandenburg, Y.E. Kim, A. Tubis
     Phys. Rev. C12 1368 (1975)
10. A. Lagergne, C. Gignoux
     Nucl. Phys. A203 597 (1973)
11. W. Glöckle, R. Offermann
     Phys. Rev. C16 2o39 (1977)
12. M.M. Giannini, D. Drechsel, H. Arenhövel, V. Tornow
     preprint, Mainz 1979
13. R.A. Brandenburg, S.A. Coon, P.U. Sauer
     Nucl. Phys. A294 305 (1978)
14. A. Reid
     Ann. Phys. (NY) 50 411 (1968)
15. R. de Tourreil, D.W.L. Sprung
     Nucl. Phys. A201 193 (1973)
16. J.G. Zabolitzky, W. Ey
     Nucl. Phys. (in print)
17. K. Emrich, J.G. Zabolitzky, K.H. Lührmann
     Phys. Rev. C16 1650 (1977)
18. J.C. Owen
     preprint 1978
19. B. Day
     private communication
20. K. Holinde
     NORDITA preprint 1979
21. K. Kotthoff, R. Machleidt, D. Schütte
     Nucl. Phys. A264 484 (1976)
22. D.W.E. Blatt, B.H.J. McKellar
     Phys. Rev. C1 614 (1974)

PRESENT STATUS OF HARTREE-FOCK CALCULATIONS OF NUCLEAR BINDING

ENERGIES USING EFFECTIVE INTERACTIONS

P. Quentin

Institut Laue-Langevin, 156X
38042 Grenoble Cedex, France

ABSTRACT

At the AMCO-5 conference, some accounts of successful descriptions of binding energies through Hartree-Fock calculations using
effective interactions have been given. Since then, many works have
been performed in two directions : (i) studies of the reliability
of the approximations made, having in mind particularly the problem
of extrapolations to unknown nuclear regions, (ii) attempts to
improve the flexibility or the feasibility of such calculations.

In this paper we aim to make a survey of these developments by
reviewing the following topics :
- an assessment of the quality of effective interactions in use,
- some comments on the pairing correlation treatment,
- the evaluation of the spurious rotational energy in deformed
  nuclei,
- the methods of solutions of the Hartree-Fock equations,
- a review of some ersatz to full Hartree-Fock calculations and a
  discussion of their accuracy.

1. INTRODUCTION

Since the last AMCO-5 conference, it has been increasingly
clear that Hartree-Fock calculations using effective nucleon-
nucleon interactions are indeed capable of reproducing within
$\sim$ 2-3 MeV nuclear binding energies. But it has also appeared that
going significantly beyond that kind of agreement would necessitate
a tremendous effort on both theoretical and numerical sides.

Even though the use of the Hartree-Fock + BCS (or in some cases of the Hartree-Fock Bogoliubov) approximation with effective interactions results in relatively heavy numerical calculations, it remains of course a rather crude approximation to the full nuclear many-body problem. This possible source of inadequacy is enhanced when the effective force in use is purely phenomenological (see Ref. [1] for a comprehensive list of references) and is not partially deduced from nuclear matter G-matrix calculations within the local density approximation as in Refs [2-3]. The direct parametrization of the Hamiltonian density often referred to [4-5] as the 'energy density formalism" is subject even in its self-consistent version to the same difficulty in addition to the fact that,in this case,the contact between the output and the bare nucleon-nucleon interaction is somewhat more obscure.

Assuming for a moment that the problems related with the use of phenomenological effective interactions would be fixed up, a complete computation of nuclear binding energies would remain an almost impossible numerical task. Indeed while for spherical nuclei, present numerical methods would allow such a study, the fact that most of the nuclei are deformed implies a lot more of practical difficulties in the numerical solution of the variational equations due to the corresponding lack of symmetries. Some approximations to the full Hartree-Fock + BCS treatment are therefore needed but they entail further numerical uncertainties.

In this paper we would like to review the problems and to assess the validity of present Hartree-Fock + BCS calculations of nuclear binding energies. In Chapter 2, we will study topics related to the approach itself whereas Chapter 3 will be devoted to a discussion of the practical solution of the problem. In Chapter 4 finally, we will survey some proposals to avoid some of the numerical difficulties encountered in the solution of the Hartree-Fock + BCS variational equations.

2.   SOME THEORETICAL PROBLEMS

2.1   The Effective Force

If one has in mind to calculate systematically a large number of nuclear binding energies one is more or less forced, in the present state of the numerical art to use fully phenomenological effective forces of short range character to which we will refer in the following,as Skyrme-like forces (see e.g. Ref. [1]). We will therefore not consider here gaussian forces like the D1 Gogny force [6] or others (see e.g. Refs. [7-8]). For the same reason we will also not study the problem of partially phenomenological forces as issued from Refs. [2-3]. The so-called "energy density formalism" in its self-consistent version will not be touched upon either, since we limit the scope of this paper to calculations deriving explicitly from an effective force (the reader may find an up-to-date account of some of its results in the contribution of F. Tondeur[5] to this Conference).

The Central Part. It has been shown in Ref. [9] that one can reproduce correctly nuclear matter saturation properties (or equivalently binding energies and radii of magic nuclei) with an infinity of 5-parameter central Skyrme forces, the remaining free parameter being the relative amount of velocity- versus density- dependence in the interaction. This is illustrated in Table 1. There is however an array of more or less phenomenological arguments leading to a determination of the velocity dependence of the force in such a way that the nuclear matter effective mass should be $\sim$ 70 % of the nucleonic mass. These reasons include :

(i) a comparison between G-matrix effective forces and Skyrme-like forces through the density matrix expansion[10]

(ii) the correct reproduction of binding energies in deformed nuclei (see Fig. 1 of Ref. [1])

(iii) the correct reproduction of the single particle level density in deformed nuclei where the coupling of individual and vibrational degrees of freedom is minimal in contradiction with the case of spherical nuclei (see Figs. 3 and 4 of Ref. [1]).

Table 1.   Reproduction of the ^{208}Pb saturation properties by forces having different velocity- versus density- dependence.

Force	m^*/m	t_3	ΔE	Δr
SV	0.38	0.	0.55	- 0.03
SIV	0.47	5000.	0.08	0.01
SIII	0.76	14000.	0.12	0.07
SVI	0.95	17000.	0.62	0.09

The nuclear matter effective mass m^* is given in units of the nucleonic mass m. The standard Skyrme density-dependence parameter t_3 is given in MeV-fm^6. The difference $\Delta E = B_{HF} - B_{exp}$ where B stands for the (positive) binding energy, is given in MeV. The difference $\Delta r = r_{HF} - r_{exp}$ (where r stands for the charge distribution radius) is given in fm.

(iiii) an interpretation of the systematics of isoscalar quadrupole giant resonance region in terms of a RPA sum rule approach[11].

Consequently the uncertainty in the determination of the
central part of the force is not as dramatic as it may seem a priori.
Nevertheless it should be concluded that a major ambiguity cannot be
removed upon calculating only binding energies for spherical (magic)
nuclei.

Usual Skyrme forces[9] include a linear dependence of the
central force on the density, resulting in a rather high value of
the incompressibility modulus K ranging from 300 to 350 MeV. While
this seems not to affect too much binding energies it is of course
of primary importance for the energy of the giant monopole resonance.
The introduction of a fractional density dependence leading to a
more reasonable value for K while maintaining the quality of the fit
to other saturation properties is highly desirable and is currently
undertaken.

The Tensor Part. Even though there is some reasons to think
that the effective tensor interaction should play some role in
nuclear static properties calculated within the Hartree-Fock approximation (see e.g. Ref. [12]) it has not been as thoroughly studied
as other parts of the effective force. With very few exceptions (as
in Ref. [7]) it has not been included in phenomenological forces;
the reason for this being that the tensor force mainly influences
the spin-orbit splitting of spin unsaturated nuclei which can be
reasonably well accounted for, simply by phenomenological spin-orbit
forces. This has led the authors of Refs [13-14] to the conclusion
that it is phenomenologically not necessary to include a tensor
part into the effective interaction.

The Spin-Orbit Part. In most cases the spin orbit force is
considered within its zero-range limit, leaving the only one overall
parameter to be fitted. This is generally done [9,15] either by
fitting some magic nuclei spin-orbit splittings or by adjusting as
best as possible the calculated single-particle level sequence to
what can be deduced from odd nuclei in the vicinity of magic nuclei.
The problem with this fitting procedure is that to retain the
computational advantage of spherical nuclei one places oneself in
the poorest case of comparison between calculated and "experimental"
single particle spectra. As already noted above, such odd nuclei
spectra need to be deconvoluted for the dressing of the individual
degrees of freedom by the core vibrational modes. It should be
concluded from the latter that a sensible way of adjusting the spin-
orbit strength would be to fit experimental energy spectra in well
deformed nuclei (as rare-earths or actinides) which is indeed what
S.G. Nilsson always did (see e.g. [16]).

In the way the parameters have been fitted so far, one must have in mind a possible error bar of $\sim$ 20 % on the spin-orbit strength. The effect of such an uncertainty for example on the super-heavy stability problem is tremendous. Indeed the closure at N=184, Z=114, N=228 etc.. is mostly due to a specific position of some high-ℓ level(s) with respect to the Fermi surface. A wild variation of the spin-orbit strength W is in fact capable of completely destroying the level scheme. This has been investigated in the $^{354}_{228}126$ case calculated with the SIII force where decreasing (increasing) W by $\sim$ 20 % resulted in decreasing (increasing) the fission half-life by 13(17) orders of magnitude [17]! Less dramatic but still important effect on nuclear binding energies of so far unknown magic nuclei, may also be expected. There is obviously here room for further improvements.

2.2. The Treatment of Pairing Correlations

Given an effective force, a fully self-consistent way of treating pairing correlations within an independent (quasi-) particle approximation is known as the Hartree-Fock Bogoliubov formalism. Practical solutions of the corresponding variational equations have been obtained by Gogny[6]. As one of the most striking results of subsequent work from the same group, one has found for most nuclear static and dynamic properties (including thus binding energies) no significant difference between the results of full Hartree-Fock-Bogoliubov and Hartree-Fock + BCS calculations. As an example B. Grammaticos[18] has shown for ^{152}Sm over a whole curve of quadrupole deformation energy that : (i) the corresponding energy curves are undistinguishable (ii) the pairing "gaps" identified approximately as the minimal quasi-particle energies differ only by $\lesssim$ 0.1 MeV. One may therefore conclude that for binding energies the full complexity of the Hartree-Fock-Bogoliubov approach is not necessary and one is entitled to use the Hartree-Fock + BCS approximation.

It is essential for the preceding conclusion to hold, that one uses the same interaction for all (Hartree-Fock-Bogoliubov, Hartree-Fock, BCS) calculations involved. In this respect, it is important to note that calculations using the Skyrme forces do not fulfill this last condition. Indeed even though there are indications[19-20] that upon slightly changing the parametrization, Skyrme-like forces may lead to good pairing matrix elements, present standard [9] versions of this force are not satisfactory in this respect.

2.3   Some Deficiencies of Single-Determinantal Wave Functions with Bearing on Nuclear Binding Energies.

Even though the effect on binding energies of <u>some</u> correlations in the nuclear ground state are taken care of, by use of an effective force, this cannot be true for all correlations. However in purely phenomenological approaches one fits the binding energies of magic (spherical) nuclei with an accuracy of ~ 0.5 MeV (see Table 1), leaving thus no room for the energy gain due to all possible RPA-type correlations. This does only make sense if one is able by using the same force to reproduce with a comparable accuracy the binding energies of non-magic nuclei.

The latter include rigidly deformed nuclei. When the spherical symmetry is broken, it is well known that a Slater determinant is not a good angular momentum state, but rather describes a so-called intrinsic state i.e. an admixture of ground-band states. This admixture generates therefore a binding energy lack. In the ideal case of a pure rotational band, this lack in energy is given, for even-even nuclei, by

$$\Delta B = B_{0+} - B_{int.} \; = \; \frac{h^2}{2J} \; < \psi \mid \vec{J}^2 \mid \psi > \tag{1}$$

where J is the moment of inertia and $< \psi \mid \vec{J}^2 \mid \psi >$ denotes the expectation value of the square of the total angular momentum in the intrinsic state $\mid \psi >$. Given a Hartree-Fock + BCS solution $\mid \psi >$, one may compute $< \psi \mid \vec{J}^2 \mid \psi >$ and also J within the Inglis Cranking approximation. Upon assuming the validity of Eq. (1) one gets an estimate of ΔB. In Table 2 the results for B obtained in some deformed nuclei are reported from calculations[21] using the Skyrme SIII force. As a consequence of such calculations, one sees that the binding energies in deformed nuclei are reproduced within ~ 2 MeV. These figures however are merely estimates. Indeed, contrary to what was the case in spherical nuclei, the extraction of binding energies from Hartree-Fock + BCS calculations suffer from some uncertainties. One is due to the estimate (1) of ΔB since : (i) not a single actual nucleus is a pure rotor (ii) should the rotational collective motion be as expected adiabatic, one would still have to deal with Thouless-Valatin corrections to the Inglis cranking value for J. Another uncertainty is due to the estimate of the truncation error (if the Hartree-Fock single particle states are expanded on a given truncated basis) or to the estimate of the errors stemming from the kinetic energy operator (if the Hartree-Fock equations are solved in a meshed $\vec{r}$— configuration space). Nevertheless the qualitative result remains valid : it is indeed possible to include phenomenologically the effects of expected correlation energy gains for both magic and deformed nuclei, through

Table 2.   Binding energies of some deformed nuclei near the actinide region. Calculated energies[26] (in MeV) have been corrected for truncation and 0+ projection effects and are compared with experimental values[27].

Nucleus	Calculated (13 shells)	Corrected for truncation	Corrected for 0+ projection	Experimental
^{224}Ra	1708.1	1715.0	1716.6	1720.4
^{230}Th	1743.2	1750.7	1752.6	1755.2
^{236}U	1778.7	1786.8	1788.7	1790.5
^{244}Cm	1823.9	1832.8	1834.7	1835.9
^{248}Cm	1847.7	1857.0	1859.0	1859.3
^{258}Fm	1900.3	1910.6	1912.9	—

Note the inadequacy of the 0+ projection approximation (Eq. (1)) in the ^{224}Ra case.

the effective force fitting procedure. To appreciate fully this
result it may be interesting to keep in mind that such a phenomenological successful adjustment was not attainable for the single
particle level density : upon reproducing it in magic nuclei (which
would correspond for the paramatrization of Skyrme forces studied
in Ref. [9] to $m^{\ast}/m \gtrsim 1$) one would fail to do it in deformed nuclei.

3.   SOLUTION OF THE HARTREE-FOCK + BCS EQUATIONS

The Hartree-Fock equations are coupled integro-differential
non-linear equations. In the case of Skyrme-like forces they
constitute only a set of partial derivative equations. Solving at
the same time the BCS problem amounts only to add to the set, two
equations (gap equation and particle number conservation condition).
To solve such equations there is mostly[(+)] two attitudes :

A) project the single particle states onto eigenfunctions of
a one-body Hamiltonian chosen for its congruence with the physical
problem under study;
B) solve the differential equation in $\vec{r}$-space.

In method A one is bound to truncate the expansion basis,
introducing thus a spurious dependence of the calculated quantities
with respect to the basis parameters. Since one is to solve
approximately a variational problem for the nuclear ground state,
one optimizes the basis parameters, i.e. chooses those leading to
the lowest energy compatible with numerical computation facilities
which are available.

In method B, the symmetries retained for the HF state have an
important bearing on the practical solution of the problem. Should
it be spherical, then one has to solve a one-dimensional differential
equation, which is achieved rather easily as shown in Ref. [15].
When this is not the case, one uses a finite-difference method taking
advantage of the fact that the Hartree-Fock Hamiltonian to be
diagonalized has a large number of vanishing matrix elements. A
practical method of solution has been proposed [23] which is based
on the Lanczos algorithm. Another one has appeared as a by-product of
TDHF calculations, the so-called imaginary time technique which is
detailed in Ref. [24].

We will compare now, binding energies calculated with both
methods (restricting our discussion for the method B to the calculations of Ref. [25]).The problem with method A calculations is the

---

(+) We only refer here to methods which have been widely used in
     practical applications, leaving thus very interesting approaches
     as the variational method of Ref. [22].

lack of binding energy introduced by the truncation as illustrated
on Table 3. With method B, inaccuracies arise in the finite difference treatment of the kinetic energy operator in a discretized
space mesh. Due to the self-consistent character of the solution,
errors in the kinetic energy will generate also error in the potential energy as shown on Table 4. Indeed an underestimation of the
kinetic energy leads to too small a nucleus causing in this case
a gain of 9.91 MeV from the attractive part (pure δ-force) of the
force which is not compensated by the respulsive terms (of which
the largest contributor is the δ density-dependent term with
4.61 MeV). At the very high price of improved approximations for
differential operators (e.g. by using a 9-point Laplacian formula)
one obtains[25] for a Hamiltonian including a spin-orbit two-body
potential, solutions which are overbound by $\sim$ 0.5 MeV in light
nuclei.

    Method B calculations avoid the delicate and sometimes costly
basis parameter optimization problem encountered in method A. On the
other hand for effective forces including spin-orbit terms they are
far more expensive. They may constitute a practical improvement for
light nuclei (especially for triaxial shapes or, in general, whenever
a 3-parameter optimization is required). This is certainly not so
for heavier nuclei (say A > 100) where in the present state of the
art, method B calculations should be preferred.

4.   SOME HARTREE-FOCK WITHOUT HARTREE-FOCK APPROXIMATIONS

    In view of the numerical complexity of an exact solution of the
Hartree-Fock equations, a number of approximate methods have been
proposed, the accuracy of two of them we will discuss now.

Table 3.   Difference in the spherical ^{40}Ca nucleus between
           exact (solution in $\vec{r}$-space according to the method of
           Ref. [15]) and approximate (solution by truncated basis
           projection according to the method of Ref. [28]) binding
           energies (in MeV).

N	4	6	8	10
ΔB	3.24	2.05	1.79	0.81

The force SII[15] has been used with the oscillator
parameter $\sqrt{m\omega/\hbar}$ equal to 0.58 fm^{-1}, and are labelled
by the maximal total number of quanta N.

Table 4. Difference in the spherical ^{40}Ca nucleus between
          exact (solution in $\vec{r}$-space according to the method
          of Ref. [15])and approximate (finite difference
          method on $\vec{r}$-space with a 0.7 fm mesh size) values of
          different contributions to the total binding energy
          (in MeV).

Kinetic	Potential	Coulomb	Total
1.93	-4.26	0.23	- 2.10

The force SIII[9] has been used.

4.1 Semiclassical Approximations and Shell Corrections

The Strutinsky shell correction method[29] to compute binding
energies, is based on a truncated expansion of the Hartree-Fock
density matrix ρ around $\bar{\rho}$ its semi-classical approximation. Upon
computing self-consistently $\bar{\rho}$, it has been shown[30] that the terms
neglected in the above truncation were contributing by less than
0.5 MeV to the total binding energy. However the self-consistent
calculation of $\bar{\rho}$ done in Ref. [30] was in practice as complicated
as the exact calculation of ρ. It was therefore needed to find,
given an effective two-body Hamiltonian, a direct way to compute $\bar{\rho}$.

Attemts based on a Wigner-Kirkwood expansion truncated to lowest
powers in $\hbar$, have failed due to difficulties associated with the
classical turning point. Following a proposal of Bhaduri [31] it
has been possible to overcome this by making a partial resummation
of the Wigner-Kirkwood series including exactly derivatives (up to
some order) of the Hartree-Fock potential[32].A quadratic expansion
(including up to second derivatives) leads to the same semi-classical
energy as in the Strutinsky method for harmonic oscillator potentials[33], which is not surprising. For Saxon-Woods potentials the
corresponding difference is of the order of 10 MeV for heavy
spherical nuclei[33]. Finding the self-consistent $\bar{\rho}$ corresponding
to a given effective force requires to solve the above problem
iteratively. Work is currently in progress towards that aim[34].It
would then remain to compute the first order shell correction
corresponding to the Hartree-Fock potential built from $\bar{\rho}$.

Once this program will be achieved, it is our opinion that even
though absolute differences with actual Hartree-Fock binding energies
might be sor ·hat large (e.g. 10 MeV) relative trends could be quite
nicely reproauced.

4.2 A One Iteraction Approximation to Hartree-Fock

It has been recently proposed[35] to compute binding energies in the following way. Let us assume, for the sake of conciseness [+] that we have only density-independent 2-body forces in our Hamiltonian and use a standard notation for the total energy $E(\rho)$ associated with the (one-body reduced) density matrix ρ :

$$E(\rho) = \text{tr } K\rho + \frac{1}{2} \text{ tr tr}\rho V\rho \tag{2}$$

Starting from an ansatz ρ_o, the exact Hartree-Fock energy may be written as

$$E(\rho) \simeq E(\rho_o) + \text{tr } (K + U_o)\delta\rho \tag{3}$$

at first order in $\delta\rho = \rho - \rho_o$ and where U_o is the Hartree-Fock field associated to ρ_o :

$$U_o = \text{tr } \rho_o V \tag{4}$$

The whole assumption made in Ref. [35] is to replace in $\delta\rho$ the actual Hartree-Fock ρ by its value ρ_1 as obtained after the first iteration of the Hartree-Fock code (when ρ_o is taken for the starting point). Thus one gets :

$$E(\rho) = E(\rho_o) + \text{tr } (K + U_o) (\rho_1 - \rho_o) \tag{5}$$

In practice given a starting ρ_o, one has simply to compute the various parts of the total energy for ρ_o, make one Hartree-Fock iteration and evaluate the sum of occupied level energies after the first iteration, namely :

$$\sum_{\text{occ.}} \epsilon_1 = \text{tr } (K + U_o) \rho_1 \tag{6}$$

Calculations have been performed in Ref. [35] with the SIII Skyrme force[9]. An harmonic oscillator ansatz was chosen for ρ_o whose parameters (one oscillator length for each isospin state) were varied to yield the lowest total energy $E(\rho_o)$, under the constraint that the central density should be equal to the nuclear matter one. As a result (see Table 5) binding energies are given within ± 1-2 MeV for ^{16}O, ^{40}Ca and ^{208}Pb nuclei.

(+) A generalization to 3-body or 2-body density-dependent forces would be obvious.

Table 5. Binding energy difference between approximated and
          exact Hartree-Fock binding energies (in MeV) as
          extracted from Ref. [35].

Nucleus	^{16}O	^{40}Ca	^{208}Pb
ΔB	$-\ 2.28$	$-\ 0.83$	0.86

Restricted,as it is now,to spherical nuclei such an approximation was not exactly necessary in view of the small computing
time involved in exact calculations. Its extension to deformed
nuclei would be of great interest to show whether or not relative
variation in binding energies are correctly reproduced in a given
nuclear region. To achieve this goal it is probable that a careful
study of the optimization procedure for ρ_0 will prove of primary
importance.

5. CONCLUSION

We have concentrated here our study on binding energies
calculated within the Hartree-Fock approximation and using phenomenological effective forces. It is worth noting that the same forces
which reproduce correctly binding energies do reproduce other static
nuclear properties (single-particle properties, nuclear shapes etc.).
They have also met with success in the description of dynamical
nuclear properties in the giant resonance region and for ion
collisions at 1-2 MeV center of mass incident energy per nucleon.
It is therefore our opinion that even though both theoretical and
practical reasons prevent this approach to enjoy the same quality
of fit to binding energies as encountered for instance in liquid
drop type approaches, it constitutes an essential link between
atomic mass evaluations and the rest of the Nuclear Physics.

REFERENCES

[1]   P. Quentin and H. Flocard, Ann. Rev. Nucl. Part. Sci. 28:523
          (1978).
[2]   J.W. Negele, Phys. Rev. C1:1260 (1970).
[3]   X. Campi and D.W.L. Sprung, Nucl. Phys. A194:401 (1972).
[4]   M. Beiner and R.J. Lombard, Ann. Phys. (N.Y.) 86:262 (1974);
          M. Beiner, R.J. Lombard and D. Mas, Nucl. Phys. A249:1
          (1975) and At. Data Nucl. Data Tables 17:450 (1976).
[5]   F. Tondeur, Nucl. Phys. A303:185 (1978) and these proceedings.
[6]   D. Gogny, in "Nuclear Self-consistent fields" eds. G. Ripka
          and M. Porneuf, p. 333 (Amsterdam, North-Holland, 1975).

[7] B. Rouben, J.M. Pearson and G. Saunier, Phys. Lett. B42:385
     (1972).
[8] R.K. Lassey, M.R.P. Manning and A.B. Volkov, Can. J. Phys:2522
     (1973).
[9] M. Beiner, H. Flocard, Nguyen Van Giai and P. Quentin, Nucl.
     Phys. A238:29 (1975).
[10] J.W. Negele and D. Vautherin, Phys. Rev. C11:1031 (1975).
[11] O. Bohigas, A.M. Lane and J. Martorell, Phys. Reports 5:267
     (1979).
[12] D.W.L. Sprung, Adv. Nucl. Phys. 5:225 (1972).
[13] F. Stancu, D.M. Brink and H. Flocard, Phys. Lett. B 68:108
     (1977).
[14] B. Rouben, F. Brut, J.M. Pearson and G. Saunier, Phys. Lett.
     B70:6 (1977).
[15] D. Vautherin and D.M. Brink, Phys. Rev. C5:626 (1972).
[16] C. Gustafson, I.-L. Lamm, B. Nilsson and S.G. Nilsson, Ark. Fys.
     36:613 (1967).
[17] M. Brack, P. Quentin and D. Vautherin, in "Superheavy elements"
     ed. M.A.K. Lodhi, p. 309 (New York, Pergamon, 1978).
[18] B. Grammaticos, "Applications de la théorie du champ self-
     consistant en physique nucléaire", thesis Orsay # 1921
     (1977).
[19] H. Flocard, private communication (1977).
[20] M. Waroquier, J. Sau, K. Heyde, P. Van Isacker and H. Vincx,
     Phys. Rev. C19:1983 (1979).
[21] D.W.L. Sprung, S.G. Lie, M. Vallières and P. Quentin, Nucl.
     Phys. A, in press (1979).
[22] A. Bonaccorso, M. Di Toro and G. Russo, Phys. Lett. B72: 27
     (1977).
[23] P. Hoodboy and J.W. Negele, Nucl. Phys. A288 : 23 (1977).
[24] J. Cugnon, H. Doubre, H. Flocard and M.S. Weiss to be published.
     This method in fact is directly deduced from the method
     used to solve TDHF equations described in H. Flocard,
     S.E. Koonin and M.S. Weiss, Phys. Rev. C17 : 1682 (1978).
[25] P.H. Heenen, P. Bonche and H. Flocard, in "Time-dependent
     Hartree-Fock Method" eds. P. Bonche, B. Giraud and
     P. Quentin, p. 175 (Editions de Physique, Orsay, 1979).
[26] J. Libert, P. Quentin and H. Flocard, to be published.
[27] A.H. Wapstra and N.B. Grove, Nucl. Data A9 : 269 (1971).
[28] H. Flocard, P. Quentin, A.K. Kerman and D. Vautherin, Nucl.
     Phys. A203:433 (1973).
[29] V.M. Strutinsky, Nucl. Phys. A95 : 420 (1967); A122 : 1 (1968).
[30] M. Brack and P. Quentin, Phys. Lett. B56 : 421 (1975).
[31] R.K. Bhaduri, Phys. Rev. Lett. 39:329 (1977).
[32] M. Durand, M. Brack and P. Schuck, Z. für Phys. A286 :381 (1978).
[33] J. Bartel, P. Schuck, M. Durand, R.K. Bhaduri and M. Brack,
     to be published.
[34] J. Bartel and M. Vallières, private communication
[35] A.K. Dhutta, R.K. Bhaduri, M.K. Srivastava and M. Vallières
     Physics department preprint, McMaster University (1979).

HARTREE-FOCK APPROACH TO MASS FORMULA

J.M. Pearson, M. Farine, J. Côté, B. Rouben,
and G. Saunier

Laboratoire de Physique Nucléaire, Univ. de Montréal
Montréal, P.Q., Canada

One of the problems of the semi-empirical mass formula
is that functional forms differing only slightly, and fitted
to the same data, will give significantly different extrapolations
to the astrophysically interesting but experimentally inaccessible
region far from the stability line; see, for example, the contribution of M. Arnould to this conference (1). A part of the
difficulty stems from the so-called two-part macroscopic-microscopic
approach, characterized by the separation into droplet-model
(DM) and shell-model terms. Some synthesis of the two is
desirable, and the Hartree-Fock (HF) method obviously suggests
itself, since it takes shell-model effects into account auto-
matically and self-consistently. Even better would be Hartree-
Fock-Bogolyubov (HFB).

The ideal way of exploiting HF (or HFB) would be to take
an effective interaction with, say, 10 or 15 parameters, and
fit it directly to all the mass data, just as with current
mass formulas. Such a procedure is out of the question with
the present generation of computers, and our point of view
is that two-part mass formulas will be around for a little
longer. (They are more likely to be displaced by the density-
functional method, as described by Tondeur (2) at this conference.)

However, it is still possible to use HF to tie down some
of the ambiguities we have mentioned, while on the other hand
it could also serve to check the various mass formulas for
internal consistency. We do this by fitting finite-range
density-dependent HF forces to just a few key nuclei (doubly-
magic), and then using these forces in two ways, as follows.

341

   i) Perform HF calculations on infinite and semi-infinite
      nuclear matter in order to obtain the DM coefficients.
  ii) Extract shell systematics.
The first and the last topic presented here fall into the first
category, the second into the second.

Surface-Symmetry Properties

     Recent fits of the droplet-model mass formula (3,4) indicate
a volume-symmetry coefficient of $J \simeq 37$ MeV. However, these
fits allow some latitude in J, provided a suitable compensation
is made in the other symmetry coefficients (5), and there is
some theoretical indication of rather lower values of J (6,
7, 8, 9). In an attempt to resolve this question we have devised
(10) two finite-range density-dependent forces, labelled P1
and P2, which give comparable fits to the known doubly-magic
nuclei, although their J-coefficients are quite different:
30 and 37.5 MeV, respectively (note that the extrapolations
away from the stability line also are quite different). Our
objective is to calculate the higher-order volume-symmetry
coefficients L and M, and also the surface-symmetry "stiffness"
coefficient Q for each of these forces, and thus establish
constraints which should be satisfied in any mass formula fit.
L and M are calculated in Ref. 10, and here we consider Q.

     The HF calculation of Côté and Pearson (11) for symmetric
semi-infinite nuclear matter has been generalized to the asymmetric
case. For each of our two forces P1 and P2 we perform completely
self-consistent calculations for different values of the asymmetry
parameter $I = (N-Z)/A$ ranging from zero out to the drip line
($I \simeq 0.3$). (For semi-infinite nuclear matter I is simply the
local asymmetry deep below the surface.)

     From these calculations we first extract the surface energy
per unit area, given by

$$\sigma(I) = \frac{1}{4\pi \, r_0^2} \left[a_{sf} + \frac{9}{4} \frac{J^2}{Q} I^2 \right] \tag{1}$$

where r_0 is the charge-radius constant (= 1.13 fm for both
these forces) and a_{sf} is the usual surface coefficient. Results
for a_{sf} and Q are shown in Table 1, along with the values of
L and M obtained in Ref. 10. We note that Q, like L, is strongly
correlated with J. It will be seen that the results obtained
with our force P2 are in general agreement with the mass-formula
fits of Refs. 3 and 4, although our quoted value of Q is a
little high (see, however, below). However, although these
mass-formula fits, and the higher value of J, are given some
microscopic basis, being consistent with HF, smaller values

of J are not, of course, eliminated. Nevertheless, any mass-formula fit obtained with a smaller value of J should respect the constraints on L, M and Q implied by force P1, shown in Table 1.

Table 1.   Droplet parameters (in MeV) for HF interactions P1 and P2, and for mass-formula fits of Myers (3) and the Darmstadt group (4).

	J	L	M	Q	a_{sf}
P1	30	12.2	1.73	52	20.5
P2	37.5	92.9	0.06	~26	20.8
Myers	36.8	100	0	17	20.7
Darms-stadt	38.2	100	0	17.7	20.8

We now look at our results in more detail. In Fig. 1 we plot $a_{sf}(I^2) = 4 \Pi r_0^2 \sigma(I)$ as a function of I^2. For P1 this plot is effectively linear, but for P2, the force with the higher J, there is a significant contribution from terms of higher order than I^2 (von Groote (12) also found such an effect in Thomas-Fermi calculations). This casts some doubt on the adequacy of the droplet model, so that the accord obtained for P2 with the mass-formula fits may be somewhat questionable. (The value $Q \simeq 26$ MeV that we have quoted for P2 is an average over the entire range of I.)

In Fig. 2 we show the surface diffusivities b for the neutron and proton distributions in semi-infinite nuclear matter. For force P2 these vary considerably with I, and when I is large they are quite different for neutrons and protons (a similar effect has been observed in the Thomas-Fermi calcula-tion of Ref. 12). This result is in conflict with the assumptions of an earlier DM study of matter distribution in finite nuclei (13), but probably is not serious for masses. For force P1, with its lower value of J, the effect is much weaker; both forces give b=0.98 fm when I=0.

An independent determination of Q from semi-infinite nuclear matter is made possible by the following DM relation (14) for the neutron skin thickness:

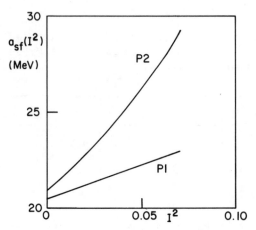

Fig. 1.   Variation of surface-energy coefficient $a_{sf}(I^2)=$
$a_{sf}+(9J^2/4Q)I^2$ with asymmetry I^2 for two forces
P1(J=30 MeV) and P2(J=37.5 MeV).

$$\theta_n = \frac{3}{2}\frac{J}{Q}\, r_0\, I \tag{2}$$

The values of Q that we can obtain in this way are 42 and 21 MeV
for P1 and P2, respectively. The agreement with the above
values of Q is reasonable, considering the difficulty of extracting
a well defined skin thickness from the computed HF matter
distributions.

Exotic Magic Numbers

     Spherical HF calculations were performed with the above
forces P1 and P2 for all possible configurations of filled
j-shells corresponding to nuclei lying between the drip lines.
All the usual numbers N and Z=8, 20, 28, 40, 50, 82, and N=126
remain magic right across from one drip line to the other.
Also no new magic numbers appear off the stability line; in

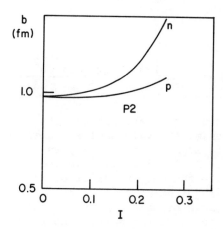

Fig. 2.   Variation of neutron and proton distribution diffusivities
          b with asymmetry I for force P2.

particular, the recent claim (15) that Z=64 is strongly magic
when N=82 is not substantiated by spherical HF calculations.
The addition of a tensor force does not change this conclusion.

What is the Density of Nuclear Matter?

A large part of the Coulomb energy of nuclei is determined
by the equivalent sharp radius of the proton distribution,
given by the droplet model for spherical nuclei as

$$R_p = r_0 A^{1/3} \{1 + \bar{\varepsilon} + \frac{1}{3}[\bar{\delta}-I]\} \tag{3}$$

where, in the usual notation,

$$\bar{\delta} = \frac{I + \frac{9}{80} \frac{e^2}{r_0 Q} Z A^{-2/3}}{1 + \frac{9J}{4Q} A^{-1/3}} \tag{4}$$

and

$$\bar{\varepsilon} = \frac{1}{K} \left[- 2a_{sf} A^{-1/3} + L \bar{\delta}^{-2} + \frac{3e^2}{5r_0} \frac{A^2}{A^{4/3}} \right] \tag{5}$$

Modern DM mass-formulas (3,4) are fitted not only to the masses
but also to the experimental values of R_p, obtained from electron
scattering and muonic atoms. In this way an essentially unique
value of r_0 emerges: 1.17 - 1.18 fm, i.e., a Fermi momentum
for symmetric infinite nuclear matter at equilibrium of k_F=1.29 -
1.30 fm^{-1}. (If the requirement of fitting R_p is dropped, a
considerably wider range of values of r_0 is possible (16),
but then there will be no quarantee of a correct distribution
of the total binding energy between Coulomb and nuclear parts,
and the extrapolation away from the stability line will be
correspondingly unreliable.)

Another way to determine r_0 is via HF effective interactions
that have been fitted to the gross properties of finite nuclei.
These show a much greater range of values of r_0 (k_F), but it
was pointed out by Blaizot et al. (17) that this was strongly
correlated with the incompressibility K, r_0 increasing with
K. The value K ≃ 220 MeV extracted from breathing-mode measurements then implies r_0=1.13 fm (k_F=1.35 fm^{-1}), in contradiction
with the above droplet-model value.

It has been suggested (18) that the same r_0 - K ambiguity
exists in the droplet model as with the HF method, and that
the quoted value of r_0 can be shifted by an appropriate change
of K, which has little effect on the masses (16). Qualitatively,
inspection of Eqns. (3) and (5) shows that this should be possible,
but we now show quantitatively that it is impossible. Consider
the case of Pb^{208}, for which $R_p A^{-1/3}$=1.13 fm experimentally
(19). Thus in order to bring r_0 into accord with its HF value
we require that the correction factor in Eq. (3) be equal to
1. Since $\bar{\delta}$-I is always <0 for heavy nuclei it follows that
$\bar{\varepsilon}$ would have to be positive. But for all reasonable values
of the droplet-model parameters appearing in Eq. (3) it is
negative, so that no matter how K is varied there is no hope
of reducing the droplet-model value of r_0 to its HF value.
In fact, there is this fundamental difference between the two
approaches that in the HF case the bulk density of Pb^{208}

corresponds to a dilation of infinite nuclear matter ($\bar{\epsilon} > 0$)
while according to the droplet model there is a compression.
[Note, however, that the compression is very small, and that
most of the difference $r_0 - R_p A^{-1/3}$ comes from the "neutron
skin" term $\frac{1}{3} (\bar{\delta}-I]$.

It could be, of course, that the breathing-mode experiments are being incorrectly interpreted and that the correct
value of K is much higher, whence the HF value of r_0 would
be closer to its DM value. However, it is to be noted that
no HF interaction with $r_0 > 1.15$ fm gives a good fit to nuclear
radii: this limit corresponds to the Skyrme-Orsay force S5
(20).

There is, furthermore, the fundamental difficulty that
the observed r_0 - K correlation of HF interactions cannot be
comprehended within the framework of the droplet model, no
matter what the correct value of K. The increase of r_0 with
K corresponds to a positive $\bar{\epsilon}$ in the droplet model, according
to Eqns. (3) and (5). But even if this were possible for heavy
nuclei, for lighter nuclei we must have $\bar{\epsilon} < 0$, since surface
tension will then dominate in Eq. (5). Thus according to the
droplet model it should not be possible for the HF forces to
maintain the fit to the radii of all nuclei with changing r_0
no matter how K is varied. A correlation between r_0 and K
is comprehensible only if $\bar{\epsilon}$ is negative for all nuclei, and
then it would go in the opposite direction to what is found
with HF forces.

Until this paradox is resolved, and the problem shown
to lie with HF, we cannot have complete confidence in the physical
basis of the droplet model. But then if HF were found to be
defective, we would have to use great care in using it to impose
constraints on the droplet model.

An unambiguous determination of the nuclear-matter density
is also of obvious interest for the nuclear many-body problem.

References

1. M. Arnould, Proceedings AMCO 6 Conference (1979).
2. F. Tondeur, Proceedings AMCO 6 Conference (1979).
3. W.D. Myers, ADNDT 17, 411 (1976).
4. H. von Groote, E. Hilf and K. Takahashi, ADNDT 17, 418
    (1976).
5. S. Ludwig, H. von Groote, E. Hilf, A.G.W. Cameron, and
    J. Truran, Nucl. Phys. A203, 627 (1973).
6. K.A. Brueckner, S.A. Coon, and J. Dabrowski, Phys. Rev.
    168, 1184 (1968).

7. P.J. Siemens, Nucl. Phys. A141, 225 (1970).

8. J. Côté, B. Rouben, and J.M. Pearson, Can. Journ. Phys.
    51, 1619 (1973).

9. M. Haensel and P. Haensel, Zeit. f. Physik A279, 155 (1976).

10. M. Farine, J.M. Pearson, and B. Rouben, Nucl. Phys. A304,
    317 (1978).

11. J. Côté and J.M. Pearson, Nucl. Phys. A304, 104 (1978).

12. H. von Groote, Proc. of 3rd International Conference on
    Nuclei far from Stability, Cargese (1976), p. 595. (CERN
    76-13).

13. W.D. Myers and H. von Groote, Phys. Lett. 61B, 126 (1976).

14. W.D. Myers and W.J. Swiatecki, Ann. of Phys. 55, 395 (1969).

15. M. Ogawa, R. Brodo, K. Zell, P.J. Daly and P. Kleinheinz,
    Phys. Rev. Lett. 41,289 (1978).

16. H. von Groote, private communication.

17. J.P. Blaizot, D. Gogny and B. Grammaticos, Nucl. Phys.
    A265, 315 (1976).

18. J.M. Pearson, B. Rouben, G. Saunier and F. Brut, Nucl.
    Phys. A317, 447 (1979).

19. W.D. Myers, Droplet Model of Atomic Nuclei (Plenum, 1977).

20. M. Beiner, H. Flocard, Nguyen von Giai and P. Quentin,
    Nucl. Phys. A238, 29 (1975).

GENERALIZED HARTREE-FOCK AND MASS RELATIONSHIPS

E. Comay and I. Kelson

Department of Physics and Astronomy
Tel-Aviv University
Tel-Aviv, Israel

ABSTRACT

The use of homogeneous mass relationships as an extrapolative device, is based on the observation that the experimentally measured deviations from them are randomly distributed around zero. These deviations,however, are more than simple statistical noise; they contain a lot of physically significant information. They are investigated and analyzed, and correlations with nuclear characteristics are found. A possible link with the Generalized Hartree-Fock approach to nuclear structure is suggested..

Homogeneous mass relationships are signficant for two basic reasons. First, they contain a specific statement about nuclear forces and binding energies. This statement is not nearly as explicit as what is provided by mass <u>formulas</u>, and is also only approximate. It stems from the cancellation – in a formal sense – of all the one-body and two-body matrix elements in the (transverse) equation.

$$\hat{O}M(Z,N) \equiv M(Z,N+1)+M(Z-1,N)+M(Z+1,N-1)-M(Z,N-1)-M(Z+1,N)-$$

$$-M(Z-1,N+1)=0 \tag{1}$$

Second, they provide a simple and generally reliable algorithm for extrapolating the mass surface to unknown regions; this algorithm is particularly dependable for small, direct extrapolations. It is important to understand the reason for the possible use of the mass equation as an extrapolative algorithm. In reality, $\hat{O}M(Z,N)$ is only approximately zero, having roughly a Gaussian

349

distribution satisfying:

$$< |\hat{OM}| > = 175 \text{ KeV} \tag{2}$$

$$<\hat{OM}> = 14 \text{ KeV} \tag{3}$$

where the average extends over all known nuclides. The $<\hat{OM}>$ average
is actually consistent statistically with zero, when the experimental uncertainties attached to the masses are taken into account.
Thus, the deviations are thought to constitute a random statistical
"noise", which may be disregarded in the application of the exact
equation (1). Following this reasoning, a number of detailed (and
different) approaches were used to obtain a complete, extrapolated
mass table.

There are, however, definite, consistently non-zero deviations
from the homogeneous relationship, as was pointed out long ago by
Jänecke and others. To demonstrate the most outstanding effect,
we plot in Fig. 1, as a function of mass number A, the experimental value of $<\hat{OM}>$ averaged over ranges of ten mass units

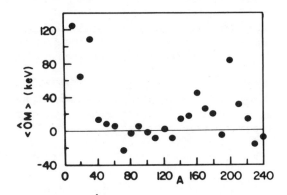

Fig.1   The value of $<\hat{OM}>$, in KeV, averaged over ranges of ten
consecutive mass numbers, as a function mass number A.

There is, obviously, not only a marked A dependence for low A, but a pronounced peak around A=200. It is instructive to note that of the various terms in the Weiszäcker semi-empirical mass formula, the Volume energy, Surface energy and Pairing energy satisfy rigorously the difference relationship (1), while the Coulomb energy and Symmetry energy do not. In fact, writing

$$E_c = \text{Coulomb Energy} = C_{coul.} \, Z^2/A^{1/3} \qquad (4)$$

$$E_{sym} = \text{Symmetry Energy} = C_{sym} \, (N-Z)^2/A \qquad (5)$$

we have

$$\hat{OE}_c = C_{coul} \, [(2/3)A - (8/9)Z]/A^{7/3} \qquad (6)$$

$$\hat{OE}_{sym} = 8C_{sym} \, (N-Z)/A^3 \qquad (7)$$

In order to demostrate the numerical significance of these two terms, we have given their values in Table 1, for various mass numbers. For each A we have selected the value Z_β, which formally minimizes the semi empirical M(Z,A) for constant A (namely, the β-stability line).

Table 1.   The deviations, in KeV, of the difference operator applied to the Coulomb and Symmetry terms along the β-stability line.

A	$Z\beta$	$\hat{OE}_{sym}$	$\hat{OE}_c$
5	2.45	165	19
10	4.86	65	8
20	9.55	25	3
50	23.00	7	1
100	43.96	2	.4
200	82.08	1	.2

It is indeed no surprise that such inhomogeneous terms are present. We recall that the most general solution of eq.(1) has the form

$$M(Z,N) = f_1(Z) + f_2(N) + f_3(Z+N) \qquad (8)$$

where f_1, f_2, f_3, are any arbitrary functions. Obviously, it is impossible to mock up complicated functions of Z and N with such a restricted form. In particular, any strong local correlation around any specific Z_o, N_o - such as seems to be the case for magic numbers - will escape the representation of eq.(8) completely.

Consequently, a procedure has been developed by a number of authors, which introduced specific inhomogeneous terms to M(Z,N) in addition to a functional of the form (8). Thus, it is argued, not only will the newly obtained deviations between experimental masses and fitted masses be truly random, but the functions f_1 and f_2 could become identical (as they should, in the absence of the Coulomb term). While the second claimed benefit is a substanial improvement, the first - in our opinion - is rather problematic. By describing some phenomenon as statistically random, one, in fact, gives up hope of uncovering any meaningful content which this phenomenon may have. Moreover, even if the phenomenon were truly random (and this is quite difficult to establish), one would like to understand <u>why</u> this is so, and be able to derive its <u>magnitude</u> in a more fundamental way.

In other words, we consider the "random" differences between the experimentally measured masses and the theoretically values obtained by the above approach, to be worthy of thorough analysis and investigation. This is,in effect, the central point that we wish to make in this presentation.

We would like to describe now, in very general terms, the framework in which we intend to carry out this investigation - the framework of the Generalized Hartree-Fock approach (GHF).

The GHF to which we refer is a particular generalization of the Hartree-Fock description of nuclear states. In the HF theory, the system is described by a single anti-symmetrized independent particle wave function ϕ. In the GHF the system is described by a linear combination of such states,

$$\psi = \sum_{i=1}^{n} a^{(i)} \phi^{(i)} \tag{9}$$

with each $\phi^{(i)}$ being a determinantal state in its own independent representation. This multitude of single particle representations may formally be regarded as a rearrangement of the degrees of freedom necessary to describe ψ . However, it may provide some significant advantages when one realize that these representations may be interrelated in a basic way, reflecting a fundamental underlying physical picture of the system. Thus, for example, approximate angular momentum projection can be effected formally by a function of this form, which provides a discrete point approximation to the well known projection integration relation.

$$\psi_M^J(\underline{r}) = \sum_K \int D_{MK}^J (\Omega \underline{rr}') \chi_K(\underline{r}') d\Omega \underline{rr}' \tag{10}$$

where $\underline{r}$ and $\underline{r}'$ are the laboratory and intrinsic frames respectively. However, since we have a hierarchy of approximations, which - morphologically - contains other model descriptions as well, it is hoped to be applicable to all nuclei, regardless of collectivity, deformation, magicity or angular momentum characteristics.

It will obviously be naive to believe in the practicality of carrying out the GHF program for all nuclei. After all, the simple HF - with realistic interactions - is a formidable task in itself. However, if we recall that what we are after, are the <u>differences</u> of the results of such an approach applied to sets of neighboring nuclei, we intuitively realize that massive cancellations of terms and contributions are likely to occur. Thus we propose to apply the GHF (and particularly the HF) directly to the differences $\delta M(Z,N)$, rather than separately to the individual nuclei appearing in the difference.

MICROSCOPIC ENERGY DENSITY MASS FORMULA

François Tondeur

Physique Nucléaire Théorique - CP 229, Université Libre
de Bruxelles, Campus Plaine, Bd. du Triomphe
B 1050   Bruxelles - Belgium

and   Institut Supérieur Industriel - Bruxelles

INTRODUCTION

Some years ago, the semi-empirical mass formulae were sometimes characterized by a quality factor, obtained by the product
of the r.m.s. deviation by the number of parameters. Two formulae
have a good quality factor : the forty-year-old Bethe-Weizsäcker
formula (five parameters, deviation better than 3 MeV) , and the
formula of Myers and Swiatecki [2] (seven parameters, r.m.s. deviation better than 1.5 MeV).

The concept of a quality factor seems to have disappeared for
ten years, probably because all improvements of the precision of
the mass formulae have been obtained with an important increase
of the number of free parameters, leading to very bad quality factors. For example, the high precision formulae reaching r.m.s. deviations lower than 0.3 MeV need more than hundred parameters [3,4],
whereas the most precise droplet model formulae need thirty seven[6]
or fifty[5] parameters to obtain a r.m.s. deviation around 0.7 MeV .
In these last formulae, a majority of the free parameters is used
in the shell corrections . Any attempt to reduce the number of free
parameters should thus first look into the shell corrections.
This seems uneasy as long as the shell model is considered separately from the macroscopic model of the nucleus. A calculation of
the shell correction with a five-parameter -Woods-Saxon potential
and the Strutinski method did not give satisfactory results [7]. A
good fit to experimental data with this type of potential probably
also needs a much greater number of parameters (see e.g. the potential used by Tanaka et al. [8]).

It is possible to go beyond this difficulty by using a model

355

where the calculations of the macroscopic and shell energies are
unified, like in the folding model of Nix and his coworkers [9,10,
11], where the shell model potential is related to the droplet
model with a few free parameters. This model has been applied to
heavy nuclei for several years [10], but the attempt to build a mass
formula with it is recent [11] and shows encouraging results.

The mass formula described here is of another type, based on
the self-consistent version of the energy density formalism. This
model is close to the approach of Beiner et al. [31] and to calculations with the Skyrme interaction [32].

SELF-CONSISTENT MODEL FOR SPHERICAL NUCLEI [12, 15]

The binding energy of the nucleus is obtained by

$$B = \int \mathcal{H} \cdot dV + \iint \frac{e^2}{8\pi\varepsilon_0} \frac{\rho_p(\vec{r})\rho_p(\vec{r}')}{|\vec{r}-\vec{r}'|} dV \, dV'. \qquad (1)$$

The second term in (1) is the direct Coulomb energy. The energy
density $\mathcal{H}$ is approximated by

$$\mathcal{H} = \frac{\hbar^2}{2m_n} \tau_n + \frac{\hbar^2}{2m_p} \tau_p + a\rho^2 + b\rho^\delta + c\rho^{5/3}\left(\frac{\rho_n - \rho_p}{\rho}\right)^2$$

$$+ d \cdot \vec{J} \cdot \vec{\nabla}\rho + \eta |\vec{\nabla}\rho|^2 - \frac{3}{4}\left(\frac{3}{\pi}\right)^{1/3} \frac{e^2}{4\pi\varepsilon_0} \rho_p^{1/3}. \qquad (2)$$

The last term in (2) is an approximate Coulomb exchange energy
density. The densities ρ_q, τ_q and $\vec{J}_q$ are defined as usually by :

$$\rho_q = \sum_i v_i^2 \left| \phi_i(\vec{r},\sigma,q) \right|^2 \qquad (3)$$

$$\tau_q = \sum_i v_i^2 \left| \vec{\nabla}\phi_i(\vec{r},\sigma,q) \right|^2 \qquad (4)$$

$$\vec{J}_q = -i \sum_{i,\sigma,\sigma'} v_i^2 \, \phi_i^*(\vec{r},\sigma,q) \, \vec{\nabla}\phi_i(\vec{r},\sigma',q) \times \langle\sigma|\vec{\sigma}|\sigma'\rangle \qquad (5)$$

where the ϕ_i are the individual wave functions of the nucleons, σ
and q are spin and isospin coordinates, and the sums extend to all
levels of given q. In (2), $\rho = \rho_n + \rho_p$ and $\vec{J} = \vec{J}_n + \vec{J}_p$. The
v_i^2 in (3,4,5) are occupation numbers calculated in the BCS
pairing method. The pairing equations for the gap parameters Δ_i :

$$\Delta_i = \sum_j \frac{G_{ij} \Delta_j}{\sqrt{(e_j - \lambda)^2 + \Delta_j^2}} \qquad (6)$$

and for the nucleon number

$$N \ (\text{or } Z) \ = \sum_i v_i^2 = \sum_i \frac{1}{2}\left(1 - \frac{e_i - \lambda}{\sqrt{(e_i - \lambda)^2 + \Delta_i^2}}\right) \quad (7)$$

are solved separately for neutrons and protons. The gap matrix G_{ij} is calculated for a delta-interaction $v = V_0\ \delta(\vec{r} - \vec{r}\,')$. However, in a preliminary version of the model [15], the constant G approximation has been used, with approximate formulae for G_n and G_p.

The variations of B (1) with respect to the ϕ_i lead to Schrödinger equations for each nucleon is a central plus spin-orbit potential, which is related to the densities ρ_n, ρ_p and $\vec{J}$. These equations are solved in a iterative calculation, until the self-consistent solution is reached.

APPROXIMATE METHOD FOR DEFORMED NUCLEI

Computers are not yet fast enough to perform a self-consistent calculation for thousands of deformed configurations in a reasonable time. A useful time-saving approximation is the expectation value method (EVM) developed by Brack and his coworkers [13], [14], who use the wave functions ϕ_i obtained in a realistic deformed potential instead of the self-consistent ones. This method cannot give directly a good value for the energy B_d of deformed nucleus, but gives rather accurate values for the deformation energies $\Delta B = B_d - B_s$. This quantity can be added to the self-consistent energy B_s obtained for spherical symmetry.

The choice of the deformed potential is of course an important point of the EVM. Brack [14] uses a Woods-Saxon potential fitted on the spherical self-consistent potential. We have followed two other methods. The first one, which is valid only for small deformations, has been used in [16] to calculate equilibrium deformations : the deformed potential is directly obtained by deforming the self-consistent spherical potential. The second method uses the relations between the potentials and the densities to obtain the deformed potential with deformed densities of the Fermi type , fitted on the self-consistent spherical densities. This method has been used to calculate fission barriers.

A further approximation is to follow Strutinski's method. But, instead of calculating the smooth component of the binding energy with a macroscopic model, we obtain them for a few tens of nuclei and for each deformation by introducing smoothed occupation numbers when calculating the densities (3,4,5) with the ϕ_i of the EVM. The smooth energies vary smoothly with N and Z for each given deformation , and can be interpolated in the (N,Z) plane together with the deformed EVM single-particle levels used in the calculation of the shell corrections.

Till now, only the constant G approximation has been used for pairing in deformed configurations.

PRECISION OF THE MASS FORMULA

A complete calculation of all known binding energies has not yet been performed. The most extended calculation concerns 424 nuclei (including 108 deformed nuclei) with the constant G approximation for pairing, giving a r.m.s. deviation of 0.96 MeV. Only 142 nuclei have been considered with the delta pairing interaction. The r.m.s. deviation is then 0.90 MeV , against 0.97 MeV for the same set of nuclei with the constant G approximation.

We believe that a better precision can still be obtained after a careful analysis of the remaining systematic deviations of the formula and a subsequent refit of the parameters. However, a few unsolved problems like the calculation of fission barriers, must first be examined.

EXTRAPOLATION OF THE FORMULA

The self-consistent basis of the formula probably gives more reliable predictions of the nuclear properties far from the stability region. A few interesting results have already been obtained, which are summarized hereafter.

Magic numbers far from the stability line

The magic numbers are characterized by all Δ_i = 0 in the pairing calculations. With this criterion , we have determined the limits of magicity of the magic numbers [12, 15]. The results are given in table 1. They depend on the method followed to obtain the gap matrix G_{ij}. The better PD (or ND) indicate that the magic numbers is still magic at the proton drip (or neutron drip) line. With both methods for pairing, the most important variations of the magic numbers are found for neutrons in the neutron-rich region between N = 28 and N = 58. With the delta pairing interaction, the usual magic numbers N = 28 and 50, are replaced near the neutron drip line by N = 32, 34, 40 and 58. The great sensibility of these results to the pairing strength, as well as their possible consequences for astrophysical problems, show that it would be interesting to make self-consistent pairing calculations in this region of the (N,Z) plane.

Table 1

Limits of magicity of magic numbers, according to the method used in pairing calculations.

Magic number	Constant G		δ interaction	
	from	to	from	to
N = 16	Z = 16	ND	Z = 15	ND
N = 20	PD	ND	PD	ND
N = 28	PD	Z = 24	Z = 28	Z = 27
N = 32	Z = 20	Z = 15	Z = 21	ND
N = 34	Z = 15	ND	Z = 17	ND
N = 40	-	-	Z = 19	ND
N = 50	PD	Z = 28	PD	Z = 22
N = 58	-	-	Z = 25	ND
N = 82	PD	Z = 38	PD	ND
N = 126	PD	ND	PD	ND
Z = 16	N = 11	N = 16	PD	N = 16
Z = 20	N = 17	ND	PD	ND
Z = 28	N = 25	N = 56	PD	ND
Z = 50	PD	N = 109	PD	ND
Z = 82	PD	N = 162	PD	ND

Superheavy nuclei with N = 184

The stability of nuclei with 184 neutrons has been examined [12, 17]. This number is found to be magic near the β-stability line, as well as Z = 126 for N = 184. The most stable isotone with N = 184 is either $^{294}_{110}x_{184}$ or $^{292}_{108}x_{184}$, according to the method used to calculate the gap matrix. With the delta pairing interaction, $^{294}_{110}x_{184}$ is the most stable, and is expected to be α-unstable with a half-life of approximately three years.

Astrophysical applications

The mass formula has also been used in the study of neutron star matter [18] and in a very rough analysis of the r-process [7]. Those questions will not be examined here, by lack of place.

Even-odd mass differences far from the stability line

The even-odd mass differences P obtained with the delta pairing interaction do not follow the usual empirical expressions in $A^{-1/2}$ when leaving the stability line towards the neutron-rich or proton rich regions [12]. For an unpaired neutron, this difference indeed increases with A at fixed N. For an unpaired proton, it has a maximum value when A varies at fixed Z.

RECENT RESULTS

Surface thickness of nuclei

Apart from binding energies and related quantities, the model also gives full information about the nuclear densities.

Nuclear radii are in rather good agreement with experimental values, and are similar to those predicted by the droplet model, but include shell effects neglected by this model [7, 15].

Recently [19], the variations of the surface thickness of nuclear densities and potential have been studied with the self-consistent model described in sect. 2. The values obtained for the surface thickness of neutron and proton densities along the stability line are given in fig. 1. They show shell effects as well as long range variations of the surface thickness.

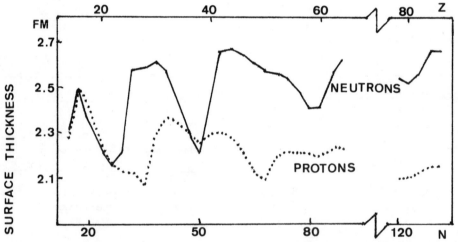

Fig. 1 : Surface thickness of neutron and proton self-consistent densities for spherical nuclei along the stability line.

It is interesting to notice that these results contradict those of Myers and von Groote [20], who predict that the neutron surface thickness increases more slowly and is lower than the proton surface thickness. The origin of this discrepancy is perhaps the Thomas-Fermi model used by these authors, which is not expected to give a very satisfactory description of the nuclear surface.

Other interesting features are found when the surface thickness of the self-consistent densities and potentials is divided into two parts (inside and outside the half-density radius).It is then found that the two parts are not very different for the densities, but that the internal thickness is about 1.6 times the external thickness for the potentials. This result disagrees with the common description of the potential by a Woods-Saxon potential, for which the two parts are equal.

α stability of nuclei with $N > 184$

Although the search for superheavy elements has mainly been focused on the vicinity of $N = 184$, several "superheaviologists" have followed Sobiczewski et al . [21] in more remote regions of the valley of stability. These authors predicted as magic numbers in a Woods-Saxon well $Z = 164$ and $N = 228$, 308 and 406. A self-consistent calculation has been done by Kolb [22] till $A = 500$. He confirms the magic character of $Z = 164$ and $N = 228$, 308. However, other candidates could be $Z = 120$ [28], 126 [17, 30] , 138 [29] and $N = 218$ [30], 258.

The model we have described enables a study of the nuclear stability in this region of the (N,Z) plane. Fission, α and β decay should be examined. On a first step, we shall not consider fission, which needs very long numerical calculations.

Except near magic numbers, α - stability is expected to increase with the neutron excess. We have thus examined the nuclei along the neutron-rich limit of the β-stability zone. The binding energies have been calculated with the self-consistent model for spherical symmetry. Using the same criterion of magicity as for Table 1, only $N = 308$ and $Z = 164$ are found to be magic. The doubly magic $^{472}_{164}X_{308}$ is predicted to be β -stable. On the other hand, some shell effects seem to be spread between $Z = 120$ and 126, and between $N = 228$ and 258. These regions do unfortunately not overlap, except in a small region near $^{354}_{126}X_{228}$.

Fig. 2 shows the calculated values of the α -decay energies Q_α . Dots under the main line indicate Q_α values for nuclei which have a predicted Q_β-value lower than 0.10 MeV. The Q_α value for $^{472}_{164}X_{308}$ is indicated by a circle, and is smaller than values obtained for more neutron-rich isotopes due to the shell effects. The figure also shows the half-lives as predicted by the empirical formula of Viola and Seaborg.

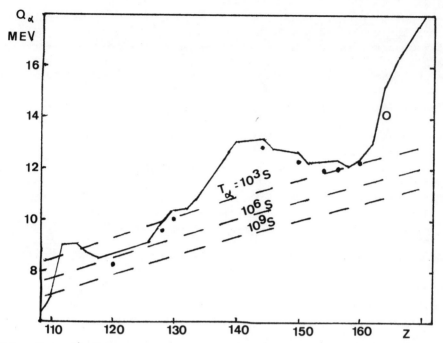

Fig. 2 : α -decay energies of superheavy nuclei on the neutron-
rich limit of the β -stability zone and α -decay half-
lives predicted by the formula of Viola and Seaborg.

Two favoured regions are found with half-lives of a few days
near Z = 120 (including $^{354}_{126}X_{228}$) and of a few minutes near Z =
160. The rapid increase of Q_α for Z $>$ 160 gives no chance of finding an island of α -stability for heavier nuclei. Of course, the
above results give only rough indications, because deformation
has been neglected and because the extrapolation of the model as
well as of the formula of Viola and Seaborg is uncertain. Moreover,
the stability against fission must also be examined.

Nuclear level densities

A few results about the calculation and the extrapolation of
nuclear level densities have been presented previously [25]. New
calculations , taking into acccount the delta pairing force,
confirm the important discrepancies with the back-shifted Fermi
gas model [26] far from the stability line. Those results will be
detailed elsewhere [27].

AN UNSOLVED PROBLEM : THE FISSION BARRIERS

Microscopic models have not yet succeeded in reproducing

experimental fission barriers [23], whereas good results have been obtained with the macroscopic-microscopic method based on the droplet mass formula (see e.g. [24]).

Calculations with the energy density functional described in sect. 2 have been performed for ^{244}Pu with the EVM (without the smoothing procedure used in [16] for deformed nuclei) and the constant G approximation for pairing. The fission barrier has been calculated in the (c,h) plane [33] with h = o. A double humped barrier is found, but the second hump is too high ($\sim$ 20 MeV) above the ground state. Although a small part (1-2 MeV) of the height is probably due to the numerical inaccuracy of the program for c $>$ 1.6, this indicates a major inadequacy either of the model or of the parameter set.

A much better barrier can be obtained with other values for the parameters. For example, we have obtained a second hump 13 MeV above the ground state by setting the compressibility modulus equal to 200 MeV instead of 235 MeV, and by a modification of the symmetry energy density term in (2), now taken as $c.\rho.(\rho_n - \rho_p)^2$, the parameter c being refitted. Taking into account the inaccuracy of the numerical calculation and the neglect of the h degree of freedom, this is in rather good agreement with the results of the microscopic-macroscopic method [24].

This is the result of a preliminary study, and we do not know if a good fit of the masses and of the fission barriers is possible with the same parameters. This problem is still being investigated.

CONCLUSION

The microscopic mass formula we have described represents a progress compared to the earlier work of Beiner et al. [31] with the same type of model, and to the droplet model mass formulae.

The mass formula of Beiner et al. [31] does reach a very good precision, although using more parameters than we do. Moreover, those authors do not include the effects of deformation in their model.

Our mass formula has a better precision than liquid drop formulae with a similar number of parameters. Its self-consistent basis probably enables better extrapolations, and leads to the prediction of new effects, e.g. in the extrapolation of shell effects or of even-odd mass differences.

REFERENCES

1. H. Kümmel, J. Mattauch, W. Thiele and A. Wapstra, Nucl. Phys. 81 : 129 (1966).
2. W. Myers and W. Swiatecki, Nucl. Phys. 81 : 1 (1966).
3. S. Liran and N. Zeldes, Atomic Data and Nucl. Data Tables

17 : 428 (1976).

4. J. Jänecke , Atomic Data and Nucl. Data Tables 17 : 455 (1976).
5. H. von Groote, E. Hilf and K. Takahashi, Atomic Data and Nucl. Data Tables 17 : 418 (1976).
6. P. Seeger and W. Howard, Atomic Data and Nucl. Data Tables 17 : 428 (1976).
7. F. Tondeur, Thèse de doctorat. Université Libre de Bruxelles (1977) unpublished.
8. Y. Tanaka, Y. Oda, F. Petrovitch and R. Sheline, Phys. Lett. 83B : 279 (1979).
9. M. Bolsterli, E. Fiset, J. Nix and J. Norton, Phys. Rev. C5 : 1050 (1972).
10. P. Möller, S. Nilsson and J. Nix, Nucl. Phys. A229 : 292 (1974).
11. H. Krappe, J. Nix, A. Sierk, Preprint LA-UR-79-885.
12. F. Tondeur, Nucl. Phys. A315 : 353 (1978).
13. C. Ko, H. Pauli, M. Brack and G. Brown, Nucl. Phys. A236 : 269 (1974).
14. M. Brack, Phys. Lett. 71B : 239 (1977).
15. F. Tondeur, Nucl. Phys. A303 : 185 (1978).
16. F. Tondeur, Nucl. Phys. A311 : 51 (1978).
17. F. Tondeur, Z. Phys. A288 : 97 (1978).
18. F. Tondeur, Astron. Astroph. 72 : 88 (1979).
19. F. Tondeur, Journ. of Phys. G5 : 1189 (1979).
20. W. Myers and H. von Groote, Phys. Lett. B61 : 125 (1976).
21. A. Sobiczewski, T. Krogulski, J. Blocki and Z. Szymanski, Nucl. Phys. A168 : 519 (1971).
22. D. Kolb, Z. Phys. A280 : 143 (1977).
23. M. Brack, Proc. of the Intern. Symp. on Physics and Chemistry of Fission, Jülich (1979) IAEA.
24. M. Junker and J. Hadermann, Z. Phys. A282 : 291 (1977).
25. M. Arnould and F. Tondeur, Annual Nuclear Physics Meeting of the German Physical Society, Heidelberg (1978).
26. J. Holmes, S. Woosly, W. Fowler and B. Zimmerman, Atomic Data and Nucl. Data Tables 18 : 305 (1976).
27. M. Arnould and F. Tondeur, in preparation.
28. B. Rouben, J. Pearson and G. Saunier, Phys. Lett. 42B : 385 (1972).
29. B. Rouben, F. Brut and J. Pearson, Phys. Lett. 70B : 6 (1977).
30. M. Vallières, D. Sprung, Phys. Lett. 67B :253 (1977).
31. M. Beiner, R. Lombard and D. Mas, Atomic Data and Nucl. Data Tables 17 : 450 (1976).
32. M. Beiner, H. Flocard, Nguyen van Giai and Ph. Quentin, Nucl. Phys. A238 : 29 (1975).
33. M. Brack, J. Damgaard, A. Jensen, H. Pauli, V. Strutinski and C. Wong, Revs. Mod. Phys. 44 : 320 (1972).

THE PHENOMENOLOGICAL SHELL-MODEL

AND THE SYSTEMATICS OF NUCLEAR MASSES*

J. J. E. Herrera and M. Bauer

Centro de Estudios Nucleares and Instituto de Física
Universidad Nacional Autónoma de México
A. P. 20-364, México 20, D.F. MEXICO

INTRODUCTION

The purpose of the present work is the understanding of the
systematics of nuclear masses from the shell model point of view
at a consistent phenomenological level. With respect to previous
work[1], the fundamental modifications contained in this paper are:
first, the use of the renormalized Brueckner-Hartree-Fock theory
(RBHF) as a guideline, second, the link with the optical model based
on a unified phenomenological description of bound and unbound single
particle states and finally the specification of the single particle
energies through an analytic expression arising from the square well
potential rather than an harmonic oscillator with a Nilsson $D\ell^2$ term.
We proceed to discuss these points.

THE THEORETICAL GUIDELINE

In the RBHF theory, the rearrangement energies are in principle
minimized and the analogue of Koopman's theorem is then approximately
valid [2,3]. This is achieved by incorporating third order contributions of the two-body G-matrix into the self-consistent potential,
which is so "renormalized". The total energy is then given by

$$E = \frac{1}{2}\sum_{\lambda \leq F}\left\{e_\lambda + \langle\lambda|T|\lambda\rangle\right\} +$$

$$+ \frac{1}{2}\sum_{\lambda \leq F}\langle\lambda|U|\lambda\rangle[1 - \rho(\lambda)] \tag{1}$$

*Work supported in part by the Instituto Nacional de Investigaciones
Nucleares and by C.O.N.A.C.y T., México, through grant PNCB-00025.

Table 1. Single particle energies in ^{40}Ca.

State	η_λ	$e_\lambda = \frac{m}{m^*}\eta_\lambda$	E^{th} (a)		E^{exp}		
			BHF	RBHF	^{n}(b)	(c)^p	(d)
1s	41.5	61.0	68.2	54.8		56	50±11
1p	29.3	43.1	$\begin{cases}45.3 \\ 41.4\end{cases}$	$\begin{cases}35.1 \\ 32.4\end{cases}$		41	34±6
1d	16.4	24.1	$\begin{cases}24.7 \\ 18.8\end{cases}$	$\begin{cases}17.5 \\ 13.4\end{cases}$	$\begin{cases}21.3 \\ 15.8\end{cases}$	14.9	$\begin{cases}15.5 \\ 8.3\end{cases}$
2s	15.2	22.4	21.8	5.8	18.2	11.2	11.6

(a) ref. 8, (b) ref. 10, (c) ref. 11, (d) ref. 12.

where $e_\lambda \equiv \langle\lambda|T|\lambda\rangle + \langle\lambda|U|\lambda\rangle$ is the single particle energy and $\rho(\lambda)$ is the occupation probability of the level λ. The renormalization implies that $\rho(\lambda) < 1$ even if the summation only goes up to the Fermi level F. Thus RBHF yields formally an additional attractive term to the Hartree-Fock energy.

Furthermore, the single particle energies e_λ are formally equal to the mean removal energies which are the quantities that are determined experimentally, through pick up, (p, 2p) and (e,e'p) reactions[3]. It is because of the relation of the total energy with removal energies that we adopt the RBHF as a theoretical guideline.

THE SINGLE PARTICLE LEVEL SCHEME

The self-consistent potential is expected to be nonlocal and energy dependent[4]. Taking into account the nonlocality of the nuclear potential in the so-called effective mass approximation, it can be shown[5] that a single-one-body Schrödinger equation yields the increased spacing of the bound energy levels relative to that predicted by a local potential and the energy dependence of the local potential that describes the scattering. The following correspondence between the pehnomenological shell and optical models is established at this level of approximation. If the elastic scattering cross-section is fitted with an energy dependent local optical potential whose real part is given by

$$\mathcal{V}(r;E) = -[V - \alpha E]\, f(r),\qquad (2)$$

then the shell model bound state energies e_λ are obtained from the eigenvalues $\eta(\lambda)$ of the potential $\mathcal{V}(r;E=0)$ by the relation

$$e_\lambda = (m/m^*)\,\eta_\lambda \tag{3}$$

where the scaling effective mass factor m/m^* is given by

$$1 - \left(\frac{m}{m^*}\right)^{-1} = \alpha = + \frac{d}{dE}\,V(r;E). \tag{4}$$

Thus $m/m^* = (1-\alpha)^{-1} > 1$ and an increase in spacing follows. In Table 1 we show the neutron single particle spectrum of ^{40}Ca obtained according to eqs. (3) and (4) when the Saxon-Woods potential that fits elastic scattering is used. The parameters are[6] $V = 55.5$ MeV, $\alpha = 0.32$, $r_0 = 1.3$F, $a = 0.65$F and the effective mass is then $m^* = 0.65\,m$.

In the phenomenological models, the Saxon-Woods potential is the one used more often to represent the self consistent potential. However it does not admit an analytical solution. In previous work[1] we took the well-established Nilsson approach to single-particle energies, which uses the harmonic oscillator plus a $D\ell^2$ term to break the degeneracy and simulate the effect of flat bottom potential. However, the exact summation of oscillator levels does not yield a surface term ($\sim A^{2/3}$) in the mass formula[7]. The surface term was obtained because a Strutinsky-type averaging procedure was used instead of a straightforward summation.

To obviate this problem and obtain a more realistic representation, we now use an analytic expression arising from the solution of the infinite square well potential. In this case, the eigenvalues are given by the zeros of the spherical Bessel functions, for which there exists an infinite series expression that converges very rapidly[6]. Using the first two terms of the series, the single-particle energies of eqs. (1) and (3) are written as

$$e_{n\ell j} = (m/m^*)\left\{ \langle T \rangle_{n\ell j} + \langle U \rangle_{n\ell j} \right\} \tag{5}$$

where the (local) kinetic and potential energies are given by:

$$\langle T \rangle_{n\ell j} = \frac{\hbar^2}{2mR^2}\left\{ \left(\frac{n+2}{2}\right)^2 \pi^2 - \ell(\ell+1) \right\} \tag{6}$$

$$\langle U \rangle_{n\ell j} = -V - C\left\{ j(j+1) - \ell(\ell+1) - 3/4 \right\} \tag{7}$$

Here n is the principal quantum number. We have included a spin-orbit coupling term but no deformation effects so that the mass formula will apply only to spherical nuceli.

Furthermore, in order to take into account the fact that the nuclear potential is finite and has a certain surface diffuseness, we allow the radius R to increase as one goes from the deeply bound to the least bound states. Thus we take

$$R \equiv R(n) = r_0 A^{1/3} \left(1 + \sigma A^{-p/3} n \right)$$

(8)

where σ is an adjustable parameter and p (integer) $\geqslant$ 1, because the n of the Fermi level goes as $A^{1/3}$. The spin-orbit coupling parameter is expected to have an $A^{-2/3}$ dependence[1] so that we write $C = K A^{-2/3}$. Finally, we know from the phenomenological optical model that the nuetron and proton potential well depths are different, namely[6]

$$V = V_0 + \tau V_1 \left(\frac{N-Z}{A} \right) + \left(\frac{1+\tau}{2} \right) \alpha_c Z/A^{1/3}$$

(9)

where $\tau = +1$ for protons and $\tau = -1$ for neutrons.

THE MASS FORMULA

Numerical calculations[8] have yielded the result that the occupation probability $P(\lambda)$ of the RBHF theory is rather independent of the particular state. We thus assume that all depletion factors $d(\lambda) = 1 - P(\lambda)$ have the same value and represent them by a single parameter d . Then the RBHF ground state energy, eq. (1), can be written as

$$E = \frac{m}{m^*} \sum_{n\ell j \leqslant F} \left\{ \langle T \rangle_{n\ell j} + \frac{1}{2}(1+d)\langle U \rangle_{n\ell j} \right\} N_{n\ell j}$$

(10)

where $N_{n\ell j}$ is the number of nucleons in the $(n\ell j)$ level, and the total number of particles N is given by

$$N = \sum_{n\ell j \leqslant F} N_{n\ell j}$$

(11)

As in reference 1, the nucleus is considered to be a system of N neutrons and Z protons moving independently, the energy and number of particles of each species, given by eqs. (10) and (11), where m/m^* is different for each, due to its well depth dependence. Instead of the Strutinsky-type average procedure used in reference 1,

the terms in equation (10) are summed up exactly. Since we consider
the spherical single particle spectra, only summation over closed
shells is carried out. The expression that is thus obtained yields
and energy per nucleon adequate for doubly closed shell nuclei.
Extended to all values of N and Z , the formula leaves out small
contributions reflecting partial shell filling and deformation[1].

Adding to the result thus obtained the Coulomb energy E_c as
given in reference 9, i.e.:

$$E_c = a_c \, \frac{Z^2}{A^{1/3}} \left[1 - 0.7636 \, Z^{-2/3} - 1.641 \, A^{-2/3} \right] ,$$ (12)

the final form of the binding energy is given as

$$E = \sum_{j=-2}^{3} \sum_{k=0,2} a_{jk} \, A^{j/3} \left(\frac{N-Z}{A} \right)^k + E_c$$ (13)

where the coefficients a_{jk} are given by

$$a_{jk} = \sum_{i=1}^{4} c_{ijk} \, t_i$$ (14)

where the c_{ijk}'s are numerical coefficients that depend on the parameters
σ and K and the t_i's are nonlinear functions of the parameters
V_o , V_1 , r_o and m/m^* (see Appendix I).

THE NUMERICAL FIT

Once the parameters σ , K and a_c are fixed, eq. (13) is fitted
to experimental values by least mean squares taking the functions
t_i as adjustable parameters. Solving them for the single particle
parameters, this yields the values of V_o/V_1 , m/m^* , r_o^2 and
$V_o(1+\alpha)$, which should fall within physically acceptable ranges.
This does not happen in general (negative values for r_o^2 may be
obtained for instance), but only for certain values of σ , K and
a_c .

Table 2 shows the results of two different reasonable fits. Fit
I is an example in which p =3. The difference between the theoretical
and experimental binding energies is shown in figure 1. As it may be
seen the fit resembles the one of the Bethe–Weiszäcker formula, as
expected, since no deformation effects have been included so far.
Although the form of eq. (13) may also resemble that of the Bethe–
Weiszäcker formula, it should be observed that the additional terms
that appear in it are not negligible in comparison with the usual
ones. Thus, it is interesting to find out what happens if only the

Table 2. Sets of parameters that fit experimental masses.

	(a)				(b)	
	Shell Model Parameters			k	a_{jk} (MeV)	
	Fit I		Fit II	j	0	2
p	3		2		Fit	I
σ	2.05		.45	3	-17.54	28.53
a_d	.76	MeV	.69 MeV	2	42.61	-1.612
K	8.0	MeV	8.19 MeV	1	-13.71	-116.2
d	.35		.30	0	339.0	-66.96
V_0	43.85	MeV	42.95 MeV	-1	459.4	-27.38
V_1	13.29	MeV	17.36 MeV	-2	-2822.0	370.9
r_0	1.30	F	1.285 F		Fit	II
m/m^*	1.644		1.784	3	-15.16	30.49
S	2.59	MeV	2.57 MeV	2	17.17	-58.11

a) Sets of shell model parameters for the results shown in figures 1 and 2. b) Values of the a_{jk} coefficients as given by eq. (14).

powers j =3 and 2 are kept. This is the case of fit II. As seen in figure 1 there is no significant difference between both fits, but the coefficients a_{jk} are now found to be closer to those usually quoted[13,14]. The energy levels which are obtained from the values of the single-particle parameters compare favorably with those of Hartree-Fock calculations[15] with Skyrme interactions as seen in figure 2.

The following conclusions have been drawn out of this work:

1) It is possible to build up a formula for the binding energies of nuclei starting from the single-particle picture, and write it as an expansion in terms of powers of $A^{1/3}$.

2) If deformation effects are not included, a root-mean-square of the order of that of the Bethe-Weiszäcker-Bacher formula (2.7-2.8 MeV) is obtained.

3) The values of the single-particle parameters V_1/V_0, m/m^*, r_0 and $V_0 (1 + d)$ are of the order of the physically acceptable values.

4) The values of the coefficients of the liquid drop formula are reproduced only if a comparable number of terms is kept.

5) It is not clear whether the additional terms predicted in this work are meaningful, and that should be studied by adding information (e.g. fission barriers) once the formula is further developed.

Figure 1.   Difference between theoretical and experimental
            binding energies as function of N .

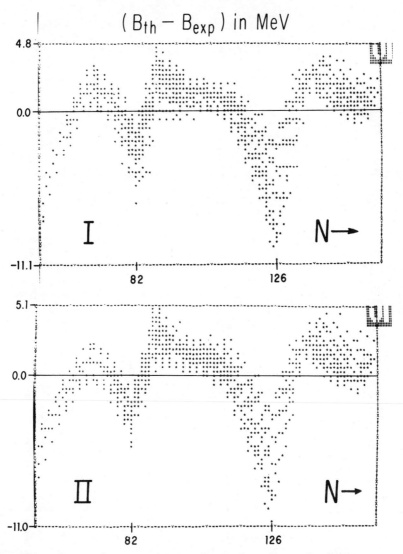

$$(B_{th} - B_{exp})\ \text{in MeV}$$

The fits shown above correspond to the sets of parameters given
in table 2.

APPENDIX

$$t_1 = -\tfrac{1}{2}\,\tfrac{m}{m^*}\,V_0\,(1+d)$$

$$t_2 = t_1\,\tfrac{V_1}{V_0}\left[\left(\tfrac{m}{m^*}-1\right)\tfrac{V_1}{V_0}-\tfrac{m}{m^*}\right]$$

$$t_3 = \tfrac{\hbar^2}{2m r_0^2}\,\tfrac{m}{m^*}$$

$$t_4 = t_3\,\tfrac{V_1}{V_0}\left(\tfrac{m}{m^*}-1\right)$$

$$c_{130} = 1.$$

$$c_{232} = 1.$$

Figure 2. Single Particle energy levels for ^{208}Pb.

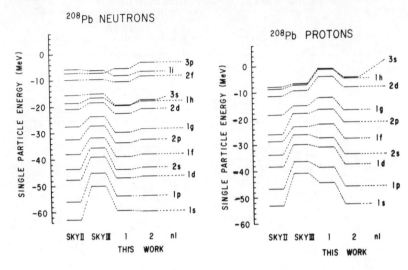

The first two columns in each figure show some values obtained through Hartree-Fock calculations[15]. Columns 1 and 2 correspond to the values obtained in this work with the parameters obtained from fits I and II to nuclear masses.

$$b_1 = \left(\tfrac{3}{2}\right)^{1/3}$$

$$b_2 = \left(\tfrac{3}{2}\right)^{2/3}$$

$$b_3 = \tfrac{\pi^2}{20} - \tfrac{1}{10}$$

$$b_4 = \tfrac{\pi^2}{16}$$

$$b_5 = \tfrac{\pi^2}{8} - 2$$

$$b_6 = \tfrac{13}{48}\pi^2 - 2 - 2\Lambda$$

$$b_7 = -\tfrac{37}{60}\pi^2 + \tfrac{389}{90} + 2\Lambda$$

$$b_8 = -\tfrac{25}{6}\pi^2 - \tfrac{40}{3} + \tfrac{40}{3}\Lambda$$

$$b_9 = \tfrac{\pi^2}{24} - \tfrac{1}{12}$$

$$b_{10} = -\tfrac{\pi^2}{40} + \tfrac{3}{20}$$

$$b_{11} = -\tfrac{\pi^2}{16} - \tfrac{3}{4}$$

$$b_{12} = \tfrac{17}{24}\pi^2 + 2 - 2\Lambda$$

$$b_{13} = -\tfrac{49}{48}\pi^2 + \tfrac{57}{12} + 4\Lambda$$

$$\Lambda = .6522560$$

$$c_{330} = 3\,b_2\,b_3$$

$$c_{320} = 3\,b_1\,b_4 - 9\sigma\,b_9$$

$$c_{310} = b_5 - 6\sigma\,b_2\,b_{10}$$

$$c_{300} = 2\,b_2\,b_6 - 6\sigma\,b_1\,b_{11}$$

$$c_{3-10} = 2\,b_1\,b_7 - 6\sigma\,b_{12}$$

$$c_{3-20} = b_8 - 4\sigma\,b_2\,b_{13}$$

$$c_{332} = \tfrac{5}{3}\,b_2\,b_3$$

$$c_{322} = \tfrac{2}{3}\,b_1\,b_4 - 9\sigma\,b_9$$

$$c_{312} = -\tfrac{10}{3}\,\sigma\,b_2\,b_{10}$$

$$c_{302} = -\tfrac{2}{9}\,b_2\,b_6 - \tfrac{4}{3}\,\sigma\,b_1\,b_{11}$$

$$c_{3-12} = -\frac{2}{9}\, b_1 b_7 \qquad\qquad c_{3-22} = \frac{4}{9}\, \sigma\, b_2 b_{13}$$

$$c_{432} = -5\, b_2 b_3 \qquad\qquad c_{422} = -4\, b_1 b_4 + 18\, \sigma\, b_9$$

$$c_{412} = -b_5 + 10\, \sigma\, b_2 b_{10} \qquad\qquad c_{402} = -\frac{4}{3}\, b_2 b_6 + 8\, \sigma\, b_1 b_{11}$$

$$c_{4-12} = -\frac{2}{3}\, b_1 b_7 + 6\, \sigma\, b_{12} \qquad\qquad c_{4-22} = \frac{8}{2}\, \sigma\, b_2 b_{13}$$

REFERENCES

1. M. Bauer, <u>At</u>. <u>Nuc</u>. <u>Data</u> <u>Tables</u> 17:442 (1976).
2. R.L. Becker, K.T.R. Davies and M.R. Patterson, <u>Phys</u>. <u>Rev</u>. C9:1221 (1974).
3. D.S. Koltun, <u>Phys</u>. <u>Rev</u>. C9:484 (1974).
4. C. Mahaux, Nuclear Matter Approach to the Nucleon-Nucleus Optical Model, <u>in</u>: "Microscopic Optical Potentials", H. von Geramb, ed., Springer-Verlag, Berlin (1979).
5. M. Bauer, J.J. E. Herrera and J. Quintanilla, preprint IFUNAM.
6. F.D. Bechetti and G.W. Greenlees, <u>Phys</u>. <u>Rev</u>. 182:1190 (1969).
7. M. Bauer and V. Canuto, <u>Nucl</u>. <u>Phys</u>. 72:33 (1965).
8. K.T.R. Davies and R.J. Mc. Carthy, <u>Phys</u>. <u>Rev</u>.C1:81 (1971).
9. S.G. Nilsson, Chin Fu Tsang et al., <u>Nucl</u>. <u>Phys</u>. A131:1 (1969).
10. A. Bohr and B.R. Mottelson, "Nuclear Structure", Benjamin, N.Y. (1969).
11. J. Mougey, M. Bernheim et al., <u>Nucl</u>. <u>Phys</u>.A262:461 (1976).
12. R.W. Manweiler, <u>Nucl</u>. <u>Phys</u>. A240:373 (1975).
13. W.D. Myers, <u>At</u>. <u>Nuc</u>. <u>Data</u> <u>Tables</u> 17:411 (1976).
14. P.A. Seeger and W.M. Howard, <u>At</u>. <u>Nuc</u>. <u>Data</u> <u>Tables</u> 17:428 (1976).
15. C.M. Ko, H.C. Pauli, M. Brack and G.E. Brown, <u>Nucl</u>. <u>Phys</u>. A236:269 (1974).

ATOMIC MASSES IN ASTROPHYSICS

Marcel Arnould[+]

Institut d'Astronomie et d'Astrophysique
Université Libre de Bruxelles, Belgium

1. INTRODUCTION

Very neutron-rich or neutron-deficient nuclei, even up to the drip lines (and sometimes beyond!) are expected to be produced or to exist in certain astrophysical circumstances. A correct description of these events requires in particular a reliable prediction of the masses of the involved highly unstable nuclei.

The aim of this paper is to review briefly some astrophysical problems in which atomic mass predictions enter more or less crucially, and to discuss possible consequences of the divergences which may appear between several mass laws in the relevant regions of the nuclidic chart.

The production of very neutron-rich nuclei is related in particular to the Type II supernova phenomenon, whose description is one of the most important current problems in astrophysics. Such an event corresponds to the catastrophic destruction of a star whose mass is at least about ten times larger than the solar mass. It is characterized by the explosive ejection of a certain amount of more or less nuclearly processed material into the interstellar medium and, in certain cases at least, by the formation of a central neutron star (or even black hole) remnant. We briefly analyse the role possibly played by the predictions of the masses of very neutron-rich nuclei in the description of the following pre- to post-supernova phases: hydrostatic silicon burning (Sec. 2), iron core collapse (Sec. 3), explosive ejection of a certain fraction of the collapsed core

---

[+]Chercheur Qualifié F.N.R.S. (Belgium)

(Sec. 4), structure and composition of the neutron star remnant
(Sec. 5) and possible decompression of cold neutron star matter
(Sec. 6).

On the other hand, explosions of supermassive stars and of
hydrogen-rich supernova envelopes might lead to the formation of very
neutron-deficient nuclei. The role of mass laws in such circumstances
is shortly discussed in Sec. 7. Finally, some conclusions are drawn
in Sec. 8.

2. PRE-SUPERNOVA SILICON BURNING

Silicon burning has received much attention recently[1,2], as it
is predicted to be the last stage in the hydrostatic evolution of
massive stars. During such a phase, the iron group nuclei are expected
to be produced and to be partly transformed into more or less neutron-
rich species by captures of free electrons originating from the ionization of the atoms[2].

Nuclear binding energies and electron capture Q-values constitute
basic ingredients for the calculation of the composition of the
Si-burning zones. As these data may be lacking experimentally for
certain of the relevant neutron-rich nuclei in the vicinity of the
iron group, some theoretical estimates may still be unavoidable (see
also ref.[3]). It is not known if differences in these estimates arising
from the use of different mass formulae can have a significant influence on the modeling of the Si-burning evolutionary stage.

3. IRON CORE COLLAPSE

With the end of Si burning, the nuclear energy reservoir of the
central parts of a star goes to exhaustion (the produced iron-group
nuclei indeed possess the highest binding energy per nucleon). In such
conditions, the further evolution of those zones is dominated by gravitational contraction associated to a temperature and density increase (for details about these features, see e.g. ref.[4]). For temperatures in excess of about $5 \ 10^9$ K, the iron-group nuclei start being
photodisintegrated into α-particles and neutrons. Such an important
energy sink (in addition to secondary processes of energy loss by
neutrinos produced during the neutronization of the iron-group nuclei)
is at the origin of a stage of catastrophic gravitational collapse
during which temperatures well in excess of 10^{10} K can be reached,
while densities might increase dramatically up to, and even beyond,
nuclear density. This collapse phase is considered to be at the origin
of the supernova event.

Recent calculations emphasize the sensitivity of the collapse
models upon the nature and abundance of the heavy nuclei expected to
be replenished at some stage of the collapse following the iron photo-
disintegration. The presence of such nuclei in a hot and dense

material also composed of free nucleons, α-particles, electrons and
neutrinos might have a drastic influence both on the equation of
state (relationship between pressure, density and temperature) and
on the neutrino mean free path, which is related to the maximum degree
of neutronization during the collapse (see e.g. ref.[5] for a detailed
discussion of such questions and further references).

The problem of the nuclear abundances during collapse is by far
the easiest to solve when the stellar core can be assumed to be composed of a non-interacting Fermi gas of free electrons, nucleons and
possibly neutrinos coexisting with a classical gas of heavy nuclei in
nuclear statistical equilibrium (i.e. equilibrium with respect to
strong and electromagnetic interactions) or in β-equilibrium (i.e.
equilibrium with respect to weak, as well as strong and electromagnetic interactions). Such a situation is referred to in the following
as the "non-interacting gas regime". The abundance problem becomes
much harder to solve when the nuclear forces between nucleons begin
to play a significant role, or when the effect of the external
nucleons on the nuclei and the Coulomb forces between nuclei cannot
be neglected anymore. Such a situation is referred to in the following as the "interacting fluid regime". It is estimated on very qualitative grounds[6] that the non-interacting gas regime remains valid
as long as $\rho \lesssim 50T$, ρ being the density in gcm^{-3} and T the temperature
in K.

3a. Non-interacting Gas Regime

In such conditions, the composition of the matter can be calculated as a function of temperature, density and electron concentration (when a nuclear statistical equilibrium is achieved) or neutrino
chemical potential (in β-equilibrium situations) if the nuclear
binding energies and partition functions are known (see e.g. ref.[6]
for details about the relevant equations).

For certain possible, although still rather uncertain, collapse
conditions which satisfy the non-interacting gas requirements, it is
expected that rather heavy (up to A≳100) and very neutron-rich nuclei
(up to the neutron drip line) constitute an important component of
the matter. This of course implies that certain of the nuclear binding energies, which are a basic ingredient of the abundance calculations, have to be predicted theoretically. A systematic comparison
between the abundance distributions in likely collapse conditions
resulting from the adoption of various mass formulae is not available.
However, some qualitative trends can be extracted from Fig. 1, where
several mass excess predictions are compared for N=2Z-10 and N=2Z
nuclei. Such a comparison clearly illustrates the dramatic differences which may exist between the various estimates in the very neutron-
rich region. Schematically, and as tightly bound nuclei are favoured
in the considered equilibrium situations, Fig. 1 indicates that, for
given physical conditions, the mass formula of von Groote et al.[7]

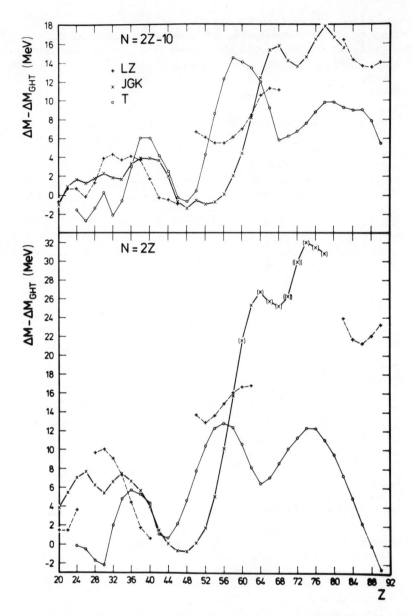

Fig. 1.   Comparison between mass excesses ΔM derived from various
          mass laws for N=2Z-10 and N=2Z, even Z nuclei. The conside-
          red mass formulae are GHT (ref.[7]), LZ (ref.[8]), JGK (ref.[9])
          and T (ref.[10]). Data are represented in single (double)
          parentheses when at least one nucleus with N≤2Z-10 or N≤2Z
          is unstable with respect to single (double) neutron emission.

has a general tendency of predicting a stronger neutronization than the mass laws of Liran and Zeldes[8] or Jänecke[9], as well as, although to a lesser extent, of Tondeur[10]. Such a tendency already appears in the $Z \lesssim 50$ range, but develops most strongly for higher Z values which, however, are most likely to be of no relevance in the problem considered here.

The predicted location of the neutron drip line or the expected electron concentration pertaining to β-equilibrium situations may also vary more or less markedly from one mass formula to the other. The former problem has been examined by Tondeur[11] in the $8 \leq Z \leq 44$ range, whereas certain aspects of the latter one have been discussed by Bethe et al.[5].

3b. Interacting Fluid Regime

In such conditions, the conventional mass formulae as well as the equations governing the abundance calculations in non-interacting gas conditions become useless for estimating the nature and concentrations of nuclei in collapsing matter. In fact, the qualitative features of the two nuclear phases (nuclei and external neutron fluid) are determined primarily by the properties of uniform bulk matter (see e.g. ref.[12] and references therein). A more recent calculation includes in addition surface and Coulomb contributions to the nuclear energy[13]. On grounds of the simplifying assumption of the existence of only one type of nuclei (this assumption appears in fact to be invalidated by the calculations), such a study reaches several important conclusions: (i) nuclei or nuclear clusters can survive to high temperatures ($\approx 10^{11}$ K) and up to densities approaching the nuclear matter density along all likely collapse trajectories, and (ii) the collapse is halted only at nuclear densities or greater, so that a reliable description of hot matter beyond nuclear densities appears to be required for describing the last phases of the collapse (such results are in qualitative agreement with those of Bethe et al.[5]).

Such high densities are reached about one second after the start of the collapse. At that point, a bounce probably occurs because of the exceedingly stiff equation of state above nuclear matter density, and an outgoing shock wave is generated. This, possibly combined with the effect of neutrino transport outwards, leads to the rapid decompression of part of the imploded material. This fraction of the matter is finally dispersed into the interstellar medium (Sec. 4), while a central neutron star (Sec. 5) or even black hole remnant is formed.

4. SUPERNOVA EXPLOSIONS AND THE r-PROCESS

While propagating, the outgoing shock wave generated deep inside the star by the bounce of the core leads to a temporary heating and

nuclear processing of the crossed layers. Such a processing ("explosive nucleosynthesis") is however limited by the rapid expansion of
the material which is accelerated beyond its escape velocity (for a
review, see e.g. ref.[14]).

The heavy (A≳70) stable even-A nuclei located on the neutron-
rich side of the valley of β-stability, a certain fraction of the
odd-A nuclei situated at the bottom of that valley as well as the
transbismuth elements are considered to be produced during the explosive nucleosynthesis episod by rapid neutron captures taking place
in the very vicinity of the bottom of the decompressing material.
Such a mechanism is referred to in the following as the canonical
r-process.

Both astrophysical and nuclear physics uncertainties affect the
study of that process. In particular, as no fully consistent supernova model is available to date, the initial temperatures, densities
and r-process seed nuclei abundance distributions are still rather
uncertain. As these distributions are generally calculated as discussed in Sec. 3a, further uncertainties arise, related in particular to
mass predictions. A limited evaluation of the level of such uncertainties has been performed by El Eid[15].

It is beyond the scope of this contribution to provide a detailed analysis of the canonical r-process along likely expansion paths
(for a review, see ref.[16]). In fact, detailed calculations indicate
that the gross features of the r-process abundances can be accounted
for by a very simple model, referred to in the following as the static r-process model. In such a framework, the r-process is assumed to
take place in conditions of temperature, density, neutron concentration and time scale τ such that charged particle reactions are too
slow to be of interest, whereas an equilibrium between neutron captures (n,γ) and photodisintegrations (γ,n) can be established for each
isotopic chain. The flow of material to higher and higher Z is due to
β⁻-decays, which of course have to be slower than the above mentioned
reactions in order to preserve the isotopic equilibrium. Such an evolution towards higher and higher Z values is most probably stopped by
neutron-induced fissions, which lead to a cycling back of a portion
of the material to lower Z elements. After the time interval τ, all
nuclear reactions are assumed to be suddenly frozen. At such a stage,
β⁻-cascades, as well as α-decays (in the A≳210 region), spontaneous
or β-delayed fissions and single or multiple β-delayed neutron emissions drive the matter towards β-stable nuclei[17].

The exact location of the high-Z cutoff, as well as the probability and relative importance of all the post-freezing processes
depend more or less dramatically, among other things, upon mass formula predictions. Such questions are not examined here. We concentrate instead on some gross effects mass laws might have on the location
of the r-process path in the (N,Z) plane and, ultimately, on the abi-

lity of models to reproduce the dominant features of the solar system
r-nuclei abundance distributions.

In static r-process conditions, the neutron capture path mainly
goes through nuclei whose neutron separation energy is close to[17]

$$S_n^o = T_9(34.075 - \log n_n + 1.5\log T_9)/5.04 \text{ MeV}, \qquad (1)$$

where T_9 is the temperature in 10^9 K and n_n is the neutron number
density in cm^{-3}. For processes susceptible of explaining the observed
solar system r-nuclei abundance distribution, S_n^o is such that the
r-process path lies in a region where no experimental S_n data are
available (see below for details about this statement). Thus, mass
formulae have to be called for, which implies a certain degree of
uncertainty in the location of the r-process path for given T_9 and
n_n. In particular, a comparison between various theoretical S_n values
for N=2Z-10 and N=2Z nuclei (Fig. 2) indicates that, for a given S_n^o,
the r-process path in the Z≳50 range is predicted to be pushed, on
average, deeper into the neutron-rich region if Tondeur's[10] mass evaluations are used instead of the other considered models[7,8,9]. The
only exception to this concerns nuclei just after closed shells and
when the comparison is made with the formula of von Groote et al.[7].
Paths much closer to the stability line are predicted with the mass
laws of Liran and Zeldes[8], and especially of Jänecke[9]. In this latter
case, N=2Z nuclei in the Z≳50 range are even predicted to already lie
beyond the neutron drip line. In the Z≲50 region, the situation is
less clear-cut, and no qualitative conclusions are attempted to be
drawn.

In fact, the solar system r-nuclei abundance distribution[18] puts
strong constraints on possible r-process paths, particularly through
its two rather sharp peaks centered around A≃130 and 195. An important problem is to determine whether or not these two peaks can be
generated in the same astrophysical conditions[17]. Here, we simply
want to examine qualitatively if the answer to such a question depends
upon the adopted mass formula.

Such abundance peaks are attributed to the N=50 and N=82 shell
closures, which implies that ^{130}Cd and ^{194}Er and/or ^{196}Yb are their
progenitors (the reason for such an assignment is discussed in e.g.
ref.[16]). In order for ^{130}Cd to be the most abundant Z=48 isotope and,
in addition, to be more abundant than ^{132}Sn (for avoiding a displacement of the abundance peak to A=132), the conditions $S_n(^{132}$Cd$) < S_n^o$
$< S_n(^{134}$Sn$)$ have to be fulfilled. In the same way, A=194 or 196 is
expected to constitute an abundance peak if $S_n(^{196}$Er$) < S_n^o <$
$S_n(^{198}$Yb$)$ or $S_n(^{198}$Yb$) < S_n^o < S_n(^{200}$Hf$)$, respectively. Table 1 indicates that, for all considered mass formulae (with the possible ex-
ception of LZ and BLM for which insufficient information is available), the correct position of the A=130 peak requires higher S_n^o
values than the one of the A=195 peak. It appears particularly diffi-

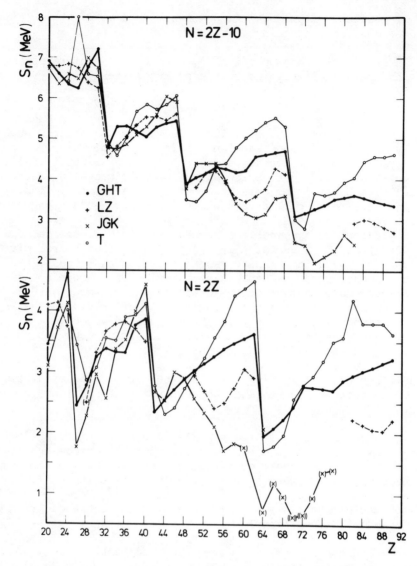

Fig. 2.   Same as in Fig. 1, but for neutron separation energies S_n.

cult to reproduce these two peaks in the same astrophysical conditions with the JGK, CK, JE and T mass formulae, whereas the incompa-
tibility is the smallest with the M and SH predictions, the GHT formula leading to some intermediate situation.

Such qualitative conclusions based on the static r-process model
probably keep a certain level of validity when more realistic r-process models are considered. Even if these latter models achieve a

Table 1. Values of S_n (in MeV) predicted by various mass formulae
          for the indicated nuclei[a]

Nucleus	M	GHT	SH	LZ	JGK	CK	JE	BLM	T
^{132}Cd	4.16	4.13	3.87	----	4.27	4.34	3.87	2.97	3.71
^{134}Sn	4.73	4.73	4.47	3.50	4.74	4.97	4.52	3.47	4.27
^{196}Er	3.06	2.80	2.77	----	2.06	2.10	2.28	----	2.40
^{198}Yb	3.46	3.21	3.27	----	2.50	2.66	2.75	----	2.71
^{200}Hf	3.88	3.63	3.57	----	3.06	3.31	3.20	----	2.73

[a]The symbols used to identify the mass laws are as follows: M: ref.[19];
GHT: ref.[7]; SH: ref.[21]; LZ: ref.[8]; JGK: ref.[9]; CK: ref.[22]; JE: ref.[23];
BLM: ref.[24]; T: ref.[10].

fair degree of success in reproducing the solar system r-nuclei abundance distribution, the main difficulty they encounter lies in the
predicted positions of the abundance peaks[16].

Two further problems encountered in r-process models and more or
less closely related to mass predictions concern a broad bump around
A=165 in the observed abundance distribution, as well as the possibility of superheavy element production. The bump is generally interpreted as a deformation effect, and the extent of the agreement between observation and theory is considered as a measure of the ability
of a mass law to account for nuclear deformation. As far as superheavy elements are concerned, the exact nature of the most stable
ones and their half-lives remain debated questions. In contrast, it
seems generally agreed to-day that such nuclei cannot be produced in
the canonical r-process (see e.g. ref.[16]), such a conclusion being
relatively independent of the adopted mass formulae. However, it has
been suggested that superheavies might have existed in nature, particularly in the early solar system or the early Moon[25,26]. If this
indeed turns out to be confirmed by further studies, many of the present ideas about the r-process in the transbismuth region would have
to be dramatically revised.

It is also worth noting that sites for the r-process different
from the canonical one have been searched for recently. In particular, it has been shown[27] that a strong neutron irradiation of r-process type might occur during the explosive decompression of a supernova helium-burning shell at T≲8 10^8 K and ρ≲10^4 gcm^{-3}. However, the
derived results, based on the mass law of von Groote et al.[7], fail to
mimic the typical solar system r-process abundance curve, due in particular to a huge overproduction of A=78 nuclei coming from the processing of the iron-group nuclei. Such calculations cannot reproduce

correctly the A=130 and 195 peaks either. Another possible non-cano-
nical r-process scenario is described in Sec. 6.

A final word concerns the r-process chronology, that is the determination of the time scales for the r-process events on grounds of
the observed or implied relative abundances of certain radioactive
nuclei emerging from such a mechanism (see e.g. ref.[28] for a review).
Part of the uncertainties encountered in that field are related to
mass predictions, as pointed out by Seeger and Schramm[29].

5. STRUCTURE AND COMPOSITION OF COLD NEUTRON STARS

Neutron stars formed at very high temperatures as a result of
the iron core collapse (Sec. 3) are expected to cool down due to neutrino losses over time scales of the order of a million years or less.
After that period, it is reasonable to assume that the neutron star
matter is in its ground state (i.e. at zero temperature). Schematically, such a cold neutron star can be divided into three regimes[30].
The crust at densities in excess of about 10^9 gcm^{-3} is mainly composed of degenerate electrons (coming from the ionization of the atoms)
and neutron-rich nuclei (produced by free electron captures). This
regime is referred to in the following as the (A,e) phase. When the
density exceeds about 10^{11}-10^{12} gcm^{-3}, electron captures start forming nuclei beyond the neutron drip line. With increasing density,
the dripped neutrons become more and more abundant and start in turn
to form a degenerate gas. An equilibrium is established between neutron captures and emissions. This regime is referred to in the following as the (A,n,e) phase. Finally, at densities in excess of about
10^{14} gcm^{-3}, all nuclei get diluted, and matter is essentially composed of neutrons (or heavier baryons above roughly 10^{15} gcm^{-3}) with a
small proportion of protons, electrons and muons. The dilution of the
nuclei may occur through several more or less complex mechanisms
which remain to be investigated in detail (see e.g. ref.[13] for some
comments about certain aspects of this question). In the following,
we briefly discuss the extent to which the modeling of the (A,e) and
(A,n,e) phases may depend upon the adopted mass formulae.

5a. (A,e) Phase

In such a phase, differences in mass predictions are not expected to sensitively affect the derived equation of state, except perhaps near the transition to the (A,n,e) phase. In contrast, they may
play a central role in the determination of the compostion and of the
density at which dripped neutrons start to appear[31,32].

For example, calculations based on the energy density method[32]
indicate that nuclei close to the N=50 or 82 shell closures can be
the most abundant ones in certain density ranges, whereas the use of
a droplet-type mass formula[33,34,35] only allows for closed shell
nuclei. Such a difference results from the fact that the adopted

energy density formalism[32] predicts a weakening of the N=50 and 82 shell closures for very neutron-rich nuclei. Concerning the density at which dripped neutrons start to appear, some differences are also encountered. In particular, the critical density for double neutron emission is predicted to be 5.66×10^{11} gcm^{-3} by Tondeur[32], whereas Baym et al.[33] and Heintzmann et al.[35] find 4.3×10^{11} and 4.65×10^{11} gcm^{-3}, respectively.

5b. (A,n,e) Phase

Just as in the interacting fluid regime at high temperature (Sec. 3b), classical mass formulae become useless for the description of such a phase, due to the interactions between the various constituents of the matter.

The energy density formalism has been applied to this problem by several authors[36,37,38]. On the other hand, a compressible liquid drop model making use of a Skyrme interaction and semi-empirical shell corrections[39] has also been used[40]. A comparison between these various works first indicates that shell corrections play a very important role[36,40]. Differences also appear between calculations including shell corrections. In particular, Negele and Vautherin[37] predict essentially Z=40 or 50 nuclei, depending upon density, whereas essentially Z=28 and 50 nuclei are found to be present by Tondeur[38]. Differences are also encountered, at least in the $\rho < 10^{13}$ gcm^{-3} range, between the compositions predicted by Tondeur[38] and Lattimer et al.[40]. It may be somewhat difficult to exactly trace back the very origin of such discrepancies. Shell effects have of course to be invoked. It has also to be noticed that the Coulomb lattice energy has been neglected by Tondeur[38], whereas it has been taken into account in the calculations of Lattimer et al.[40]. It is known that such an energy tends to favour heavier nuclei.

Differences are also encountered in the calculated equations of state. They seem to be essentially due to divergences in the predicted pressure of the neutron gas.

6. DECOMPRESSION OF COLD NEUTRON STAR MATTER

In addition to the supernova scenario thought to be responsible for the decompression of very hot and dense matter (Sec. 4), some astrophysical processes have been investigated recently in which cold neutron star matter (Sec. 5) is suddenly expanded to much lower densities. This could namely result from neutron star volcanos[41] or from the tidal disruption of a neutron star by a black hole in (still hypothetical) evolved neutron star - black hole binary systems[40].

The initial conditions for such a decompression process are provided by a cold neutron star model (Sec. 5). As the matter expands, the density of the dripped neutrons decreases, even if neutrons

rapidly leave the nuclei. When the density has dropped to low enough
values ($\rho \approx 10^{11}$ gcm^{-3}; ref.[40]), β-decays can effectively take place.
In order for the matter to remain in equilibrium, it can be shown
that those decays have to be accompanied by neutron captures. Thus,
such an expansion phase is characterized by an increase both in Z and
N of the nuclei, as well as by a temperature increase due to the li-
berated β-decay energy. Gradually, the produced nuclei become more
and more unstable with respect to fission, which finally has time to
take place. This leads to a cycling back of the nuclei to lower Z and
N values, as well as to a temperature increase up to about 10^{10} K
after the first fission. In such conditions, a nuclear statistical
equilibrium is rapidly reached, and the methods mentioned in Sec. 3
may be used to evaluate the composition of the matter. Such a situation, coupled with the relatively high dripped neutron density, may
supply the ingredients for a subsequent r-process phase. No study of
this kind of r-process has been performed to date. It is however speculated[40] that such a mechanism might be more favourable than the
canonical r-process for the production of superheavy elements.

Most of the aspects of the decompression scenario are sensitive,
among other things, to mass predictions. This is particularly so for
the composition of the matter in the initial stages of the decompression (Sec. 5), for the point at which β-decays and fissions start to
take place, as well as in the final phases of the decompression, when
an r-type process might develop (Sec. 4).

7. NUCLEI IN THE NEUTRON-DEFICIENT REGION

In certain astrophysical circumstances, namely the explosion of
certain supermassive stars[42] or of hydrogen-rich supernova envelopes[43], neutron-deficient nuclei are expected to be produced by proton
captures and/or (γ,n) reactions at temperatures in excess of 10^9 K.
In such conditions, successive proton captures probably have time to
take place and possibly drive material relatively close to the proton
drip line, at least in the A$\leq$100 mass range. However, such a flow
towards the proton drip line can be hindered by (γ,p) photodisintegrations and possibly β^+-decays[44]. In fact, for given temperatures,
densities and proton concentrations, the exact location in the (N,Z)
plane of the main flow of material depends very much upon proton
separation energies and β^+-decay Q-values. This implies a possible
sensitivity of the calculated abundances upon mass predictions if
material can indeed flow close enough to the proton drip line. A qualitative inspection of the probable location of the main nuclear flow
in the (N,Z) plane under possible astrophysical conditions, as well
as a comparison between available mass predictions lead us to think,
however, that drastic changes in the calculated abundances of
neutron-deficient nuclei due to different choices of mass formulae
are rather unlikely. In fact, uncertainties of such a nature are
most probably less important than those due to the estimated β-decay
lifetimes or nuclear reaction cross sections of interest.

8. CONCLUSIONS

Mass predictions appear to be of key importance for the description of the iron core collapse, for the understanding of the r-process, as well as for the modeling of cold neutron stars and of the possible decompression of a fraction of their material. However, the conventional mass formulae become useless when the nuclear (and Coulomb) interactions between the various components of the matter play a significant role. This is the case namely in the interacting fluid regime of the iron core collapse (Sec. 3b) or in the deep interior of neutron stars (Sec. 5). In such situations, other methods have to be developed for predicting the properties and abundances of nuclei embedded in a dense neutron fluid. It may be hoped that the study of such methods would help improving the reliability of mass formulae aimed at describing isolated very neutron-rich nuclei.

REFERENCES

1. W. D. Arnett, Astrophys. J. Suppl. 35:145 (1977).
2. T. A. Weaver, G. B. Zimmerman and S. E. Woosley, Astrophys. J. 225:1021 (1978).
3. C. N. Davids, this Conference.
4. J. P. Cox and R. T. Giuli, "Principles of Stellar Structure", Gordon and Breach, New York (1968).
5. H. A. Bethe, G. E. Brown, J. Applegate and J. M. Lattimer, preprint (1978).
6. M. J. Murphy, preprint (submitted to Astrophys. J.) (1978).
7. H. von Groote, E. R. Hilf and K. Takahashi, Atomic Data Nucl. Data Tables 17:418 (1976).
8. S. Liran and N. Zeldes, Atomic Data Nucl. Data Tables 17:431 (1976).
9. J. Jänecke, Atomic Data Nucl. Data Tables 17:455 (1976).
10. F. Tondeur, Nucl. Phys. A303:185 (1978).
11. F. Tondeur, Z. Phys. A288:97 (1978).
12. J. M. Lattimer and D. G. Ravenhall, Astrophys. J. 223:314 (1978).
13. D. Q. Lamb, J. M. Lattimer, C. J. Pethick and D. G. Ravenhall, Phys. Rev. Lett. 41:1623 (1978).
14. V. Trimble, Rev. Mod. Phys. 47:877 (1975).
15. M. F. El Eid, private communication (1979).
16. W. Hillebrandt, Space Sci. Rev. 21:639 (1978).
17. T. Kodama and K. Takahashi, Nucl. Phys. A239:489 (1975).
18. A. G. W. Cameron, Space Sci. Rev. 15:121 (1973).
19. W. D. Myers, Atomic Data Nucl. Data Tables 17:411 (1976).
21. P. A. Seeger and W. M. Howard, Atomic Data Nucl. Data Tables 17:428 (1976).
22. E. Comay and I. Kelson, Atomic Data Nucl. Data Tables 17:463 (1976).
23. J. Jänecke and B. P. Eynon, Atomic Data Nucl. Data Tables 17:467 (1976).

24. M. Beiner, R. J. Lombard and D. Mas, Atomic Data Nucl. Data Tables 17:450 (1976).

25. E. Anders, H. Higuchi, J. Gros, H. Takahashi and J. Morgan, Science 190:1262 (1975).

26. L. M. Libby, W. F. Libby and S. K. Runcorn, Nature 278:613 (1979).

27. F.-K. Thielemann, M. Arnould and W. Hillebrandt, Astron. Astrophys. 74:175 (1979).

28. W. A. Fowler, in: "Proceedings of the Robert A. Welch Foundation Conference on Chemical Research XXI - Cosmochemistry", W. O. Milligan, ed., Robert A. Welch Foundation, Houston (1978).

29. P. A. Seeger and D. N. Schramm, Astrophys. J. 160:L157 (1970).

30. G. Baym and C. J. Pethick, Ann. Rev. Nucl. Sci. 25:27 (1975).

31. F. Tondeur, Astron. Astrophys. 14:453 (1971).

32. F. Tondeur, Astron. Astrophys. 72:88 (1979).

33. G. Baym, C. J. Pethick and P. Sutherland, Astrophys. J. 170:299 (1971).

34. K. Koebke, M. El Eid and E. Hilf, Z. Phys. 271:21 (1974).

35. H. Heintzmann, W. Hillebrandt, M. El Eid and E. Hilf, Z. Naturforsch. 29A:933 (1974).

36. Z. Barkat, J. Buchler and L. Ingber, Astrophys. J. 176:723 (1972).

37. J. Negele and D. Vautherin, Nucl. Phys. A207:321 (1973).

38. F. Tondeur, Thesis, Université Libre de Bruxelles (unpublished) (1977).

39. W. D. Myers and W. J. Swiatecki, Nucl. Phys. 81:1 (1966).

40. J. M. Lattimer, F. Mackie, D. G. Ravenhall and D. N. Schramm, Astrophys. J. 213:225 (1977).

41. F. J. Dyson, "Neutron Stars and Pulsars", Fermi Lecture at the Scuola Normale Superiore di Pisa, Accademia Nacionale dei Lincei, Roma (1971).

42. W. Ober, in: "Ninth Texas Symposium on Relativistic Astrophysics" (to appear), Munich (1978).

43. J. Audouze and J. W. Truran, Astrophys. J. 202:204 (1975).

44. M. Arnould, unpublished (1976).

ALPHA DECAY OF NEUTRON-DEFICIENT ISOTOPES

STUDIED AT GSI DARMSTADT

Ernst Roeckl

GSI Darmstadt
6100 Darmstadt
Federal Republic of Germany

INTRODUCTION

Heavy-ion beams of ^{40}Ar, ^{48}Ti, ^{58}Ni and ^{84}Kr, available now from accelerators such as the UNILAC, combined with experimental facilities, such as the velocity filter SHIP or the GSI on-line mass separator, enabled a considerable extension of alpha decay studies during recent years. The present review describes the status of various experiments underway in this field at GSI Darmstadt. The experimental techniques involved are summarized and the results are discussed in terms of systematics of alpha decay energies and widths. In a final section, an outlook on some present limitations and future potentialities is given.

EXPERIMENTAL TECHNIQUES

Velocity Filter SHIP

The velocity filter SHIP[1], combining 2 electric and 4 magnetic dipole fields with 2 quadrupole triplets, allows an efficient separation between projectiles and evaporation residues. Within the entrance aperture of 3°, windows of 10% in velocity and of 20% in ionic charge are focussed for reaction products, while the primary beam is suppressed by factors up to 10^{11} depending on the projectile-target combination. An overall efficiency of (14±1)% was obtained for ^{150}Nd(^{40}Ar,6n) reactions. The separation time is determined by the time-of-flight of the ions on their 11-m flight path. Since this time is of the order of 10^{-6} s, very short half-lives are accessible using as a time reference the UNILAC macro structure and/or the arrival time of an ion at the exit of SHIP.

Schmidt et al.[2,3] applied a secondary electron detector for generating fast time signals, and a 450 mm² surface barrier detector for the energy analysis of both the incoming ion and its subsequent alpha decay. Using fusion reactions between ^{40}Ar projectiles and targets of ^{176}Hf, ^{181}Ta and ^{184}W, alpha lines were time-analyzed with respect to the time of arrival of the reaction product and to correlations between pairs of lines (see Fig. 1). The identification of the new protactinium and thorium alpha-emitters was established on the basis of time correlations with the respective daughter-isotopes.

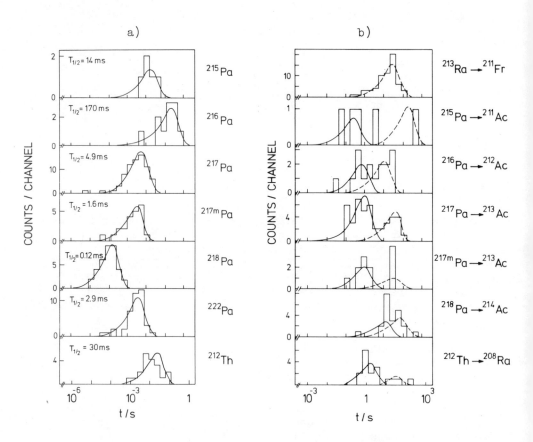

Fig. 1. Time analyses for alpha lines from neutron-deficient iso-
         topes of protactinium, thorium and actinium. Experimental
         distributions are given as histograms. Fits assuming radio-
         active decay or random time differences are shown as solid
         and dashed curves, respectively. Note the logarithmic
         scale. a) Distribution of the delay times between the pulse
         of the impinging reaction product and its subsequent alpha
         decay. b) Distribution of time differences between
         correlated (except ^{213}Ra → ^{211}Fr) pairs of alpha lines.

Another identification tool is the use of position correlations occurring at the detector of SHIP. Hofmann et al.[4] applied a position-sensitive surface-barrier detector for determining correlations both in position and time between pairs of alpha lines as exemplified in Fig. 2. The width of the position distribution (Fig. 2b) is governed by the position resolution (400 μm FWHM) which is large compared to the range (about 7 μm) of the fusion products, to the range (about 0.4 μm) of the recoiling nuclides after alpha emission, and to the range (25-40 μm) of the alpha particles themselves. The time spectrum (Fig. 2d) represents the decay of the daughter nucleus, 110 ms ^{157}Hf.

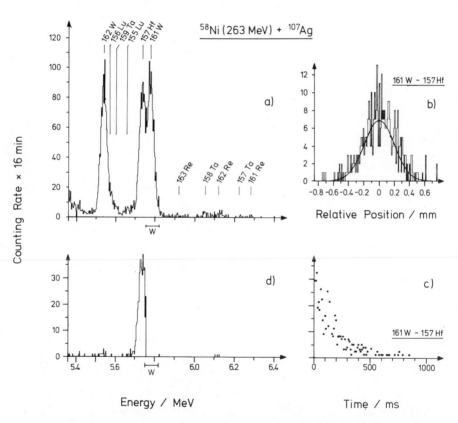

Fig. 2. Position and time correlations between the alpha lines of ^{161}W and ^{157}Hf. a) Alpha singles spectrum obtained by implanting evaporation residues from ^{58}Ni + ^{107}Ag reactions into a position-sensitive detector of 47x8 mm^2 area. b,c) Position and time distributions generated with the energy window W on the ^{161}W alpha line. d) Alpha spectrum of ^{157}W obtained from (a) by setting windows on the position (b) and time (c) distributions.

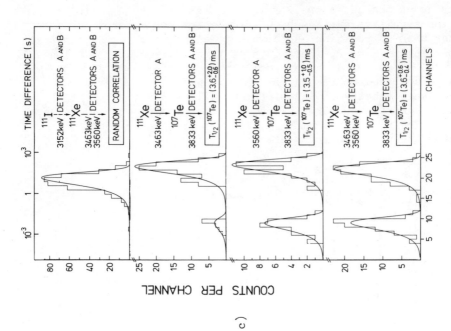

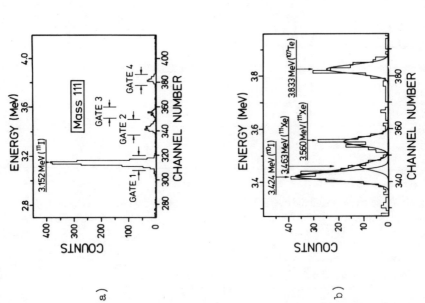

Fig. 3. Time correlation analysis of mass-111 alpha activity. a) Alpha spectrum accumulated in detector A of the windmill system. The gates indicated were used for the correlation analysis. b) High-energy part of spectrum (a) together with Gaussian fits. The ^{110}I contamination is collected due to the movement of the windmill wing through the mass-111 beam. c) Time correlation analysis.

On-Line Mass Separator

The GSI on-line separator facility[5] falls short in separation efficiency and separation time in comparison with SHIP, but it offers the advantages of unambigious mass assignment and higher source purity. Reaction products recoiling out of the target are stopped in a hot catcher inside the ion source, from where they are extracted and accelerated. After having passed a magnetic analyzer, the mass-separated beams are collected for decay spectroscopy. Except for the study of ^{114}Cs from a thermal ion source, the FEBIAD ion source has been used exclusively so far. The overall efficiency for polonium and mercury isotopes, for example, is 20% for half-lives of the order of minutes and 10% for half-lives of the order of seconds. Alpha decay spectroscopy of the mass-separated samples is performed by means of single surface-barrier detectors or telescope arrays.

In order to clarify the Z assignment of the alpha lines observed from the mass-separated sample, additional measurements of β-, γ- and X-rays were performed. In the case of the mass-111 alpha activity, the time correlation method developed by Schmidt et al.[3] was used as illustrated in Fig. 3. The mass-111 beam from the separator was implanted into a thin carbon catcher mounted on one of the four wings of a windmill system[7]. After a 90° rotation of the wings, the activity came to rest in between two surface-barrier detectors for a counting interval. The dominant mass-111 alpha line (see Fig. 3a) is ascribed to the decay of ^{111}I (Refs. 7,8) on the basis of a common half-life found in alpha particle and X-ray[9] measurements. The assignment of the higher-energy alpha lines to ^{111}Xe and ^{107}Te (see Fig. 3b) is deduced from the time correlation analysis (Fig. 3c).

RESULTS

Table 1 summarizes the heavy-ion reactions involved and lists the alpha emitters studied: more detailed information can be found in publications of Schmidt et al.[2,3], Hofmann et al.[4] and Schardt et al.[7]. Included are also preliminary results from recent work at the on-line separator, which led to identification of the new isotopes (E_α of main alpha line in MeV) ^{183}Tl (6.32), ^{183}Pb (6.69), ^{184}Pb (6.65) and ^{188}Bi (6.82).

The most short-lived isotopes identified directly so far are 0.12 ms ^{218}Pa for SHIP experiments and 0.57 s ^{114}Cs for on-line separator measurements (the lightest known alpha emitter, 3.6 ms ^{107}Te, was detected as daughter product of 0.9 s ^{111}Xe).

Table 1.   Summary of alpha particle measurements at the velocity
           filter SHIP and at the on-line separator

| Projectile | | Targets | Alpha emitters studied | Ref. |
Isotope	Energy (MeV)			
^{40}Ar	165–202	^{176}Hf; ^{181}Ta; ^{184}W	^{212}Th; $^{215-218}$Pa; ^{222}Pa	2,3
^{84}Kr	386–419	^{107}Ag	^{188,189}Bi	–
^{48}Ti	244–317	^{142,146}Nd	$^{183-188}$Pb; ^{183}Tl	–
^{58}Ni	215–275	^{103}Rh; ^{108,110},natPd; ^{107,109}Ag	$^{161-164}$Re; $^{160-166}$W; $^{157-161}$Ta; $^{156-160}$Hf	4
^{58}Ni	290	^{58}Ni	^{114}Cs; $^{111-113}$Xe $^{110-113}$I; $^{107-110}$Te	7,8

ALPHA-DECAY ENERGIES

     Alpha decay energies (Q_α) are a measure of the slope of the
mass-energy surface and indicate, if plotted in a systematic way[10]
the onset of shell-closure and deformation effects. The new decay
energies for neutron-deficient isotopes of hafnium through rhenium
complete the pattern above the neutron shell closure N=82 (see
Ref. 4). The observation of a new island of alpha emission above tin
is closely related to the expected double shell closure at Z=N=50.
A comparison of experimental Q_α values with predictions from current
mass formulae[11,12] shows best overall agreement with the three available droplet-model mass formulae, in particular with the one of Hilf
et al.[12]. As can be seen from Fig. 4, the latter formula shifts the
onset of deformation effectively to ^{117}Cs and reproduces the regular
Q_α pattern rather well including ^{114}Cs. The three droplet-model
versions differ mainly in their shell correction terms, which also
determine the onset of deformation. The Q_α values of tellurium
through cesium are particularly sensitive to this correction term.
In the case of the tellurium isotopes the values predicted by Hilf
et al.[12] derive about 90% from the shell correction term and only

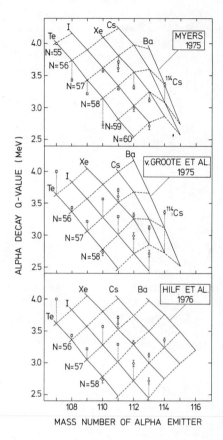

Fig. 4.   Comparison of experimental Q_α values with predictions from
the mass formulae of Myers[11], von Groote et al.[11] and Hilf
et al.[12].   The experimental results are plotted as open
circles linked by dotted lines to the corresponding
predictions.

about 10% from the droplet term. Agreement between predicted and
experimental Q_α values within an error much smaller than the shell
part of Q_α is therefore a strong indication for the existence of the
expected double shell closure at Z=N=50.

Instead of Q_α values (or mass differences), differences ΔQ_α of
Q_α values (that is double mass differences) are also useful to study
the mass surface. Schmidt et al.[2,13] pointed to the mutual support
of proton and neutron magicity around Z=82 and N=126, and similar
systematics were discussed recently[14] around Z=N=50.

In testing mass formulae, experimental mass differences such as Q_α or double differential quantities such as ΔQ_α are not so valuable as the experimental mass-excess values themselves (if only for the purpose of determining less pairing-dependent differences such as two neutron separation energies). There are two recent examples of linking alpha decay energies mentioned above to known masses. The direct mass measurements performed by the Orsay group[15] for neutron-deficient francium isotopes establishes many such links down to bismuth and up to the new protactinium isotopes. Another example for a successful linkage between known masses and very neutron-deficient ones is presented by Płochocki et al.[14] at this conference.

ALPHA WIDTHS

The reduced widths W_α for s-wave alpha transitions between ground states of even nuclei are known[16] to form a basis for the systematics of alpha decay probabilities. W_α is defined as λ_α/P, where λ_α is the partial transition probability for alpha decay determined from the measured half-life of the isotope and its branching ratio for alpha decay. P is the barrier penetrability for s-wave alpha particles calculated with the diffuse-surface potential used by Rasmussen[17]. The W_α systematics shown in Fig. 5 is taken from Ref. 16, but includes recent experimental data from Ref. 18 and in particular from Refs. 7 (^{108}Te), 4, 19, 20 (alpha emitters with Z=62-72) and 21 (^{192}Pb). The main pattern, wich is understood in

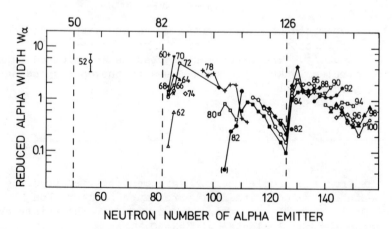

Fig. 5.   Reduced alpha W_α, normalized to ^{212}Po, as a function of the neutron number N of the alpha emitter. The proton number of the emitter is indicated.

terms of single-particle models (see e.g. Ref. 22), is a sharp
minimum occurring at the shell closure Z=82, N=126, maxima occurring
two or four particles beyond this shell closure and the one at N=82,
and a regular behavior in between. The first (and only) value
measured above to the expected Z=50, N=50 shell closure continues
the regular trend. Additional minima due to subshell closures occur
at N=152 (Ref. 16) and Z=64. The latter effect, predicted earlier
by BCS calculations[23], was established experimentally as a shallow
kink in the N=84 alpha widths[4],[19]. The lead anomaly seems to
persist even though the original data[16] are not reproduced by more
recent investigations (Ref. 21 and preliminary GSI results on
neutron-deficient mercury and lead isotopes).

OUTLOOK

      Heavy-ion induced fusion reactions have proven to offer unique
conditions for studying extremely neutron-deficient isotopes and
for their production at present seem to have a lead over high-energy
proton reactions[24]. As far as GSI work is concerned, the recent
alpha decay experiments may be understood as a _first step_. What
is important is not so much the identification of series of new
alpha emitters, but rather the improved production capabilities in
general, which will allow considerably improved spectroscopy experiments.

      A few examples may enlighten _second-step_ extensions of the
above alpha-particle experiments and their limitations. It is
interesting to note (see Fig. 6) that the limit, where the
predicted[25] production cross section for optimum (but realistic)
heavy-ion fusion reactions falls below 1 mb, has hardly been passed
yet. From cross section considerations it should therefore be
possible to further map the chart of nuclides by a few units of
neutron number towards the neutron-deficient side. Preliminary
results from very recent SHIP experiments[26] show indeed evidence
for lighter alpha emitters of osmium, iridium and platinum isotopes.
Even towards very heavy elements, where cross sections are only of
the order of 1 nb, identification of new neutron-deficient isotopes
seems to be feasible[27] using (heavy ion, xn) reactions. Deep-
inelastic reactions between targets and projectiles such as xenon,
lead and uranium might represent a novel tool for producing neutron-
rich isotopes of heavy elements at a moderate cross-section level
(see e.g. Ref. 28).

      So far no alpha decay has been observed "through" or from the
N=82 shell closure or from emitters just below this neutron number.
The experimental difficulty becomes evident considering the features
of the limits of cross section and alpha-decay half-lives on the one
hand and of the proton drip line on the other. To the left of the

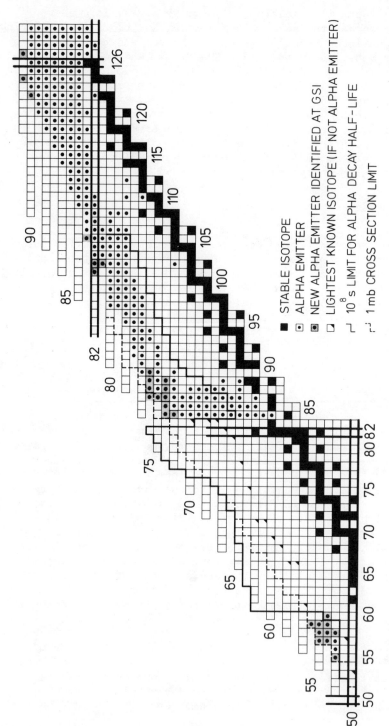

Fig. 6. Section of the chart of nuclides. Stable nuclides are given as full squares. Alpha emitters are marked by points, those identified first in GSI experiments are additionally hatched.
The lightest known isotopes (if not alpha emitters) are labelled by triangles. The chart ends
at the neutron-deficient side at the proton drip line predicted by Hilf et al.[12] The limits for
fusion cross sections of 1 mb and for alpha-decay half-lives of 10^8 s are given as dashed and
solid histograms, respectively (see text).

solid histogram-line shown in Fig. 6 the calculated alpha-decay half-lives become smaller than 10^8 s assuming Q_α predictions from Ref. 12.

Examples for more realistic second-step experiments are the investigation of alpha decay to excited states, for instance to the (unknown) excited states of ^{182,184}Hg in the region of the sudden shape change observed for light mercury isotopes, or studies of alpha emission from high-spin isomers.

It is interesting to note (Fig. 6), that the lightest alpha emitters of cesium, bismuth and of the odd-Z elements between tantalum and gold reach out now to the predicted proton drip line or lie even beyond it.

REFERENCES

1. G. Münzenberg, W. Faust, S. Hofmann, P. Armbruster, K. Güttner, and H. Ewald, Nucl. Instr. and Meth. 161 (1979) 65.
2. K.-H. Schmidt, W. Faust, G. Münzenberg, H.-G. Clerc, W. Lang, K. Pielenz, D. Vermeulen, H. Wohlfahrt, H. Ewald and K. Güttner, Nucl. Phys. A318 (1979) 253.
3. D. Vermeulen, H.-G. Clerc, W. Lang, K.-H. Schmidt and G. Münzenberg, to be published.
4. S. Hofmann, W. Faust, G. Münzenberg, W. Reisdorf, P. Armbruster, K. Güttner and H. Ewald, Z. Physik A291 (1979) 53.
5. K. H. Burkard, W. Dumanski, R. Kirchner, O. Klepper and E. Roeckl, Nucl. Instr. and Meth. 139 (1976) 275.
6. R. Kirchner and E. Roeckl, Nucl. Instr. and Meth. 133 (1976) 187.
7. D. Schardt, R. Kirchner, O. Klepper, W. Reisdorf, E. Roeckl, P. Tidemand-Petersson, G. T. Ewan, E. Hagberg, B. Jonson, S. Mattsson and G. Nyman, Nucl. Phys. A (in print).
8. E. Roeckl, R. Kirchner, O. Klepper, G. Nyman, W. Reisdorf, D. Schardt, K. Wien, R. Faß and S. Mattsson, Phys. Lett. 78B (1978) 393.
9. R. Kirchner, O. Klepper, G. Nyman, W. Reisdorf, E. Roeckl, D. Schardt, N. Kaffrell, P. Peuser and K. Schneeweiß, Phys. Lett. 70B (1977) 150.
10. A. H. Wapstra and K. Bos, Atomic Data and Nucl. Data Tables 19 (1975) 277.
11. The 1975 Atomic Mass Predictions, S. Maripuu, Ed., Atomic Data and Nucl. Data Tables 17 (1976).
12. E. R. Hilf, H. von Groote and K. Takahashi, in: Proc. 3rd Int. Conf. on Nuclei far from Stability, Cargèse, CERN 76-13 (1976) p. 142.
13. K. H. Schmidt and D. Vermeulen, contribution to this conference.

14. A. Płochocki, G. M. Gowdy, R. Kirchner, O. Klepper, W. Reisdorf,
     E. Roeckl, P. Tidemand-Petersson, J. Żylicz, U. J. Schrewe,
     R. Kantus, R.-D. von Dincklage and W.-D. Schmidt-Ott,
     contribution to this conference.
15. G. Audi et al., contribution to this conference.
16. P. Hornshøj, P. G. Hansen, B. Jonson, H. L. Ravn, L. Westgaard
     and O. B. Nielsen, Nucl. Phys. A230 (1974) 365.
17. J. O. Rasmussen, Phys. Rev. 113 (1959) 1593.
18. Table of Isotopes, Seventh Edition, C. M. Lederer and
     V. S. Shirley, Eds., New York (1978).
19. W. D. Schmidt-Ott and K. S. Toth, Phys. Rev. C13 (1976) 2574.
20. E. Hagberg, P. G. Hansen, J. C. Hardy, P. Hornshøj, B. Jonson,
     S. Mattsson and P. Tidemand-Petersson, Nucl. Phys. A293
     (1977) 1.
21. K. S. Toth, M. A. Ijaz, C. R. Bingham, L. L. Riedinger,
     H. K. Carter and D. C. Sousa, Phys. Rev. C19 (1979) 2399.
22. H. J. Mang, Annu. Rev. Nucl. Sci. 14 (1964) 1.
23. R. D. Macfarlane, J. O. Rasmussen and M. Rho, Phys. Rev. C134
     (1964) 1196.
24. P. G. Hansen, Annu. Rev. of Nucl. and Particle Sci. (in print).
25. W. Reisdorf, in: Proc. Int. Workshop on Gross Properties of
     Nuclei and Nuclear Excitations VII, Hirschegg, INKA-Conf.-
     79-001-040 (1979) p. 93.
26. S. Hofmann, private communication.
27. G. Münzenberg, private communication.
28. J. V. Kratz, W. Brüchle, G. Franz, M. Schädel, I. Warnecke,
     G. Wirth and M. Weis, to be published.

MASSES OF VERY NEUTRON DEFICIENT NUCLIDES IN THE TIN REGION

A. Płochocki[*], G. Gowdy, R. Kirchner, O. Klepper
W. Reisdorf, E. Roeckl, P. Tidemand-Petersson, J. Żylicz[*]

GSI Darmstadt, 61 Darmstadt, Fed. Rep. of Germany
U. Schrewe, R. Kantus, R. v. Dincklage, W. Schmidt-Ott
Universität Göttingen, 34 Göttingen, Fed. Rep. of Germany

INTRODUCTION

This paper describes a determination of mass-excess values for several neutron-deficient nuclides far from stability, in the $Z \approx 50$ region, and discusses the results in terms of semi-empirical mass formulae.

The new mass-excess values for nuclides indicated in Fig. 1, have been deduced from the following experimental information:
(i) known masses of ^{108,106}In and ^{104m}Ag, Ref.[1,2],
(ii) earlier mass-difference data from studies of α-decay[3], β-delayed protons[4] and positron spectra[5],
(iii) electron-capture decay energies (Q_{EC}) measured in this work for the ^{108,106}Sn and ^{104}Cd decays.

DETERMINATION OF Q_{EC} VALUES

Our experiments were carried out using γ-ray spectroscopy techniques. From the results of singles and coincidence measurements, the $\beta^+/(EC + \beta^+)$ probability ratios were determined for electron-capture plus positron transitions to individual levels of the daughter nuclei. A comparison of these ratios with the theoretical ones allowed us to deduce the positron end-point energies and, in consequence, the Q_{EC} values. Before calculating the theoretical ratios, the EC/β^+ values taken from Ref.[7] were multiplied by

---

[*] On leave of absence from the Institute of Experimental Physics, Warsaw University, 00-681 Warsaw, Poland

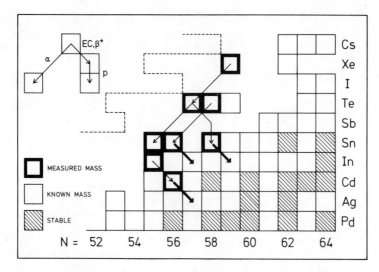

Fig. 1. A section of the nuclide chart illustrating the studies of
         the decay energies and masses. Heavy-line arrows indicate
         those (EC + β$^+$) decays for which the Q_{EC} values were
         determined in this work. The proton-drip line according to
         Jänecke and Eynon[6] is shown by the broken line.

k = 0.90 ± 0.09 (see also Fig. 2) in order to account approximately
for some discrepancies between the theory and recent[8,9] experiments
for a few medium heavy nuclides.

A determination of the β$^+$/(EC + β$^+$) ratios was facilitated by
the simplicity of the ^{108,106}Sn and ^{104}Cd decay schemes (known from
Refs.[10-12] and supported by our measurements). In each case, the
β-decay is practically limited to few transitions with very low
log ft values. For a selected level in a daughter nucleus we could
establish an intensity balance between the β- and γ-transitions
feeding this state, and γ-transitions de-exciting it, and deduce in
this way the required probability ratio. The intensity of the annihilation quanta served as a measure for the intensity of the positron
radiation (a small correction for the annihilation of positrons in
flight could be taken from Ref.[13]). In order to ensure that the
solid angles for detection of the annihilation radiation and γ-rays
were nearly the same, positrons had to be stopped close to the
source.

The way from the experimental β$^+$/(EC + β$^+$) ratios to the Q_{EC}
values is presented in Table 1.

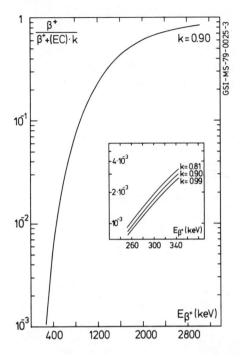

Fig. 2.   Theoretical dependence of the $\beta^+/(EC + \beta^+)$ ratio upon the
positron end-point energy for a mother nucleus with Z = 50.
Inset: Section of this dependence with its uncertainty
limits. (See text for comments).

SOME DETAILS OF THE EXPERIMENTS

The tin activities were studied at GSI using the mass-
separator[14] on-line with the UNILAC. They were produced in the
reaction between [58]Ni projectiles (5.0 MeV/u; 10^{11} ions/s) and [58]Ni
target nuclei (see Ref.[15] for the estimate of production rates for
similar experimental conditions). Mass separated samples were
collected on a tape and transported periodically to the counting
position inside a copper "annihilator". The radiation was measured
using Ge(Li) detectors having efficiencies of 18 and 15%. These
detectors were placed "face to face", about 1 cm apart from each
other. This very compact geometry of the large detectors was needed
here because of low yields of activities studied, but in the data
evaluations the effect of summing had to be carefully taken into
account. The total measuring time was 10 h for [108]Sn and 13 h for
[106]Sn. A PDP-11 computer was used for list-mode acquisition of the

Table 1. Determination of Q_{EC} values from $\beta^+/(EC + \beta^+)$

Decay	Daughter level (keV)	$\beta^+/(EC + \beta^+)$	End-point energy E_{β^+} (keV) [a]	Q_{EC} (keV) this work	Q_{EC} (keV) literature
$^{108}Sn \rightarrow ^{108}In$	669	$(351 \pm 64) \times 10^{-5}$	$357 \pm (17 + 11)$	2048 ± 28	$\gtrsim 2300$, Ref. [10]
$^{106}Sn \rightarrow ^{106}In$	1190	0.110 ± 0.031	$923 \pm (85 + 30)$	3135 ± 115	
	864	0.253 ± 0.021	$1240 \pm (40 + 40)$	3125 ± 80 [b]	
$^{104}Cd \rightarrow ^{104m}Ag$	84	$(189 \pm 50) \times 10^{-5}$	$290 \pm (19 + 7)$	1396 ± 26	1560 ± 50, Ref. [12]

a) The first contribution to the error is of purely experimental origin; the second one results from the uncertainty of the theoretical EC/β+ ratio.

b) Adopted value.

Table 2. New mass-excess data compared to predictions from mass formulae

Nuclide	M_{exp} - A (keV)	M_{calc} - M_{exp} (keV)				
		$M^{a)}$	$LZ^{a)}$	$CK^{a)}$	$JE^{a)}$	$HGT^{b)}$
^{104}In	- 76 600 ± 200	- 710	870	340	480	620
^{104}Cd	- 83 720 ± 30	- 340	- 220	- 310	- 150	620
^{105}Sn	- 73 245 ± 130	- 1690	920	70	- 280	- 120
^{106}Sn	- 77 460 ± 90	- 1170	800	30	- 30	310
^{108}Sn	- 82 050 ± 90	- 910	320	- 120	- 50	340
^{109}Te	- 67 620 ± 130	- 1460	980	- 430	- 320	- 280
^{110}Te	- 72 325 ± 100	- 720	1020	20	250	390
^{113}Xe	- 62 100 ± 130	- 1310	590	-1240	- 70	- 370
^{117}Cs	- 63 320 ± 240 c)	- 2520	- 940	-2110	- 620	- 970

a) M - Myers, LZ - Liran and Zeldes, CK - Comay and Kelson, JE - Jänecke and Eynon, Ref. 6

b) HGT - Hilf et al., Ref. 17

c) From Ref. 16

γ-γ-time coincidence events and singles multispectrum analysis.

The cadmium activity was studied at the cyclotron laboratory
of the Göttingen University. Successively, 10 sources were produced
via the ^{102}Pd (α,2n) reaction and measured using a 13% Ge(Li)
detector and a pure Ge X-ray counter (summing could be neglected).
These sources were chemically purified both before and (continuously)
during the measurements. The ion-exchange column applied during the
measurements also served as an annihilator. Coincidence events were
stored in a PDP-11 computer.

MASS EXCESS VALUES

Table 2 contains the new mass excess values from this work
and - for ^{117}Cs, the lightest cesium isotope with known mass - from
Ref.[16]. These experimental results are compared with predictions of
selected mass formulae. The best agreement is found with the
inhomogeneous-partial-difference equations of Jänecke and Eynon[6] and
the droplet-model formula in the version of Hilf et al.[17]. For a
more stringent test of mass formulae, more extensive mass data are
needed.

STRENGTH OF THE Z = 50 SHELL

Our results allow a preliminary discussion of the strength of
the Z = 50 proton shell far from stability. This strength is
approximately given by the following difference of the α-decay
energies[18]:

$$\Delta Q_\alpha (N) = Q_\alpha (^A_{52}Te_{N+1}) - Q_\alpha (^{A-2}_{50}Sn_{N+1})$$

The systematics of the ΔQ_α values is shown in Fig. 3. The points
from N = 62 to 73 are based on mass data from a compilation in Ref.[1].
For N = 57, ΔQ_α is calculated from masses of ^{108}Sn and ^{104}Cd, and
from Q_α of ^{110}Te, Ref.[3]. It is somewhat higher than ΔQ_α for N = 62.
However, because of the magnitude of the errors, we cannot claim yet
to have an indication of increase of the proton shell strength when
approaching the doubly magic ^{100}Sn. Such an increase may be expected
by analogy with the lead region[18].

The experimental ΔQ_α values are rather well reproduced by the
empirical formula of Comay and Kelson[6]. The shell-model mass
equation of Liran and Zeldes[6], equations of Jänecke and Eynon[6],
and the formula of Hilf et al.[17] are not far from the experimental
points but fail to reproduce the bump observed at N ≈ 65, which might
be related to the N = 64 subshell closure. The droplet-model formula
of Myers[6] gives ΔQ_α values which are too high and almost independent
of the neutron number.

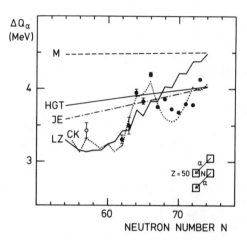

Fig. 3.   Systematics of the experimental $\Delta Q_\alpha(N)$ values in comparison
with selected mass-formulae predictions (M – Myers, LZ –
Liran and Zeldes, CK – Comay and Kelson, JE – Jänecke and
Eynon, Ref.[6]; HGT – Hilf et al.[17]).

Finally, the new mass values allow to roughly estimate the mass
excess of ^{114}Cs and to correspondingly localize the proton drip line
for cesium isotopes, which is described together with more details
on this work in a forthcoming publication[19].

REFERENCES

1. A. H. Wapstra and K. Bos, Atomic Data and Nucl. Data Tables 19
       (1977) 177.
2. H. Nutley and J. B. Gerhart, Phys. Rev. 120 (1960) 1815.
3. D. Schardt, R. Kirchner, O. Klepper, W. Reisdorf, E. Roeckl,
       P. Tidemand-Petersson, G. T. Ewan, E. Hagberg, B. Jonson,
       S. Mattson and G. Nyman, Nucl. Phys. (in print).
4. D. D. Bogdanov, V. A. Karnaukhov and L. A. Petrov,
       Yad. Phys. 17 (1973) 457.

5. H. Huang, B. P. Pathak and J. K. P. Lee,
   Can. J. Phys. 56 (1978) 936.
6. The 1975 Mass Predictions, S. Maripuu, Ed., Atomic Data and
   Nucl. Data Tables 17 (1976).
7. B. S. Dzehlepov, L. N. Zyrianova and Yu. P. Suslov,
   Beta processes, Nauka, Leningrad 1972.
8. W. Bambynek, H. Behrens, M. H. Chen, B. Crasemann,
   M. L. Fitzpatrick, K. W. D. Ledingham, H. Genz, M. Mutterer
   and R. L. Intemann, Rev. Mod. Phys. 49 (1977) 77.
9. M. Campbell, K. W. D. Ledingham and A. D. Baillie,
   Nucl. Phys. A233 (1977) 413.
10. B. J. Varley, G. S. Foote and W. Gelletly, J. Phys. G:
    Nucl. Phys. 3 (1977) 1099.
11. B. J. Varley, G. S. Foote, C. Garrett and W. Gelletly,
    J. Phys. G: Nucl. Phys. 4 (1978) 1643.
12. F. Münnich, A. Kjelberg and D. J. Hnatowich, Nucl. Phys.
    A158 (1970) 183.
13. J. Kantele and M. Valkonen, Nucl. Instr. 112 (1973) 501.
14. K. H. Burkard, W. Dumanski, R. Kirchner, O. Klepper and
    E. Roeckl, Nucl. Instr. Meth. 139 (1976) 275.
15. R. Kirchner, O. Klepper, G. Nyman, W. Reisdorf, E. Roeckl,
    D. Schardt, N. Kaffrell, P. Peuser and K. Schneeweiss,
    Phys. Lett. 70B (1977) 150.
16. M. Epherre, G. Audi, C. Thibault, R. Klapisch, G. Huber,
    F. Touchard and H. Wollnik, Phys. Rev. C19 (1979) 1504.
17. E. R. Hilf, H. v. Groote and K. Takahashi, in: Proc. 3rd
    Intern. Conf. on nuclei far from stability, Cargèse, CERN
    76-13 (1976) p. 142.
18. K. H. Schmidt, W. Faust, G. Münzenberg, H.-C. Clerk, W. Lang,
    K. Pielenz, D. Vermeulen, H. Wohlfarth, H. Ewald and
    K. Güttner, Nucl. Phys. A318 (1979) 253.
19. A. Płochocki, G. M. Gowdy, R. Kirchner, O. Klepper, W. Reisdorf,
    E. Roeckl, P. Tidemand-Petersson, J. Żylicz, U. J. Schrewe,
    R. Kantus, R.-D. von Dincklage and W. D. Schmidt-Ott,
    to be published.

ALPHA-DECAY SYSTEMATICS FOR ELEMENTS WITH 50 < Z < 83

K. S. Toth

Oak Ridge National Laboratory*

Oak Ridge, Tennessee 37830, U.S.A.

INTRODUCTION

From observing the slope of the mass-defect curve, one notes that most isotopes whose mass numbers are $\gtrsim$ 140 are unstable toward α-emission. However, with the exception of naturally occurring ^{147}Sm, α-decay was not observed for elements below bismuth until 30 years ago. This was due to the fact that the rate for α-decay is a very sensitive exponential function of the decay energy. The energy available for decay increases rapidly with mass, so that in the region above lead α-decay becomes a dominant decay mode. It was also known that, if one could produce nuclides sufficiently far to the neutron-deficient side of the β-stability line, then with nuclei Z < 83 would undergo α-particle emission. This was shown to be true experimentally in 1949 when Thompson, Ghiorso, Rasmussen and Seaborg[1] reported the discovery of α-radioactivity in proton-rich isotopes of gold, mercury and the rare earths. Since that time, with the availability of new accelerators, the number of known α-active nuclides below bismuth has steadily increased.

In this paper, we will review recent data reported on α-emitting isotopes in this mass region, compare α-decay energies with predictions of mass formulae, and discuss α-decay rates of even-even nuclei.

---

*Operated by Union Carbide Corporation under Contract No. W-7405-eng-26 with the U.S. Department of Energy.

NEW ALPHA-EMITTERS

The first big impetus for the study of α-decaying nuclides in
the medium-weight mass region came in the early 1960's, when several
groups at the Berkeley HILAC began using heavy-ion beams and the
helium gas-jet technique. Since then, experimentalists at other
laboratories have taken up these studies and a large number of α-
emitters with 50 < Z < 83 are now known. The most recent compilation summarizing information for isotopes in this general mass
region was published in 1975 by Gauvin et al.[2] Interest in the
field has not abated. In addition to improving the quality of data,
particularly with regard to α-decay rates, investigators[3-10] have
identified more than 30 new α-emitters during the past few years.
Table I summarizes their half-lives and α-decay energies.

Most of the isotopes listed in Table I were produced in bombardments with nickel and krypton ions accelerated at UNILAC:
1)$^{110-112}$I, ^{112,113}Xe, ^{114}Cs(or ^{114}Ba), Roeckl et al.[3] and Kirchner
et al.,[4] with the use of an isotope separator; 2) ^{156}Hf, ^{160}W,
$^{157-161}$Ta, $^{161-164}$Re, Hofmann et al.,[6] with the use of the velocity
filter SHIP; and 3) $^{166-168}$Re, ^{169,170}Ir, Schrewe et al.,[7] with the
use of a gas-jet system. At the accelerator ALICE, Cabot et al.[8]
utilized copper beams and the gas-jet technique to produce ^{168,169}Re,
$^{165-168}$Os, and $^{168-170}$Ir. The isotope separator facility at Oak
Ridge, UNISOR, was used[10] to identify $^{184-187}$Tl. Finally, the
ISOLDE collaboration in their investigations of rare earth and
mercury isotopes reported two new α-emitters, ^{158}Yb (Ref. 5) and
^{188}Hg (Ref. 9)

The experimental situation extant in the region from neodymium
to lead is shown graphically in Fig. 1. The figure shows isotopes
and their α-decay energies as a function of neutron and proton
numbers. For clarity, even-Z nuclides are indicated by bars while
odd-Z nuclei are represented by dots. It is seen that α-decay
energies increase both with increasing Z (and A) and with decreasing
N (as one gets further away from the valley of stability).

EXPERIMENTAL AND PREDICTED Q_α VALUES

Because the characterization of an α-emitter involves combining
a half-life with a specific α-particle group, α-decay provides us
with a convenient means of discovering new isotopes. Their identification opens the way for further, more extensive studies. In
addition, α-decay energies in many instances can be used to determine energy differences between the parent and daughter ground
states. Such measurements have therefore been used not only to
obtain estimates of masses for nuclei far from stability but also
for comparisons with mass formulae predictions.

TABLE I

New α-Emitting Isotopes

Nuclide	$T_{1/2}$ (sec)	$E\alpha$ (MeV)	References	
^{110}I	0.69 (4)	3.424 (15)	3	(4)
^{111}I	2.5 (2)	3.150 (30)	3	(4)
^{112}I	3.42 (11)	2.866 (50)	3	
^{112}Xe	2.8 (2)	3.185 (30)	3	
^{113}Xe	2.8 (2)	2.990 (30)	3	
^{114}Cs (or ^{114}Ba)	0.57 (2)	3.226 (30)	3	
^{158}Yb	99 (12)	4.069 (10)	5	
^{156}Hf	0.025 (4)	5.878 (10)	6	
^{157}Ta	0.0053 (18)	6.219 (10)	6	
^{158}Ta	0.0368 (16)	6.051 (6)	6	
^{159}Ta	0.57 (18)	5.601 (6)	6	
^{160}Ta		5.413 (5)	6	
^{161}Ta		5.148 (5)	6	
^{160}W		5.920 (10)	6	
^{161}Re	0.010 $\left(^{+15}_{-5}\right)$	6.279 (10)	6	
^{162}Re	0.10 (3)	6.119 (6)	6	
^{163}Re	0.26 (4)	5.918 (6)	6	
^{164}Re	0.9 (7)	5.778 (10)	6	
^{166}Re	2.2 (4)	5.495 (10)	7	
^{167}Re	2.0 (3)	5.33 (1)	7	
^{168}Re	2.9 (3)	5.14 (1)	7	
	5.5 (5)	5.26 (1)	8	
^{169}Re		5.05 (1)	8	
^{165}Os		6.20 (2)	8	
^{166}Os	0.3 (1)	6.00 (2)	8	
^{167}Os	0.65 (15)	5.84 (1)	8	
^{168}Os	2.0 (4)	5.66 (1)	8	(7)
^{168}Ir		6.22 (2)	8	
^{169}Ir	0.4 (1)	6.11 (1)	8	(7)
^{170}Ir	1.1 (2)	6.01 (1)	8	(7)
^{188}Hg		4.61 (2)	9	
^{184}Tl	11 (1)	6.162 (5)	10	
	11 (1)	5.988 (5)		
^{185}Tl	1.7 (2)	5.975 (5)	10	
^{186}Tl	~25	~5.76	10	
^{187}Tl	18 (3)	5.51 (2)	10	

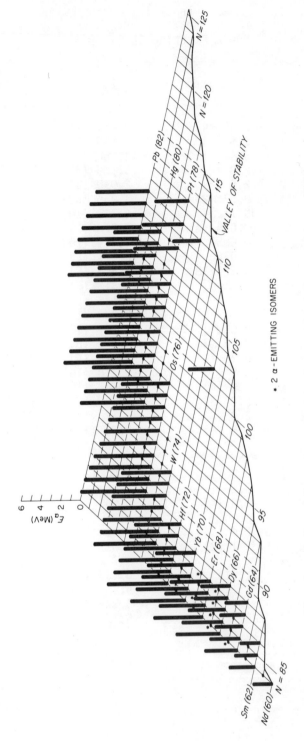

Fig. 1. Known α-Emitters in the Region from Neodymium to Lead.

We have compared experimental decay energies for α-emitters with atomic numbers between 50 and 83 with values taken from four sets of mass predictions, i.e., the shell-model formula of Liran and Zeldes,[11] the formula of Myers and Swiatecki[12] which is based on the liquid-drop model with shell corrections, the subsequent mass formula developed by Myers[13] which uses the droplet model, and finally, the Garvey-Kelson mass relations as updated by Jänecke.[14]

A detailed comparison cannot be shown in this short presentation. Instead, we have summarized in Table II the average difference between the experimental Q-values and the four sets of predictions for all 145 isotopes considered. In addition, average deviations were determined for isotopes of each element (or group of elements). The largest and smallest of these deviations are also listed in the table. The Liran and Zeldes formula[11] agrees best with data, an average difference of 152 keV, compared with 252, 373, and 630 keV for the predictions of Refs. 12, 13, and 14, respectively. Their formula also shows the least spread in differences ranging from 312 keV for isotopes in the tellurium region to 75 keV for the iridium nuclei. It is interesting to note that the droplet model[13] yields a larger deviation than the conventional liquid-drop model.[12] However, if the lead, thallium, and mercury nuclides are omitted, then the newer predictions are slightly better, a deviation of 226 keV <u>versus</u> 252 keV for the older liquid-drop model. In Table II the updated Garvey-Kelson predictions have also been broken up into three groups of elements. One sees that for Ref. 14 the deviation is greatest, 872 keV, for the middle group of elements (ytterbium → gold). The discrepancies for the remaining two groups, 137 keV (tellurium → thulium), and 300 keV (mercury → lead) are comparable with those deduced from the other three sets of predictions.

ALPHA-DECAY RATES

In α-decay, half-lives for transitions between ground states of doubly-even nuclei are taken to represent unhindered decays. The reduced widths of these s-wave transitions are considered to be standard. A rather regular behavior as a function of both neutron and atomic number is observed for s-wave α-decays. Their reduced widths are largest for nuclei two or four particles beyond a closed shell (with sharp minima occurring at the closed shell), followed by a decrease as one approaches the next closure. These trends can be understood in terms of single-particle models which have shown that the extremely sharp break at $N = 126$ is essentially a shell structure effect.

Figure 2 shows s-wave reduced widths for α-emitting nuclei with Z from 52 to 88 plotted as a function of N. In calculating

TABLE II

COMPARISON OF EXPERIMENTAL AND PREDICTED Q_α's

Mass Formula		$Q_{exp} - Q_{pred}$ (keV)
Liran and Zeldes [145]*	152	312 (Te region) 75 (Iridium)
Myers and Swiatecki [145]*	252	479 (Gadolinium) 39 (Osmium)
Myers (Droplet Model) [145]*	373	1503 (Lead) 57 (Erbium)
Myers (Without Pb, Tl, Hg) [121]*	226	431 (Gold) 57 (Erbium)
Garvey-Kelson (Updated) [145]*	630	1702 (Tantalum) 67 (Samarium
Garvey-Kelson (Te → Tm) [40]*	137	269 (Thulium) 67 (Samarium)
Garvey-Kelson (Yb → Au) [81]*	972	1702 (Tantalum) 500 (Gold)
Garvey-Kelson (Hg → Pb) [24]*	300	365 (Mercury) 156 (Thallium)

*Number of isotopes included.

these widths we have utilized Rasmussen's formalism,[15] wherein the width, δ^2, is defined by the equation: $\lambda = \delta^2 P/h$, where λ is the decay constant, h is Planck's constant, and P is the penetrability factor for the α-particle to tunnel through a barrier. One sees in Fig. 2 that, with the exception of the lead isotopes (to be discussed below in more detail), the trends mentioned above do manifest themselves. Following the sharp drop at N = 126, the widths increase as the neutron number decreases with a maximum at N = 86 due to the influence of the N = 82 closed shell. There are only two points presently available in the tin-tellurium region so that no general pattern can be discerned. However, the ^{108}Te and ^{112}Xe widths are not inconsistent with the δ^2 values for N > 82 nuclei.

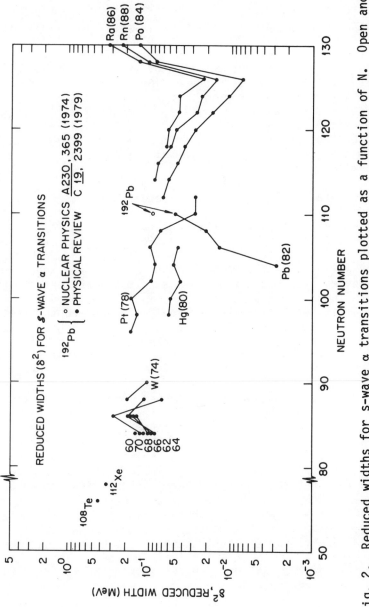

Fig. 2. Reduced widths for s-wave α transitions plotted as a function of N. Open and closed points for 192Pb are deduced from Ref. 18 and the present study, respectively.

Proton shell effects can also be noted; e.g., the widths decrease from radium to radon to polonium as Z = 82 is approached. Another consequence of systematic α-decay-rate studies, the result of recent investigations in the rare earth region, has been mounting evidence for a subshell closure at Z = 64 where the $g_{7/2}$ and $d_{5/2}$ proton orbitals are filled. The subshell was first proposed when a discontinuity in the progression of α-decay energies for N = 84 nuclides was noted at Z = 64. Macfarlane et al.[16] made calculations using a BCS treatment for the proton system of 82-neutron nuclei and were able to reproduce a discontinuity in theoretical binding energies at that atomic number. In addition, they calculated α-decay transition probabilities for the N = 84 even-A nuclei. The theoretical reduced widths indicated a significant dip at Z = 64. Contrastingly, widths determined[16] from then available data showed a general constancy in value except for a dramatic reduction for ^{150}Dy, i.e., at Z = 66.

Newer data, however, lead to a result which agrees with theory, i.e., the minimum is at Z = 64 (see Fig. 2). This point is illustrated more fully in Fig. 3, where we have plotted reduced widths for N = 84 even-even nuclei as a function of Z. Included in the figure are the calculations of Macfarlane et al.[16] Both the data and the calculations indicate a minimum at ^{150}Dy. The biggest difference between the earlier[16] and the newer sets of experimental data has to do with the ^{150}Dy α-branch. The new value of 0.36 ± 0.03, determined after investigating the nuclide's electron-capture decay scheme,[17] is a factor of two greater than the branch of 0.18 ± 0.02 (deduced from gross γ-ray counting) reported earlier. The result is a significantly larger reduced width which eliminates the dip at Z = 66.

The α-decay rates for the even-even lead isotopes, from ^{186}Pb to ^{192}Pb, have been reported[18] not to follow the pattern described above. The reduced width of ^{192}Pb is unexpectedly large. Also, the widths increase from N = 104 (^{186}Pb) to N = 110 (^{192}Pb) by a factor of about 30. The expectation is that the values should decrease as one approaches N = 126. In Ref. 18, the E.C./β$^+$ strengths were determined from K x-ray intensities. Such determinations are subject to a number of corrections. A more precise method involves a known decay scheme. With the use of the UNISOR on-line separator facility, we undertook the investigation of the E.C./β$^+$ decay properties of these lead isotopes, our purpose being to determine new values for their α-decay branches. At this time, the ^{192}Pb study is complete, while some of the data for ^{190}Pb are still being analyzed.

The open and closed points for ^{192}Pb in Fig. 2 represent the widths deduced from the data of Ref. 18 and the present study, respectively. Our value is less, by a factor of two, due primarily

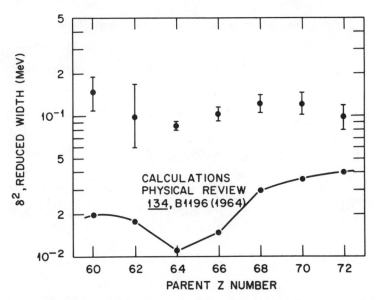

Fig. 3. Reduced widths for N = 84 even-A α-emitters. Calculations are taken from Ref. 16.

to the fact that the half-life of ^{192}Pb is 3.5 and not 2.2 min (see Table III). Nevertheless, the ^{192}Pb reduced width is still too large; it should be smaller by another factor of three to place it below the ^{188}Pt value (calculated from data[19] recently acquired at UNISOR).

In Table III we have also listed our ^{190}Pb results and compared them with the earlier data[18] for the same nuclide. Our α-decay branch, as shown in the table, is a preliminary number. Note, however, that it is about 5 times greater than the value published in Ref. 18. The resultant reduced width is 0.098 MeV, which places the ^{190}Pb point in Fig. 2 above, rather than below, the curve joining the platinum isotopes.

In contrast to Ref. 18, our results indicate that the lead reduced widths may very well follow the pattern with neutron number observed for other elements. To see if this is indeed so, we plan to remeasure the ^{190}Pb branching ratio and to extend our study to include ^{188}Pb.

TABLE III

α-Decay Properties of ^{192}Pb and ^{190}Pb

		Present Study	P. Hornshoj et al.
^{192}Pb	E_α(keV)	5112 (5)	5110
	Half-life (min)	3.5 (1)	2.2
	α Branch	5.7 x 10^{-5} (10)	6.9 x 10^{-5}
	δ^2(MeV)	0.050 (12)	0.094
^{190}Pb	E_α(keV)	5577 (5)	5590
	Half-life (min)	1.2 (1)	1.1
	α Branch	1.0 x 10^{-2} (0.4)[a]	2.1 x 10^{-3}
	δ^2 (MeV)	0.098 (39)[a]	0.021

[a]Preliminary value.

REFERENCES:

1. S.G. Thompson, A. Ghiorso, J.O. Rasmussen and G.T. Seaborg, Phys. Rev. 76:1406 (1949).
2. H. Gauvin, Y. Le Beyec, J. Livett and J.L. Reyss, Ann. Phys. (Paris) 9:241 (1975).
3. E. Roeckl et al., Phys. Lett. 78B:393 (1978).
4. R. Kirchner et al., Phys. Lett. 70B:150 (1977).
5. E. Hagberg et al., Nucl. Phys. A293:1 (1977).
6. S. Hofmann et al., Z. Physik A291:53 (1979).
7. U.J. Schrewe et al., Z. Physik A288:189 (1979).
8. C. Cabot et al., Z. Physik A287:71 (1978).
9. E. Hagberg et al., Nucl. Phys. A318:29 (1979).
10. K.S. Toth et al., Phys. Lett. 63B:150 (1976).
11. S. Liran and N. Zeldes, At. Data Nucl. Data Tables 17:431 (1976).
12. W.D. Myers and W.J. Swiatecki, Nucl. Phys. 81:1 (1966); LBL Report No. UCRL-11980 (unpublished).
13. W.D. Myers, "Droplet Model of Atomic Nuclei", Plenum Pub. Co., New York (1977).
14. J. Jänecke, At. Data Nucl. Data Tables 17:455 (1976).
15. J.O. Rasmussen, Phys. Rev. 113:1593 (1959).
16. R.D. Macfarlane, J.O. Rasmussen and M. Rho, Phys. Rev. 134:B1196 (1964).
17. K.S. Toth, C.R. Bingham and W.-D. Schmidt-Ott, Phys. Rev. C10:1550 (1974).
18. P. Hornshøj et al., Nucl. Phys. A230:365 (1974).
19. Y.A. Ellis, K.S. Toth and H.K. Carter, Phys. Rev. C18:2713 (1978).

MASSES OF NEW ISOTOPES IN THE fp SHELL

C. N. Davids

Physics Division
Argonne National Laboratory
Argonne, IL   60439

INTRODUCTION

During the past few years there has been a active program at the Argonne FN tandem accelerator aimed at the study of new isotopes far from beta stability. The particular region of the periodic table under investigation has been for 50 < A < 70, roughly the 1f-2p shell, on both sides of the valley of beta stability. Besides establishing the existence of previously unknown nuclides, the original scope of the program was to obtain information on the spectroscopic properties of parent and daughter for comparison with various nuclear models, and to measure the beta decay energies to compare with various predictions. In addition, calculations of explosive carbon burning during a supernova explosion rely on predicted masses of unknown neutron-rich nuclei in the iron region. Measurements of the masses of experimentally-accessible nuclei can help to decide which predictions are the most reliable to use in the nucleosynthesis calculations.

A total of 4 new neutron-rich isotopes have been studied, all made by heavy-ion-induced reactions on ^{48}Ca targets. These include ^{53}Ti (Ref. 1), ^{57}Cr (Ref. 2), ^{59}Mn (Ref. 3), and ^{60}Mn (Ref. 4). The bombardment of ^{58}Ni by various heavy-ion beams has resulted in the discovery of proton-rich ^{67}As (Ref. 5) and permitted extensive measurements of the superallowed decays of ^{62}Ga, ^{66}As, and ^{70}Br. These latter studies were aimed at extracting accurate ft values for Fermi decays in heavy nuclei.

EXPERIMENTAL METHODS

The isotopes with half-lives greater than 1 s were studied
using the conventional techniques of γ and β spectroscopy. Large
Ge(Li) detectors were used for γ singles and γ-γ coincidence
measurements, and in conjunction with a plastic scintillator, for
β-γ coincidence studies.

In order to perform sensitive studies of short-lived radioactivities it is necessary to transport the active product rapdily
from the production region to a low-background counting station.
This has been done by either a "rabbit" system to transport a solid
target, or using a helium-jet recoil transfer system to transport
only radioactive nuclei recoiling from the target.

The rabbit system consists of a chamber holding a carrousel
upon which are mounted 8 rabbit holders fabricated from rectangular
waveguide. The targets are mounted on lightweight rabbits made of
delrin plastic. The transfer tube is mode of the same waveguide,
and the rabbits are propelled out of the bombardment chamber by a
burst of helium. At the counting station 3 meters distant the
rabbit is stopped by a bumper made of PVC tubing. Typical transit
time is ~0.4 s. After a suitable counting interval the rabbit is
propelled back to the bombardment chamber, and a new target is
rotated into position. This allows the background activities on
each target to decay in between bombardments, greatly reducing the
count rate due to unwanted activities. During transfer and counting
periods, the bombardment chamber is isolated from the accelerator
vacuum system by a solenoid valve. Figure 1 shows a front view of
the rabbit chamber.[6]

The helium-jet system involves transferring recoils from a
target through a capillary tube in a stream of helium gas. The
recoils, after being thermalized in the gas, are thought to attach
themselves to giant (A ~ 10^7) clusters. These clusters are
presumably formed by the action of the beam on impurities which are
added to the helium stream (in the present case, water vapor is used).
At the end of the tube the products are collected on a paper tape,
while the helium is pumped away. After a suitable collection period,
the tape is moved to the detector position for counting. Since a
new area of the tape is used for collection, no buildup of long-
lived activities takes place.

For half lives less than 1 s, it is not practical to utilize
the transfer methods, and the detectors are placed next to the
target chamber, which has thin mylar vacuum windows. This configuration is undesirable in that the detectors are thus exposed to
intense radiation from the beam during the bombardment.

Fig. 1.   Front view of the multiple-rabbit chamber with cover
          removed. The numbered items are: 1) rabbit holders.
          One of the holders is removed (bottom position).
          2) Rabbit transfer tube. 3) Carrousel disk. 4) Wire
          for retaining rabbits in their holders. 5) Position-
          sensing assembly. 6) Helium inlet line.

     A crystal-controlled programmable sequence timer is used to
time the various events such as bombardment, transfer, and data
acquisition. At present a microprocessor-based system is under
development to take over this function.

     For the study of the Fermi decays along the N = Z line, a
ΔE-E plastic scintillator telescope was used to detect high-energy
positrons. Half-life and decay energy measurements were obtained
simultaneously from data taken in a multispectrum-scaling mode.

     The experiments involving heavy-ion bombardments produce a
large number of known radioactivities whose β-γ coincidence spectra
can be used to calibrate the energy scale of the β detector.
Choosing one as a standard spectrum, it is compressed or stretched
to fit as many known spectra as possible, and the linearity of the
procedure is checked by plotting the stretch factors so obtained as
a function of β endpoint energy. Stretch factors are obtained for
the unknown β spectra, and, after adding the appropriate excitation
energies, total decay energies and therefore mass excesses are

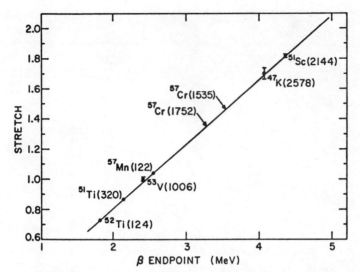

Fig. 2. Endpoint energy calibration for the known β spectra
         obtained in the β–γ coincidence experiment. The straight
         line is a linear least-squares fit to the data points.
         Also shown are the positions of the 2 ^{57}Cr endpoints.

extracted with typical uncertainties of ±100 keV. Allowance can
also be made in the computer fits for inner β branches as well as
coincident γ rays.

Using the empirical shapes from the detector system corrects
for systematic effects such as detector resolution, backscattering,
energy loss in intervening materials, and edge losses, which are the
same to first order for all β groups. The shape method also uses
nearly all of the data, instead of just the high-energy portion.
An example of the system linearity is shown in Fig. 2, taken from
the ^{57}Cr experiment.

RESULTS FOR NEUTRON-RICH NUCLEI IN THE f-p SHELL

Using 1.1 mg/cm^2 enriched ^{48}Ca metal foil targets, the
nuclides ^{53}Ti, ^{57}Cr, ^{59}Mn, and ^{60}Mn were produced using the
^{48}Ca(^{7}Li,pn)^{57}Ti, ^{48}Ca(^{11}B,pn)^{57}Cr, ^{48}Ca(^{13}C,pn)^{59}Mn and
^{48}Ca(^{18}O,αpn)^{60}Mn reactions. The measured half lives and mass
excesses for these nuclides are shown in Table I, along with mass
excess predictions. Also included are data on ^{55}V from the
Brookhaven group[7] and J^π values measured for the ground states.

An example of the spectroscopic information obtained from
these studies is the decay scheme for ^{57}Cr (Ref. 2) shown in Fig. 3.

TABLE I. Results for neutron-rich isotopes in the 1f-2p shell.

Nuclide	Ground State $J^{\pi a}$	Half-Life[a] (seconds)	$(M-A)^a_{exp}$ (MeV)	(M-A) predicted (MeV)			
				MSMME[b]	JGK[c]	LZ[d]	SH[e]
^{53}Ti	$(3/2)^-$	32.7 ± 0.9	-46.84 ± 0.10	-46.50	-46.53	-46.33	-46.4
^{55}V	$(3/2, 5/2, 7/2)^{-f}$	6.54 ± 0.15^f	-49.15 ± 0.10^f	-49.15	-49.41	-49.37	-49.4
^{57}Cr	$3/2^-$	21.1 ± 1.0	-52.39 ± 0.10	$-52.12+E_x^g$	-52.67	-52.62	-52.3
^{59}Mn	$(5/2)^-$	4.61 ± 0.15	-55.5 ± 0.11 -55.49 ± 0.04^h	-55.35	-55.94	-56.08	-55.6
^{60}Mn	3^+	1.79 ± 0.10	-52.89 ± 0.10	$-$	-53.01	-53.37	-52.2

[a] Present work unless otherwise noted.
[b] Reference 12.
[c] J. Jänecke, At. Data Nucl. Data Tables 17, 455 (1976).
[d] S. Liran and N. Zeldes, At. Data Nucl. Data Tables 17, 431 (1976).
[e] P. A. Seeger and W. M. Howard, At. Data Nucl. Data Tables 17, 428 (1976).
[f] Reference 7.
[g] The modified shell model mass equation (MSMME) used in Ref. 12 predicts the mass excess of the lowest J = 5/2$^-$ state, not the 3/2$^-$ ground state. A linear extrapolation from ^{53}Cr and ^{55}Cr puts the J = 5/2$^-$ state at E_x = 0.036 MeV.
[h] E. Kashy, W. Benenson, D. Mueller, H. Nann, and L. Robinson, Phys. Rev. C 14, 1773 (1976).

A large number of spin and parity restrictions have been made in ^{57}Mn, and the correspondence with levels in the daughter nucleus observed by charged-particle reactions is excellent. The ground-state spin and parity of ^{57}Cr turns out to be 3/2$^-$, the same as for all other known N = 33 odd-A nuclides. The simplest shell model would predict J^{π} = 5/2$^-$, but at N = 33 the binding energy gained by pairing 2 1f$_{5/2}$ neutrons is greater than that obtained from filling the 2p$_{3/2}$ orbital and allowing the 33rd neutron to reside in the 1f$_{5/2}$ orbital. Figure 4 shows the β-delayed spectrum from ^{48}Ca + ^{11}B and Fig. 5 shows the β spectra used to obtain the ^{57}Cr mass excess.

In general the production cross sections for the isotopes of interest range from 0.6 mb to ∿8 mb, as predicted by the evaporation code Alice,[8] making them always less than 1% of the total reaction cross section. The sensitivity of the detection system is such that isotopes with production cross sections in the few hundred μb region are still amenable to investigation.

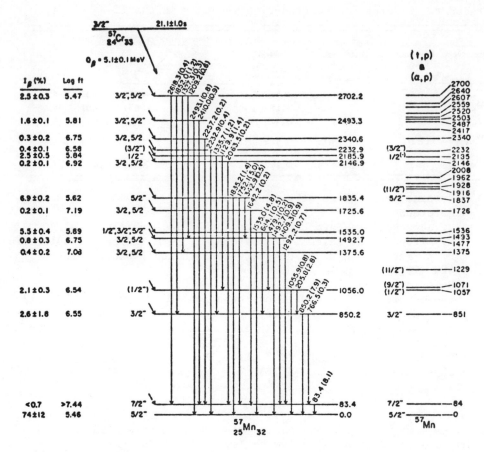

Fig. 3.   β-decay scheme for ^{57}Cr. Also shown are levels in ^{57}Mn
from the ^{54}Cr(α,p)^{57}Mn and ^{55}Mn(t,p)^{57}Mn reactions.

RESULTS FOR PROTON-RICH NUCLIDES ON THE N=Z LINE

The nuclides ^{62}Ga, (Ref. 9) ^{66}As, and ^{70}Br have been produced
via the ^{58}Ni(^{6}Li,2n)^{62}Ga, ^{58}Ni(^{10}B,2n)^{66}As, and ^{58}Ni(^{14}N,2n)^{70}Br
reactions. A decay curve for high-energy positrons from ^{62}Ga is
shown in Fig. 6. The positron spectrum obtained in the same
experiment is shown in Fig. 7. The energy calibration was
accomplished by measuring the decays of ^{46}V, ^{50}Mn, ^{54}Co, ^{58}Cu,
and ^{28}P, with Q_{EC} values ranging from 7.05 to 12.55 MeV. These
isotopes were produced by the (p,n) reaction on appropriate targets.

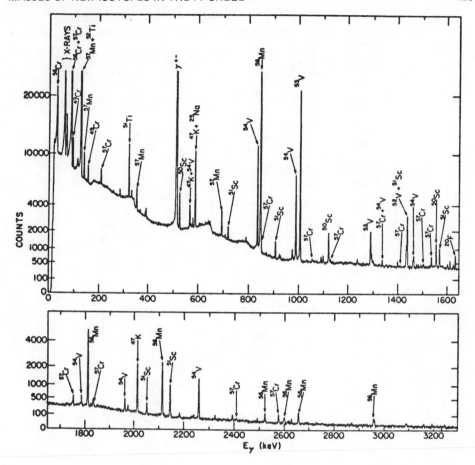

Fig. 4. Singles γ-ray spectrum from ^{48}Ca + ^{11}B during the first
         40 s following bombardment. The vertical scale is in
         square-root format.

        Preliminary values for the half-lives and decay energy Q_{EC} of
^{62}Ga, ^{66}As, and ^{70}Br are given in Table II. The resulting ft
values have been corrected for charge-dependent and radiative
effects of the order of 1— 2%, and are listed in Table II. The
prediction of the conserved vector current (CVC) hypothesis is that
the ft values for all Fermi decays will be equal. Careful evaluations of the well-studied Fermi decays between ^{14}O and ^{54}Co by
Towner and Hardy[10] and by Raman, Walkiewicz and Behrens[11] yield an
yield an average corrected ft value of 3085.5 ± 3 sec. Higher
precision in the Q_{EC} measurements is required before the present
experiments will be sensitive to any systematic deviations from
the predicted value.

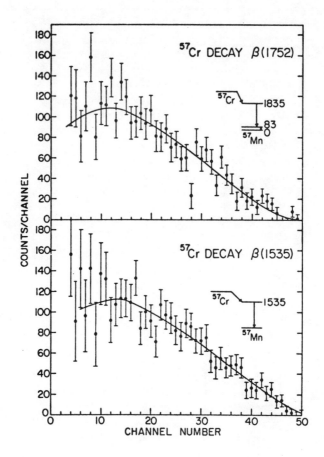

Fig. 5.   Background-corrected ^{57}Cr β spectra feeding the 1535.0-
and 1835.4-keV states in ^{57}Mn.   Solid line indicates
fit of standard shape to each spectrum.

RESULTS FOR PROTON-RICH As and Ge NUCLIDES

In addition to ^{66}As, the β+ decays of the light As isotopes
^{67}As and ^{68}As have been studied.   They were produced via the
^{58}Ni(^{14}N,αn)^{67}As and ^{58}Ni(^{12}C,pn)^{68}As reactions. ^{67}As was
previously unknown, as were the states of the daughter nucleus ^{67}Ge.
An extensive spectroscopic study of ^{67}Ge has been completed, with
early results reported in Ref. 13.   The mass of ^{67}Ge was measured
using the threshold of the ^{64}Zn(α,nγ)^{67}Ge reaction, and differs by

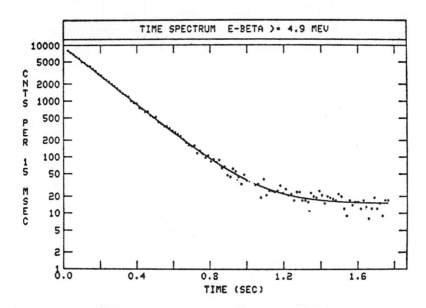

Fig. 6. Time behavior of high-energy positrons from the decay of ^{62}Ga. The solid curve is a fit to the data using an exponential decay plus a constant background.

TABLE II. Preliminary half-lives and Q_{EC} values for odd-odd N = Z nuclides in the 1f-2p shell.

Nuclide	Half-life[a] (ms)	Q_{EC} (keV)	$\mathcal{F}t$ (sec)
^{62}Ga	116.34 ± 0.35	9171 ± 26	3081.2 ± 47.1
	115.95 ± 0.30[b]		
	116.4 ± 1.5[c]		
^{66}As	96.37 ± 0.46	9550 ± 50	3062 ± 90
	95.78 ± 0.39[b]		
^{70}Br	80.2 ± 0.8[b]	9970 ±170	3118 ± 295

[a] Present work, unless otherwise noted.
[b] D. E. Alburger, Phys. Rev. C 18, 1875 (1978).
[c] R. Chiba et al., Phys. Rev. C 17, 2219 (1978).

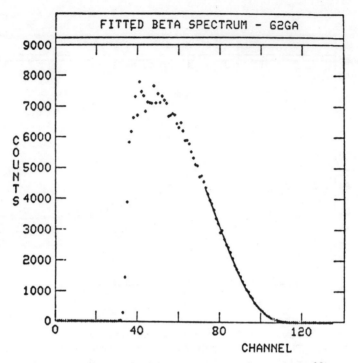

Fig. 7.   Background-corrected positron spectrum of ^{62}Ga decay. The
          solid curve is a fit obtained by stretching a standard
          shape (in this case ^{54}Co) to fit the data.

more than 200 keV from the previous value. Our work has confirmed
a spin of $1/2^-$ for the ^{67}Ge ground state. No mass has been reported
for ^{68}As, and our work has established its ground-state spin to be
$3^{(+)}$. Table III summarizes the results obtained for these nuclides.

DISCUSSION

     As can be seen from Table 1, the modified shell model mass
equation[12] does well in predicting the masses of the T_z = 9/2 nuclei
in the f-p shell. Most of the other formulations seem to over-
estimate the binding energy of these nuclides. A possible
exception is for ^{53}Ti, where all predictions yield a less tightly
bound nucleus than is observed. An independent measurement of the
^{53}Ti mass, possibly via the ^{48}Ca(^{18}O,^{13}C)^{53}Ti reaction, would help
to clear up this point.

TABLE III.   Results for proton-rich As and Ge isotopes.

Nuclide	Ground state $J^{\pi a}$	Half-Life[a]	$(M-A)_{exp}$ [a] (MeV)	(M-A) predicted (MeV)			
				MSMME[b]	JGC[c]	LZ[d]	SH[e]
^{67}Ge	$1/2^-$	1140 ± 18[f]	-62.666 ± 0.012	-62.70	-62.76	-62.71	-61.9
^{66}As	0^+	0.0964 ± 0.0005	-52.07 ± 0.05	--	--	-51.80	-50.9
^{67}As	$(5/2^-)$	42.4 ± 1.2	-56.66 ± 0.10	--	-56.54	-56.33	-56.2
^{68}As	$3^{(+)}$	151.5 ± 0.9	-58.9 ± 0.1	--	-58.74	-58.95	-58.0

[a] Present work unless otherwise noted.
[b] Reference 12.
[c] J. Jänecke, At. Data Nucl. Data Tables 17, 455 (1976).
[d] S. Liran and N. Zeldes, At. Data Nucl. Data Tables 17, 431 (1976).
[e] P. A. Seeger and W. M. Howard, At. Data Nucl. Data Tables 17, 428 (1976).
[f] Table of Isotopes, 7th Edition, Ed. C. M. Lederer et al., John Wiley and Sons, 1978, p. 214.

Large-basis shell-model calculations in the 1f-2p shell have had reasonable success in explaining low-lying levels in this group of nuclei. It is hoped that new data being provided will stimulate further calculation, especially on the nuclides lying off the stability line.

Precise measurements of the Q_{EC} values for the decay of ^{62}Ga, ^{66}As, and ^{70}Br are of interest because of the possibility of observing systematic deviations from the CVC predictions for the decay rate due to charge-dependent effects. The ability to measure the deviations caused by charge-dependent effects, which have a Z^2 dependence, would allow calculations of these effects to be checked. However, the outlook for significantly increased precision is clouded by the fact that the statistical rate function f has an E^5 dependence, making a reduction of the uncertainty in the ft value very difficult.

ACKNOWLEDGMENTS

This research was performed under the auspices of the U. S. Department of Energy. The work was done in collaboration with C. A. Gagliardi, D. F. Geesaman, W. Kutschera, M. J. Murphy, E. B. Norman, R. C. Pardo, L. A. Parks, and S. L. Tabor.

REFERENCES

1. L. A. Parks, C. N. Davids, and R. C. Pardo, Phys. Rev. C 15,
    730 (1977).
2. C. N. Davids, D. F. Geesaman, S. L. Tabor, M. J. Murphy,
    E. B. Norman, and R. C. Pardo, Phys. Rev. C 17, 1815 (1978).
3. R. C. Pardo, C. N. Davids, M. J. Murphy, E. B. Norman, and
    L. A. Parks, Phys. Rev. C 16, 370 (1977).
4. E. B. Norman, C. N. Davids, M. J. Murphy, and R. C. Pardo,
    Phys. Rev. C 17, 2176 (1978).
5. M. J. Murphy, C. N. Davids, and E. B. Norman, Bull. Am. Phys.
    Soc. 21, 968 (1976), and to be published.
6. L. A. Parks, C. N. Davids, B. G. Nardi, and J. N. Worthington,
    Nucl. Instrum. Methods 143, 93 (1977).
7. A. M. Nathan, D. E. Alburger, J. W. Olness, and E. K. Warburton,
    Phys. Rev. C 16, 1566 (1977).
8. M. Blann, University of Rochester, Nucl. Structure Lab. Report
    No. COO-3494-29 (unpublished).
9. C. N. Davids, C. A. Gagliardi, M. J. Murphy, and E. B. Norman,
    Phys. Rev. C 19, 1463 (1979).
10. J. C. Hardy and I. S. Towner, Nucl. Phys. A254, 221 (1975).
11. S. Raman, T. A. Walkiewicz, and H. Behrens, At. Data Nucl.
    Data Tables 16, 451 (1975).
12. C. N. Davids, Phys. Rev. C 13, 887 (1976).
13. M. J. Murphy, C. N. Davids, E. B. Norman, and R. C. Pardo,
    Phys. Rev. C 17, 1574 (1978).

Q_{EC}-VALUES OF SOME NEUTRON-DEFICIENT Ag ISOTOPES

J. Verplancke, D. Vandeplassche, M. Huyse, K. Cornelis
and G. Lhersonneau

LISOL, Instituut voor Kern- en Stralingsfysika
Katholieke Universiteit Leuven
Celestijnenlaan 200 D, B-3030 Leuven, Belgium

INTRODUCTION

Recently, extensive studies on the decay of neutron-deficient
Ag isotopes have been performed at LISOL, the Leuven Isotope
Separator On Line. We found evidence for the new isotopes $^{99}Ag^g$
(124 sec), $^{99}Ag^m$(15 sec), ^{98}Ag(44.5 sec) and ^{97}Ag(21 sec)[1]. These
and some heavier Ag nuclides have been studied using singles γ and
β measurements as well as γ-γ-t and β-γ-t coincidence measurements.
We report here the results for the disintegration energies of the
Ag-isotopes ranging from mass 98 to 102.

EXPERIMENTAL SET-UP

The radioactive sources were produced with a ^{14}N-beam of 110,
125 and 145 MeV, impinging on a natural Mo or an enriched ^{92}Mo
target. Ion sources and detection times were optimised in order to
have practically pure Ag spectra of a certain mass at the detector
site.

The β-detector at its working position is shown on fig. 1. It
is similar to the one used by Beck[3] at the ISOLDE facility. It
consists of 2 plastic scintillators. The first, small detector,
with a thickness of 0.5 mm acts as discriminator to filter out the
response of γ radiation. The second one, the energy detector has
a thickness of 5 cm and is shaped to have a minimum of non-active
material. It stops electrons and positrons up to about 10 MeV.

For calibrating the β telescope we used the electron sources
^{207}Bi and ^{137}Cs and the β^- sources ^{90}Sr, ^{144}Ce and ^{106}Ru with end

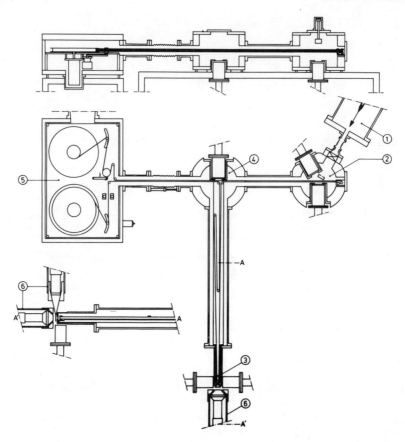

Fig. 1.   Tape collecting system. (1) separator beam, (2) implanta-
          tion station, (3) first detection station, (4) second
          detection station, (5) tape drive unit and (6) β telescope.

points close to the end points of the Ag nuclides under investigation. We fitted the last part of the pure and allowed β^+ spectra,
obtained in $\beta-\gamma$ coincidence, with a function $\int N(E)f(E,E')dE'$. $N(E)$
is the theoretical β^+ spectrum and $f(E,E')$ the response of our
detector at energy E' to a mono energetic positron beam input of
energy E. The simplest response function that fits our experimental
spectra with a good accuracy is a Gauss-function with a constant
tail down to zero energy. The fit parameters are the ratio of peak
area to total area (k) and the resolution of the detector which is
expected to be proportional to $\sqrt{E}$. All fitted spectra give the
same values for these parameters, namely k $\sim.2$ and Res. $\sim 13 \sqrt{E(keV)}$.
A result of these fits is shown on fig. 2. The effect of the
response on the Fermi-Kurie plots is given in fig. 3. However, no
unfolding of our experimental spectra has yet been used. The
results in the next paragraph are obtained by fitting the raw F.K.
plots.

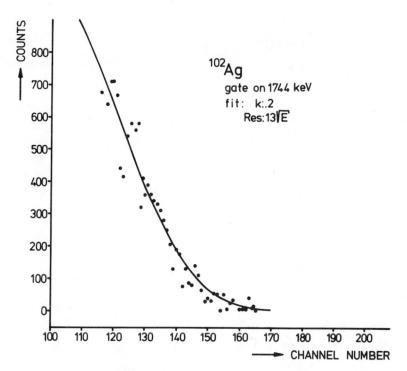

Fig. 2. Spectrum of ^{102}Ag from β-γ coincidences with γ gate on 1744 keV. The solid line is the calculated spectrum with fit parameters k and Res. (see text).

RESULTS

The results we obtained are brought together in table 1. They are compared with some values from the literature.

A. From Coincidences

For ^{102}Ag (13 min) we obtained 2.3 x 10^5 coincidences. Setting gates on 1582, 1744 and 1257 keV[8], we found a disintegration energy of 5.88 ± .11 MeV.

We stored 1.4 x 10^5 β-γ coincidences in the decay of ^{100}Ag(2min). Gates were set on 3 gamma transitions deexciting the 2921 keV level, namely 1504, 731 and 450 keV and on the 1116 keV γ-line, deexciting the 2532 keV level[5]. The resulting Q_{EC}-value for ^{100}Ag is 7.075 ± .090 MeV.

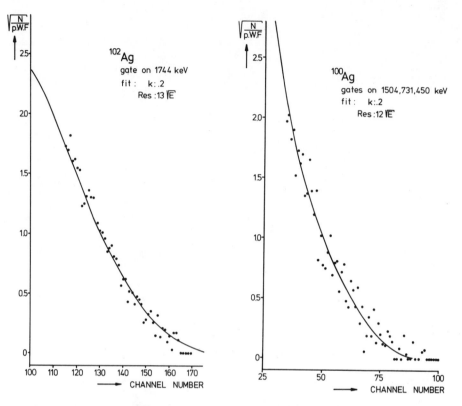

Fig. 3.   Fermi-Kurie plots. The solid line is the calculated spec-
          trum with fit parameters k and Res. (see text).

B. From Singles β Spectra

     The interpretation and the analysis method of our singles spectra
have proved to be reliable as the results for ^{102}Ag and ^{100}Ag from
the singles (5.81 ± .15 MeV and 7.27 ± .15 MeV respectively) agree
with the results of our coincident spectra.

     According to Hayakawa et al.[6], 42 % of the ^{101}Ag (11 min)
activity feeds into the 261 keV level in ^{101}Pd. By analysing the
last part of the singles β spectrum we get the end point of the β
branch to that level, yielding a Q_{EC}-value of 4.18 ± .15 MeV in
agreement with the value 4.35 ± .2 MeV reported in ref. 6.

     The Q_{EC}-values of ^{99}Ag were based on recent gamma work at LISOL[1,9].
In ^{99}Pd there is again only one level that is favourably fed by the
β,EC decay of ^{99}Ag, namely the 264 level which has a 32 % feeding.
Taking this into account we deduced a Q_{EC}-value of 5.43 ± .15 MeV.

     For ^{98}Ag the single β measurement gives a lower limit for the

Table 1. Disintegration energies in MeV.

Isotope	from singles	from coincidences	other measurements	estimated[4]
^{102}Ag (13 min)		5.88 ±.11	6.1 ±.1[10]	5.8
^{101}Ag (11 min)	4.18 ±.15		4.92±.14[7] 4.35±.2[6]	4.1
^{100}Ag (2 min)		7.075 ±.090	7.1 ±.2[5]	
^{99}Ag (124 sec)	5.43 ±.15			5.6
^{98}Ag (45 sec)	6.88 ±.15			8.6

disintegration energy of 6.88 ±.15 MeV. This is considered to be
a lower limit as it is believed that the major β,EC feeding goes
into a yet unknown high lying particle excitation level in ^{98}Pd
similar with the 2.9 MeV level in ^{100}Pd.

This work has been supported by the I.I.K.W.
We are also much indebted to our technical staff, especially B. Brijs
and J. Gentens and to the team that is in charge of the cyclotron.

REFERENCES

1. M. Huyse et al., The decay of neutron deficient ^{97}Ag, ^{98}Ag and
     ^{99}Agg,m, Z. Physik A 288 : 107 (1978).
2. G. Dumont et al., LISOL, The Leuven Isotope Separator On-Line
     at the "CYCLONE"-cyclotron., Nucl. Instr. and Meth. 153 : 81
     (1978).
3. E. Beck, A plastic scintillation spectrometer for high-energy
     beta particles., Nucl. Instr. and Meth. 76 : 77 (1969).
4. A.H. Wapstra and K. Bos, Atomic Data and Nucl. Data Tables,
     vol 19/3 (1977).
5. S.I. Hayakawa, I. Hyman and J.K.P. Lee, Decay of ^{100}Ag, to be
     published.
6. S.I. Hayakawa, I.R. Hyman and J.K.P. Lee, The decay of ^{101}Ag,
     Nucl. Phys. A296 : 251 (1978).
7. E. Beck, Beta decay of some neutron deficient nuclides, in
     Proc. of the Int. Conf. on the properties of nuclei far from
     the region of Beta-stability. Leysin, vol 1 : 353 (1970)
8. D.J. Hnatowich, F. Münnich and A. Kjelberg, The decay of 8 min

$^{102}Ag^m$ and 13 min $^{102}Ag^g$., <u>Nucl. Phys.</u> A178 : 111 (1971).

9.   M. Huyse et al. : to be published.

10.   Charoenkwan and Richardson, Decay of ^{102}Ag and levels in ^{102}Pd, <u>Nucl. Phys.</u> A94 : 417 (1967).

THE STUDY OF NUCLEAR MASS DIFFERENCE BY THE YASNAPP PROGRAMME

Henryk I. Lizurej

Laboratory of Nuclear Problems
Joint Institute for Nuclear Research
Moscow, 101000, USSR

The study of the nuclear mass difference by the YASNAPP programme (Nuclear Spectroscopy on the Proton Beam) has been carried out by means of the beta-spectra end-point energy measurement. The determination of mass difference for nuclei far from the beta-stability line presents considerable difficulties due to the complex decay schemes of these nuclei, the necessity of taking into acount the background under the measured continuum and the instrumental factors distorting the spectra (response function of the spectrometer, the scattering in the source and source backing). There is no universal solution of the problem of accounting all the effects that cause the beta-spectra distortion. Therefore it is necessary to study and to take into account all the possible effects causing the distortion of beta-spectra for each given instrument. Than,the measured positron distribution M(E) should be corrected by the introduction of the respective coefficients accounting for the above mentioned factors. This enables to obtain the real distribution N(E).

In the study of the low intensity branches of the neutron-deficient nuclei positron decay an iron-free beta-spectrometer ST-2 with a toroidal magnetic field (transmission T=20%, resolution R=1%) was used[1].

In the majority of cases we have used the carrier-free sources obtained by the chemical separation of elements from the irradiated target with the subsequent implantation of ions into an aluminized 680 $\mu g/cm^2$ mylar foil. The depth of implantation did not exceed 15 $\mu g/cm^2$. For such sources we have observed no beta-spectra distortions at the energy $E_\beta^+ > 100$ keV. In Fig.1 the positron-energy dependence of the scattering parameters $C_B(E) = N_d(E)/N_0(E)$ at different thicknesses of the source-backing is shown. The parameter C_B for the used mylar foil equals 1.035 at $E_\beta = 150$ keV.

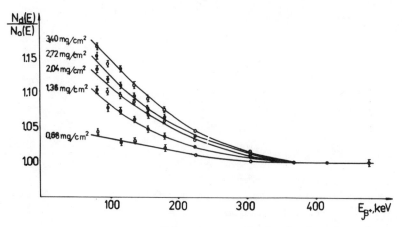

Fig.1. The positron-energy dependence of the backscattering parameters $C_B(E)$ at different thicknesses of the source-backing.

The efficiency of the detection system is constant at positron energy $E_\beta + > 100$ keV.

The response function of our spectrometer is not a universal function as in the case of an "ideal" spectrometer and the sum of the gaussian peak F_G and of the exponential tail F_{tail}, expanding towards lower energies, equals

$$F(p,p_0) = F_G(p,p_0) + F_{tail}(p,p_0).$$

The second term F_{tail} should be taken into account for every instrument. The analytical form of this function[1] has been received on the basis of the experimental distributions of the conversion electrons line K208 ^{167}Tm, K308 ^{169}Yb, K256 ^{152}Dy, K344 ^{152}Tb and K587 ^{152}Tb. The correction for the real response function is

$$C_0(p) = N(p)/M(p),$$

where $N(p)$ is the real and $M(p)$ the experimantal distribution. It turned out that the correction coefficients $C_0(p)$ for the positron-spectra in the low energy region differ significantly from the corrections $C_R(p)$ for the spectrometer resolution, when the tails of the monochromatic lines are not taken into account. It is essential to take into account the $C_0(p)$ corrections in the beta-spectra analysis, especially in the case of multicomponent beta-spectra in which the spectrum components with high E_0 values have greater intensity and in the case of single-component beta-spectra with 1-2% statistical error.

Fig.2 represents the correction coefficients $C_R(p)$ and $C_0(p)$ for positrons with different end-point energies.

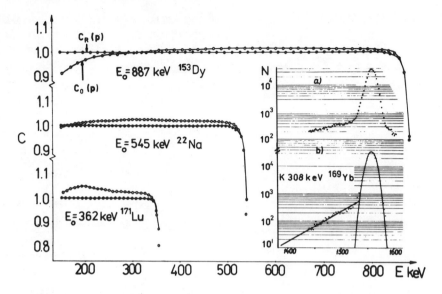

Fig.2 The correction factors $C_0(p)$ and $C_R(p)$ for positrons with
different end-point energies. The spectrum of conversion
electron: a. experimental distribution effect+background,
b. distribution "gauss+exponential" after substraction of the
background.

In the positron mode the background, generated by the spectrometer,
is energy-independent if measured at $E_{\beta^+} > 100$ keV. Thus, the measurements at the energies higher than the maximum positron energy give
directly the value of the spectrometer background . In the case
when the end-point energy E_{β^+} exceeds the maximum permissible
energy for the given spectrometer the conclusion on the background
can be drawn out of the measurements made by a source with an aluminium absorber.

The accomplished methodic investigations have demonstrated
that it is possible to detect and to study the positron decay branches of up to 10^{-5} per decay by means of the ST-2 beta-spectrometer.
The end-point energies can be measured with an accuracy up to several
keV. A significant transmission of the toroidal spectrometer (T=20%)
and the possibility of small-step-measurements (up to $\Delta E=1$ keV)
ensure the conditions for the separation of the measured complex
β-spectra, comprising 4÷5 components at energy difference
$E_{ok}-E_{o(k+1)} \geqslant 40$ keV. It is necessary to undertake the study of
coincidences for the nuclei with unknown decay scheme.

We will consider as an example the employment of the described method for the investigation of the positron decay of the rare-earth elements' deformed nuclei and for the estimation of their decay energy Q_{β^+}.

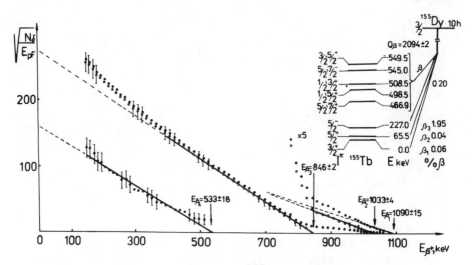

Fig.3. The Fermi-Kurie plot of the ^{155}Dy positron spectrum.

In Fig.3 the Fermi-Kurie plot of ^{155}Dy positron spectrum consisting of at least 4 components is shown. The positrons populate the 0 keV $3/2^+$ ground state and the excited states of ^{155}Tb with the energies 65.5 keV $5/2^+$, 227 keV $5/2^-$ as well as the states with the energy ~500 keV. Earlier, Persson et al[3]. did not succeed in finding the positrons responsible for the populating of the first $5/2^+$ excited state of ^{155}Tb with the energy 65.5 keV. It was demonstrated in their work[3] by means of $\beta^+ \gamma$ - coincidence method that the intensive component with E_0 =850 keV in the ^{155}Dy decay populates the excited state of ^{155}Tb with the energy 227 keV. Hence it follows that the mass difference ^{155}Dy-^{155}Tb is equal to $Q_{\beta^+}=$ 2094(2) keV.

Fig.4 shows the α-β decay cycles including the nuclei of A=151, 147. Some definite Q_{β^+} values in the decay of ^{147}Eu, ^{151}Tb and ^{147}Gd made possible a more accurate determination of the ^{151}Dy-^{151}Tb mass difference value ($Q_{\beta^+}=$2856(15) keV), and of the ^{147}Gd-^{143}Sm, (Q_α =1747(26) keV). But the mistakes of the original works[11-13] decreased the accuracy of the Q_β and Q_α determination over the α-β cycle.

In all the measurements of the positron spectra end-point energies the data on the nuclear decay scheme are obtained. This allows to determine their mass difference. The table 1 gives the experimentally found values Q_{β^+} for 18 nuclei. The experimental values

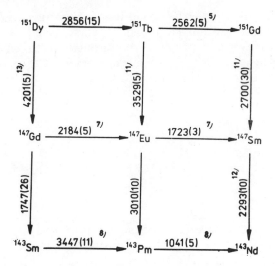

Fig.4. The α-β decay cycles including the nuclei of A=151,147.

obtained for the mass difference (column 2) are compared to the calculated data of Takahashi (column 3), Garvey (column 4) and the
summary data of Wapstra (column 5).

Table 1.   The Q_{β^+} experimental and predicted values

β^+-transition	Q_{β^+}exp.keV	9	10	8	References
$^{135}_{137}Ce-^{135}_{137}La$	2027(3)	2400	2280	2120(100)	4
$^{137}_{139}Ce-^{137}_{139}La$	1221.4(1.5)	1400	1340	1220(20)	7
$^{141}_{147}Pr-^{141}_{147}Ce$	2172(10)	1750	1960	2112(10)	14
$^{147}_{147}Nd-^{147}_{147}Pr$	1824(3)	2090	1840	1815(8)	4
$^{147}_{151}Gd-^{147}_{151}Eu$	2184(5)	2760	2410	2328(25)	7
$^{151}_{153}Eu-^{151}_{153}Sm$	1723(3)	1520	1690	1730(6)	7
$^{153}_{151}Tb-^{153}_{151}Gd$	2562(5)	2280	2220	2561(10)	5
$^{151}_{153}Tb-^{151}_{153}Gd$	1585(5)	1760	1420	1789(8)	15
$^{153}_{155}Dy-^{153}_{155}Tb$	2856(15)*	2280	3110	3007(8)	2
$^{155}_{159}Dy-^{155}_{159}Tb$	2171(2)	2900	2350	2174.0(3.0)	4
$^{159}_{159}Dy-^{159}_{159}Tb$	2094(2)	2150	2020	2099(6)	7
$^{159}_{163}Ho-^{159}_{163}Dy$	1838(3)	1580	1810	1853(9)	6
$^{163}_{165}Er-^{163}_{165}Ho$	2770(15)	2840	2930	2930(100)	7
$^{165}_{167}Tm-^{165}_{167}Er$	2440(2)	2240	2420	2400(syst)	6
$^{167}_{169}Tm-^{167}_{169}Er$	1591(2)	1300	1640	1594.5(2.4)	6
$^{169}_{171}Yb-^{169}_{171}Tm$	1954(4)	1620	1879	1954.1(3.8)	4
$^{171}Lu-^{171}Yb$	2293(3)	1940	2010	2480(25)	5
$Lu-Yb$	1479(3)	1020	1520	1480.7(2.2)	4

* Q_{β} from α-β cycles.

In a number of cases the experimental values differ from the calculated ones (up to 500 keV) and in some cases we have observed a significant difference of our experimental data on the mass differences from the data of the Wapstra summary table. This can be explained by differences in the interpretation of decay schemes.

For example if you direct the ^{169}Yb positrons populating the ground level at the excited level 191 keV, the difference from the Wapstra[8] data disappears, but the results of the study[16] and the analysis of the 191 keV γ-rays coincidence with the annihilation γ - quantum disagree with the assumption that this level is populated by the positrons with E_0=1271 keV[16]. Therefore the mass difference ^{169}Lu-^{169}Yb should be defined as Q_β+=2293(3) keV.

References

1. K.Ya.Gromov,T.Cretu,V.V.Kuznetsov,H.I.Lizurej,V.M.Gorozhankin and G.Macarie, Applied Nuclear Spectroscopy.8:59(1978).
2. A.W.Budziak,T.Cretu,V.V.Kuznetsov,N.A.Lebedev,H.I.Lizurej, Yu.V.Yushkevich,M.Janicki, JINR, P6-12403, Dubna, 1979.
3. L.Persson,H.Ryde,K.Oelson-Ryde, Nucl.Phys. 44:653(1963).
4. Proc.26 Conf. on Nuclear Physics and Nuclear Structure, Baku, L.Nauka,1976.
5. Proc.27 Conf. on Nucl.Spectroscopy and Nuclear Structure, Taszkent, Nauka, L.1977.
6. Proc.28 Conf. on Nuclear Spectroscopy and Nuclear Structure, Alma-Ata, Nauka, L.1978.
7. Proc.29 Conf. on Nucl.Spectroscopy and Nuclear Structure,Riga, Nauka, L.1979.
8. A.H.Wapatra,K.Boss, Atomic Data and Nuclear Data Tables,19:3(1977).
9. Takahashi, H.V.Groote, E.R.Hilf, JKDA 76/26.
10. G.T.Garvey,W.J.Gerace,R.L.Jaffe,I.Taimi,I.Kelson,Rev.Mod.Phys. Supl.,44:No4 (1969).
11. I.Adam,K.S.Toth,R.B.Mayer,Phys.Rev.159:985(1967).
12. M.P.Avotina,A.V.Zolotavin,Selected Nuclei with A=147,L.Nauka, 1971.
13. K.S.Toth, Phys.Rev., C10:2550(1974).
14. Ts.Vylov,I.I.Gromova,V.G.Kalinnikov,V.V.Kuznetsov,T.M.Muminov, V.A.Morozov,V.I.Fominykh,R.R.Usmanov, JINR P6-8903,Dubna,1975.
15. T.Cretu,V.V.Kuznetsov,H.T.Lizurej,V.M.Gorozhankin,G.Macarie, JINR, P6-10562, Dubna, 1977.
16. N.A.Bonch-Osmolovskaya,V.M.Gorozhankin,K.Ya.Gromov,T.Cretu, G.Macarie,A.S.Hamidov,M.Janicki, Izv.Akad.,Nauk.SSSR,ser.fiz. 41:No6, 1149(1977).

BETA-DECAY ENERGIES OF NEUTRON-RICH FISSION PRODUCTS

IN THE VICINITY OF MASS NUMBER 100*

U. Keyser, H. Berg, F. Münnich, and B. Pahlmann

Institut A für Physik, Technische Universität
Braunschweig, Germany

R. Decker, and B. Pfeiffer

II. Physikalisches Institut der Universität Giessen
Giessen, Germany

INTRODUCTION

Beta-decay energies of neutron-rich fission products have been measured by different groups using different experimental techniques and data evaluation methods at the mass separator LOHENGRIN and OSTIS installed at the high flux reactor of the Institute Laue-Langevin in Grenoble. In this communication, recent results of such experiments will be presented for light fission products; some results obtained for heavy fission products will be discussed in another contribution to this conference[1]. The decay energies have been derived from measurements with a plastic scintillation counter telescope to detect the electrons emitted in beta-decay. In most cases, the β-spectra have been measured in coincidence with γ-rays recorded with a big Ge(Li)-detector. The β-detector telescope has been calibrated by measuring β-spectra with well-known endpoint energies under the same conditions as during the main experiment. Therefore, it is not necessary for the evaluation of the experimental data to know the efficiency and response function of the detector. In some cases, β-ray singles spectra have also been recorded with a high-purity Ge-detector. More details concerning the preparation and counting of the sources and the handling of the data are given elsewhere[2].

*Supported in part by Bundesministerium für Forschung und Technologie (BMFT).

EXPERIMENTAL RESULTS FOR HEAVY Rb ISOTOPES

   The Q_β-values of neutron-rich Rb isotopes have been studied
intensively by several groups[2-6]. In addition, also direct mass
determinations of these isotopes have been performed recently at
the ISOLDE mass separator by Epherre et al.[7]. In order to compare
the results of this work with those of the decay studies, the
Q_β-values of the heavy Rb isotopes are calculated from the difference
of the mass excesses of the Rb nuclei measured at ISOLDE and of Sr
nuclei taken from the literature, as shown in Table I. These
calculated Q_β-values are then compared in Table II with the results
obtained in the present study at the mass separators LOHENGRIN and
OSTIS and with the latest values reported from experiments performed
at OSTIS[5], where a high-purity Ge-detector was used to measure the
β-spectra. The discrepancies observed between some results from
OSTIS and those from the two other studies are due to systematic
errors and will be discussed shortly.

   For the mass number 94, the result from OSTIS differs outside
the given errors from the ISOLDE value. However, in the work of
Epherre et al., an isomeric state of [94]Rb has been reported with a
mass excess of -68514±50 keV. Using this value in place of the one
given in Table I, a decay energy of 10330±50 keV is obtained in
agreement with the OSTIS result. Thus, the discrepancy between both
values might be explained by assuming that the ground state decay
of [94]Rb was not observed in the OSTIS experiments.

   For mass number 95, the Q_β-value of OSTIS is too low, due to
a wrong assignment of the levels populated by the observed
β-transitions, as can be deduced from the results obtained with the
scintillation detector telescope. With the Ge-detector, only two
coincidence measurements with low-energy γ-lines have been performed,
because the detection efficiency of this detector is very low for
β-energies above 6 MeV. In the experiment with the telescope also
performed at OSTIS, β-spectra in coincidence with 10 γ-transitions
could be evaluated, which depopulate excited levels in [95]Sr with
the following energies: 681 keV, 2264 keV, 3367 keV and 3480 keV.
From the endpoint energies of these coincidence spectra, a consistent
Q_β-value is obtained, which agrees very well with that calculated
from the direct mass determination.

   The discrepancy of the results for [97]Rb is due to the same
error in the evaluation of the OSTIS data. Also here, the coincidence
results with the Ge-detector do not allow the fixing of the level
populated by the observed β-transitions. In the experiment with the
scintillation counter, the endpoint energies of 6 coincidence
spectra with γ-transitions depopulating excited states of 986 keV,
1321 keV and 1374 keV energy in [97]Sr have been determined. The
Q_β-value based on these results agrees within errors with that
reported from ISOLDE.

Table I. Calculated Q_β-values of heavy Rb isotopes
(all values are in keV)

Mass excess[a]		Q_β-value	Mass excess
^{92}Zr −88456± 3	Y→Zr	^{92}Y 3634±16[a]	^{92}Y −84822±16
^{93}Zr −87117± 3		^{93}Y 2890±20[a]	^{93}Y −84227±20
^{94}Zr −87264± 3		^{94}Y 4903±10[b]	^{94}Y −82361±10
^{95}Zr −85663± 3		^{95}Y 4440±10[b]	^{95}Y −81223±10
^{96}Zr −85445± 4		^{96}Y 7083±40[c]	^{96}Y −78362±40
^{97}Zr −82954± 4		^{97}Y 6645±70[d]	^{97}Y −76309±70
^{98}Zr −81292±20		^{98}Y 8830±90[d]	^{98}Y −72462±92

	Q_β-value	Mass excess	Mass excess[f]
Sr→Y	^{92}Sr 1930±30[a]	^{92}Sr −82892± 34	^{92}Rb −74841±39
	^{93}Sr 4145±55[e]	^{93}Sr −80082± 59	^{93}Rb −72754±64
	^{94}Sr 3500±10[b]	^{94}Sr −78861± 15	^{94}Rb −68736±44
	^{95}Sr 6080±80[e]	^{95}Sr −75143± 81	^{95}Rb −65852±97
	^{96}Sr 5403±20[c]	^{96}Sr −72959± 45	^{96}Rb −61223±71
	^{97}Sr 7420±80[d]	^{97}Sr −68889±106	^{97}Rb −58370±81
	^{98}Sr 5870±70[d]	^{98}Sr −66590±116	^{98}Rb −54185±99

	Q_β-value
Rb→Sr	^{92}Rb 8051± 52
	^{93}Rb 7328± 87
	^{94}Rb 10125± 47
	^{95}Rb 9291±126
	^{96}Rb 11736± 84
	^{97}Rb 10520±134
	^{98}Rb 12405±152

[a]Ref.8. [b]Mean value from Refs.6 and 8. [c]Mean value from Refs.2 and 6. [d]From this work, Table III. [e]Mean value from Refs.2 and 3. [f]Ref.7.

Table II.   Experimental Q_β-values of heavy Rb isotopes
            (all values are in keV)

Isotope	Present work and Ref.2	OSTIS Ref.6	ISOLDE Ref.7
^{92}Rb	7980±100	8111± 15	8050± 55
^{93}Rb	7405±100	7485± 15	7330± 90
^{94}Rb	10185±150	10304± 30	10125± 50
^{95}Rb	9260± 70	8947±100	9290±130
^{96}Rb	11670±130	11303^{+250}_{-900}	11735± 85
^{97}Rb	10325±150	9920± 50	10520±135
^{98}Rb	12230±300	–	12405±155

The large experimental error of the Q_β-value of ^{96}Rb from OSTIS results from uncertainties concerning the response function of the Ge-detector for β-energies above 8 MeV. In the measurement with the telescope, again 6 coincidence spectra could be evaluated. The corresponding β-transitions have been assigned to populate excited levels of 1628 keV, 2567 keV, 3755 keV and 4147 keV energy in ^{96}Sr. Also here, a consistent Q_β-value is obtained from this assignment and the experimental endpoint energies.

Taking into account these explanations for the discrepancies between the Q_β-values shown in Table II, the agreement between the results of the decay studies and the direct mass determination is quite satisfactory even for the heaviest Rb isotopes studied until now.

DECAY ENERGIES OF Sr, Y, Zr AND Nb ISOTOPES: REVISED VALUES

The calibration procedure of the β-detector applied in our measurements is basically simple and straightforward, but its accuracy depends strongly on well-known endpoint energies of β-spectra from nuclei easily accessible at the LOHENGRIN separator. At the time of measurements reported earlier[2], only few such energies could be found in the literature. The value of some of these energies, which have been used as calibration standards in[2], has changed in the meantime as a result of more accurate experiments[5-7]. Therefore, it was decided to re-evaluate some of these measurements, also applying an improved data handling procedure to decompose complex β-spectra, in order to reduce the experimental uncertainties.

Table III. Q_β-values of Sr, Y, Zr and Nb isotopes

Nuclide	$T_{1/2}$	Earlier value (Ref.2)	Revised value	Other authors
	s	keV	keV	keV
$^{97}_{38}$Sr	0.4	7450±120	7420± 80	-
$^{97}_{39}$Y	3.7	6650±120	6645± 70	-
$^{98}_{38}$Sr	0.65	5880±120	5870± 70	-
$^{98}_{39}$Y^g	0.6	8750±130	8830± 90	-
$^{98}_{39}$Y^m	2.0	9780±200	9975±100	-
$^{99}_{39}$Y	1.5	-	7570± 90	-
$^{99}_{40}$Zr	2.1	4545±120	4570± 70	4500±200[a]
$^{100}_{40}$Zr	7.1	3340±130	3330± 70	-
$^{100}_{41}$Nb^g	1.5	6240±100	6215± 60	6500±300[a]
$^{100}_{41}$Nb^m	3.1	-	6745± 75	6709± 30[b]
$^{101}_{40}$Zr	3.3	-	5780±120	-
$^{101}_{41}$Nb	7.1	4570±100	4630± 70	4600±200[a]
$^{102}_{41}$Nb^g	1.3	7250±130	7200± 70	-
$^{102}_{41}$Nb^m	4.5	-	7385± 70	-

[a]Ref.9. [b]Ref.10.

The following β-endpoint energies have been adopted as standards in the new β-calibration (some calibration points are measured in coincidence with γ-rays):

^{90}Rb: E_β = 6583±13 keV ^{93}Rb: E_β = 7460±30 keV
        E_β = 5751±13 keV E_β = 7027±30 keV
        E_β = 4692±13 keV E_β = 6318±30 keV
^{96}Sr: E_β = 4471±20 keV E_β = 6075±30 keV.

The results of this re-evaluation are presented in Table III, together with the earlier values. One of these results deserves special attention: the Q_β-value of ^{100}Nb. The mass of this nuclide has been derived recently by Ajzenberg-Selove et al.[10] from the study of the ^{100}Mo(t,^{3}He)^{100}Nb reaction. A mass excess of -79840±30 keV has been determined in this experiment, yielding a Q_β-value of 6709±30 keV, which differs considerably from the value of 6240±100 keV reported by Stippler et al.[2]. In the re-evaluation of this decay energy, additional experimental data have been included from sources, where ^{100}Nb was directly produced in fission. Therefore, also the decay of the 3.1 s isomer could now be observed, as is shown in Table III. A Q_β-value of 6215±60 keV is derived from β-spectra feeding the ground state and excited states of 536 keV, 695 keV and 1064 keV energy in ^{100}Mo. The higher Q_β-value of 6745±75 keV is based on β-transitions populating levels of 1136 keV, 2087 keV, 2103 keV, 2416 keV and 2565 keV energy in ^{100}Mo. For the assignment of the β-transitions to these levels, information concerning the decay scheme of ^{100}Nb from Kocher[11] and Sadler[12] has been used. It follows from these results that the isomeric state in ^{100}Nb has the half-life of 3.1 s and an excitation energy of about 500 keV. The Q_β-value of this isomer agrees well with the value obtained by Ajzenberg-Selove et al. given above. Therefore, it must be concluded that the ground state of ^{100}Nb was not produced in the ^{100}Mo(t,^{3}He)^{100}Nb reaction studied by these authors.

Also for ^{102}Nb two Q_β-values could be obtained in the re-evaluation of the data, cf. Table III. The decay energy of the 1.3 s ground state was deduced from β-transitions to the ground state and to the excited states of 296 keV, 696 keV and 848 keV energy in ^{102}Mo. The Q_β-value of the 4.3 s isomer was calculated from the endpoints of β-spectra coincident with γ-transitions, which all depopulate the 2480 keV level in ^{102}Mo either directly or in cascade. More details concerning the Q_β-values of these two isotopes of Nb will be given in a forthcoming publication[13].

We are indebted to the technical and scientific staff of the LOHENGRIN mass separator, especially to K. Hawerkamp and H. Schrader, and to F. Blönnigen, G. Jung and F. Münzel from the OSTIS separator for their cooperation in some measurements and their help in the reduction of the data. We also want to thank Mrs. Ch. Laupheimer for her assistance in the presentation of the data and the preparation of the manuscript. The financial support of the Bundesministerium für Forschung und Technologie is gratefully acknowledged.

REFERENCES

1. U. Keyser, H. Berg, F. Münnich, B. Pahlmann, K. Hawerkamp,
    B. Pfeiffer, H. Schrader, and E. Monnand, contribution to
    this conference
2. R. Stippler, F. Münnich, H. Schrader, J. P. Bocquet, M. Asghar,
    G. Siegert, R. Decker, B. Pfeiffer, H. Wollnik, E. Monnand,
    and F. Schussler, Z. Physik A284,95(1978)
3. M. I. Macias-Marques, R. Foucher, M. Cailliau, and J. Belhassen,
    CERN-Report 70-30,321(1970)
4. I. R. Clifford, W. L. Talbert jr., F. K. Wohn, I. P. Adams,
    and I. R. McConnel, Phys. Rev. C7,2535(1973)
5. K. D. Wünsch, R. Decker, H. Wollnik, J. Münzel, G. Siegert,
    G. Jung, and E. Koglin, Z. Physik A288,105(1978)
6. R. Decker, Thesis, 1979, University Giessen
7. M. Epherre, G. Audi, C. Thibault, R. Klapisch, G. Huber,
    F. Touchard, and H. Wollnik, Phys. Rev. C19,1504(1979)
8. A. H. Wapstra, and K. Bos, At. Data Nucl. Data Tables 19,177
    (1977)
9. J. Eidens, E. Röckl, and P. Armbruster, Nucl. Phys. A141,289
    (1970)
10. F. Ajzenberg-Selove, E. R. Flynn, D. L. Hanson, and S. Orbesen,
    Phys. Rev. C19,2068(1979)
11. D. C. Kocher, Nucl. Data Sheets 11,279(1974)
12. G. Sadler, priv. comm.
13. U. Keyser, and F. Münnich, to be publ. in Phys. Rev. C

NUCLEAR Q_β-VALUES OBTAINED AT OSIRIS. A COMPARISON WITH MASS FORMULA PREDICTIONS

E. Lund, K. Aleklett, and G. Rudstam

The Studsvik Science Research Laboratory
Studsvik
S-611 82 Nyköping, Sweden

INTRODUCTION

The shape of the nuclear mass surface far out on the neutron-rich side of β-stability is of considerable interest because of many interesting applications, e.g. theories about the r-process in nucleosynthesis. Until now, direct mass determinations have been performed in very few cases[1]. Most studies of the mass surface have been carried out by the method of determining mass differences from measurements of total β-decay energies. The latter method has been used at Studsvik for a systematic investigation covering about 40 fission products in the mass ranges A = 75 - 89 and A = 120 - 135 and including isotopes of zinc, gallium, germanium, arsenic, indium, tin, antimony, and tellurium[2-6].

The Q_β-values were determined by adding the β-energy to a level in the daughter to the excitation energy of this level. This method requires good knowledge of the level structure, and in cases where such information are lacking it is only possible to give a lower limit. The accuracy of the Q_β-values determined is typically 2 - 3 %. In some cases an error below 1 % has been obtained.

EXPERIMENTAL TECHNIQUES

Using the OSIRIS facility, which is described in detail in Ref. 7, it is possible to study fission

products in the mass regions 74 to 98 and 111 to 147.
In this compilation 41 nuclides are treated.

The βγ-coincidence spectrometer used in the experiments consists of a system of Si(Li)-detectors for
β-detection and either two NaI detectors or one Ge(Li)-
detector for γ-detection. With the system used for
collecting samples it is possible to do accurate Q_β-
determinations for nuclides with half-lives down to
0.5 s. For details about the experimental procedure
see Ref. 4.

Total β-decay energies for highly unstable nuclei
are often of the order of many MeV. The energy calibration procedure, and the determination of the response
function and the efficiency function of the detection
system are described in Ref. 4. The efficiency function has been checked up to about 6 MeV by means of a
comparison between a β-spectrum of ^{86}Br measured with
the Si(Li)-detector system and the electron distribution obtained with an electromagnetic β-spectrometer.

When using two NaI-detectors for the detection of
the γ-rays the counting efficiency is high but because
of the poor energy resolution several β-branches may
contribute to the coincident β-spectrum which makes
the analysis complicated. The Ge(Li)-detector used in
this study had an efficiency of 18.5 % of an 3" NaI-
detector. In some cases this efficiency was too low
and led to unsatisfactory counting statistics. The
situation will be improved in future experiments by the
addition of a second Ge(Li)-detector.

EXPERIMENTAL RESULTS AND COMPARISON WITH MASS FORMULA
PREDICTIONS

The experimental results are collected in Table 1.
Among all published mass formulas we have chosen for
comparison the droplet model formula by W. D. Myers[8a],
the liquid drop model formula by P.A. Seeger and W.M.
Howard[8b], which pays special attention to the neutron-
rich nuclides far from stability, the semiempirical
shell-model formula by S. Liran and N. Zeldes[8c], and
finally the empirical mass relation by E. Comay and
I. Kelson[8d].

The differences between the predicted Q_β-values and
the experimental ones are shown in Figs. 1 - 4. The
error bars indicate the experimental errors in all cases

except for the comparison with the Comay and Kelson
mass data (Fig. 4). In this case the predictions include uncertainties which has been added quadratically
to the experimental errors.

Table 1. Experimental Q_β-values and differences between
predicted and experimental Q_β-values.

Nuclide	Exp Q_β-value (MeV)	Q_β-pred - Q_β-exp			
		M-S	S-H	L-Z	C-K
^{75}Zn	≥ 5.62±0.20	≤ (-0.85	-0.29	-0.02	0.15±0.75)
^{76}Zn	3.98±0.12	-1.00	-0.93	-0.06	-0.34±0.93
^{77}Zn	6.91±0.22	-0.76	-0.27	0.32	0.18±1.01
^{78}Zn	6.01±0.18	-1.64	-1.43	-0.20	-0.89±1.32
^{76}Ga	6.77±0.15	-0.74	0.17	0.15	-0.04±0.54
^{77}Ga	5.34±0.06	-1.07	-0.63	-0.32	-0.72±0.66
^{78}Ga	8.14±0.16	-0.72	0.12	0.13	-0.14±0.86
^{79}Ga	6.77±0.08	-1.12	-0.51	-0.11	-0.75±1.13
^{79}Ge	4.09±$^{0.18}_{0.07}$	-0.31	-0.04	0.37	-0.02±0.69
^{80}Ge	2.64±0.07	-0.61	-0.56	0.32	-0.61±0.93
^{80}As	5.37±0.12	-0.30	0.34	0.27	0.32±0.51
^{81}As	3.76±0.08	-0.44	-0.08	0.16	0.01±0.73
^{83}As	5.46±0.22	-0.80	-0.05	0.16	-0.40±1.48
^{85}Br	2.87±0.02	-0.45	-0.14	-0.02	-0.07±0.94
^{86}Br	7.62±0.06	0.10	-0.08	0.09	-0.04±1.26
^{87}Br	6.83±0.12	-0.86	-0.40	0.08	-0.23±1.77
^{88}Br	8.97±0.12	-0.04	0.21	0.68	-0.10±2.08
^{89}Br	8.1 ±0.5	-0.9	-0.30	0.5	-0.20±2.22

cont.

Table 1 cont.

Nuclide	Exp Q_β-value (MeV)	Q_β-pred $-$ Q_β-exp			
		M-S	S-H	L-Z	C-K
^{120}In	5.30±0.17	-0.61	-0.47	0.04	0.07±0.60
^{121}In	3.41±0.05	-0.24	-0.58	0.21	-0.02±0.64
^{122}In	6.51±0.23	-0.84	-0.60	-0.36	-0.16±0.74
^{123}In	4.44±0.06	-0.33	-0.41	0.16	-0.07±0.81
^{124}In	7.18±0.05	-0.44	-0.29	0.04	0.01±0.87
^{125}In	5.48±0.08	-0.48	-0.35	0.01	-0.20±0.94
^{126}In	8.21±0.08	-0.75	-0.24	-0.30	-0.25±0.98
^{127}In	6.49±0.07	-0.56	-0.24	-0.20	-0.32±1.02
^{128}In	9.31±0.16	-0.98	-0.44	-0.60	-0.59±1.08
^{129}In	7.60±0.12	-0.80	-0.40	-0.62	-0.60±1.16
^{127}Sn	3.201±0.024	-0.34	0.13	-0.16	-0.06±0.66
^{128}Sn	1.29±$^{0.06}_{0.04}$	0.07	0.19	0.04	0.12±0.74
^{129}Sn	3.99±0.12	-0.20	0.11	-0.03	-0.06±0.80
^{130}Sn	2.19±0.03	0.08	0.24	0.08	-0.02±0.87
^{131}Sn	4.59±0.20	0.07	0.43	0.29	-0.06±0.97
^{132}Sn	3.08±0.04	0.08	0.32	0.29	0.16±1.07
^{128}Sb	4.39±$^{0.04}_{0.06}$	-0.39	-0.11	0.11	0.16±0.47
^{130}Sb	5.02±0.10	-0.12	0.19	0.36	0.32±0.60
^{131}Sb	3.19±0.07	0.20	0.35	0.44	0.39±0.68
^{132}Sb	5.53±0.07	0.24	0.58	0.75	0.41±0.77
^{134}Sb	8.24±0.24	0.42	0.55	0.96	0.30±0.90
^{134}Te	1.56±0.09	-0.04	-0.18	-0.04	0.11±0.62
^{135}Te	5.95±0.24	-0.06	-0.27	0.28	-0.38±0.72

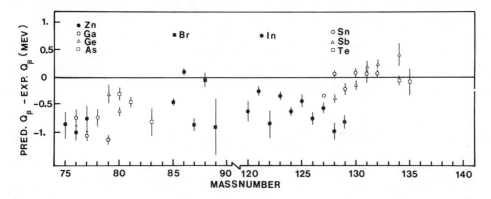

Fig. 1.   Differences between predicted and experimental Qβ-values
          concerning the mass formula by Myers[8a].

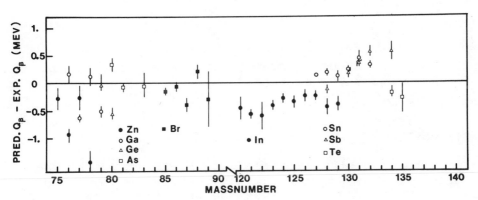

Fig. 2.   Differences between predicted and experimental Qβ-values
          concerning the mass formula by Seeger and Howard[8b].

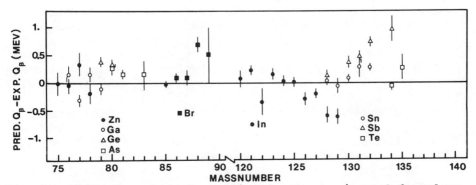

Fig. 3.   Differences between predicted and experimental Qβ-values
          concerning the mass formula by Liran and Zeldes[8c].

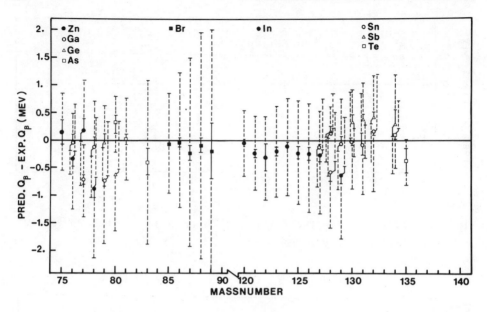

Fig. 4.   Differences between predicted and experimental
          Q_β-values concerning the mass relation by
          Comay and Kelson[8d]

DISCUSSION

     The droplet mass formula by Myers (Fig. 1) gene-
rally predicts too low Q_β-values except for the region
with $Z \geq 50$, where it gives the best fit among the predictions used in the present comparison.   Another observation is that the odd-even effect is exaggerated.

     In comparison with Myers' mass formula the one by
Seeger and Howard (Fig. 2) seems to predict Q_β-values
far from stability somewhat better as illustrated by the
range of indium isotopes (Cf. Figs. 1 and 2).

     The shell model mass formula by Liran and Zeldes
(Fig. 3) gives the most accurate predictions for all
the measured cases with $Z < 50$.   In the treatment by
Comay and Kelson (Fig. 4) the mass differences are given with errors, partly depending on the distance
between the predicted mass and those used in the extrapolation procedure.   The errors are large for the
nuclides studied here which all lie far from stability.
As shown in Fig. 4 the errors seem to be overestimated:
the predictions are better than the errors indicate.

A convenient way of comparing the precision of the various mass predictions is to use the root-mean-square deviation as a basis for the comparison. This figure takes the value 0.53, 0.35, 0.29, and 0.27 MeV for the predictions by Myers, Seeger-Howard, Liran-Zeldes, and Comay-Kelson, respectively. Thus one can say that, on the average, the Liran-Zeldes and Comay-Kelson mass predictions are the most accurate ones in the mass regions investigated here. In this connection it should be borne in mind, however, that the number of parameters used in the different mass formulas varies greatly.

It is hardly to be expected that the Myers formula with only 16 parameters or the one by Seeger and Howard with 9 parameters should yield as accurate predictions as the one by Liran and Zeldes with 178 parameters.

CONCLUSION

In summary we might conclude that all the four sets of Q$_\beta$-predictions chosen for comparison in the experimental data are reasonably accurate. In no case they are more than 1 MeV off. The comparisons indicate mass regions with systematic deviations, however, and we hope that the experimental masses presented here will be used to improve the precision of the predictions by removing these systematic deviations. Clearly, the experimental basis for this kind of work should be enlarged by providing more mass measurements. It is our intention to continue mapping the mass surface on the neutron-rich side of stability using an improved spectrometer system and also new target-ion source techniques which will enable us to reach mass regions so far not covered in the experimental programme.

REFERENCES

1. M. Epherre, G. Audi, C. Tribault, R. Klapisch,
    G. Huber, F. Touchard, and H. Wollnik, Phys.
    Rev. C19:1504 (1979).
2. K. Aleklett, E. Lund, and G. Rudstam, Nucl. Phys.
    A285:7 (1977).
3. K. Aleklett, E. Lund, and G. Rudstam, Z. Phys.
    A290:173 (1979).
4. K. Aleklett, E. Lund, and G. Rudstam, Phys. Rev.
    Cl, 18:462 (1977).
5. K. Aleklett, E. Lund, and G. Rudstam, Nucl. Phys.
    A281:213 (1977).
6. E. Lund, K. Aleklett, and G. Rudstam, Nucl. Phys.
    A286:403 (1977).

7. S. Borg, I. Bergström, G.B. Holm, B. Rydberg, L.E.
    De Geer, G. Rudstam, B. Grapengiesser, E. Lund,
    and L. Westgaard, Nucl. Instr. Meth. 91:109
    (1971).
8. Nucl. Data Tables 5-6, Vol. 17 (1976).
a)   W.D. Myers p.411 and LBL-3428 (1974).
b)   P.A. Seeger and W.M. Howard p. 428, Nucl. Phys.
        A238:491 (1975), and Los Alamos Scientific
        Laboratory Report LA 5750 (1974).
c)   S. Liran and N. Zeldes p. 431.
d)   E. Comay and I. Kelson p. 463.
9. G.T. Garvey, W.J. Grace, R.L. Jaffe, I. Talmi, and
    I. Kelson, Revs. Mod. Phys. 41:S1 (1969).

Determination of Q Values for Electron–Capture Decay of ^{172}Hf and ^{188}Pt
from X–Ray Intensities*

Y. A. Ellis and K. S. Toth

Oak Ridge National Laboratory

Oak Ridge, Tennessee 37830, USA

INTRODUCTION

Q values for electron–capture decays are not directly measurable. One must instead rely on indirect methods. One such technique utilizes x–ray intensities and/or transition intensities obtained from coincidence measurements. In these instances, the accuracy of the deduced energies depends on the correctness of the decay schemes and of theoretical capture ratios. We have applied this technique in the determination of $Q_{E.C.}$ for ^{172}Hf and ^{188}Pt.

^{172}Hf

The electron–capture decay of 1.87-y ^{172}Hf to ^{172}Lu was studied through its γ radiations. Low-energy singles and coincidence γ-ray spectra were taken with an intrinsic germanium and a Ge(Li) x-ray detector. The portion of the γ-ray spectrum below ~60 keV was also studied with a Si(Li) detector. The level scheme constructed for ^{172}Lu is shown in Fig. 1. Numbers following γ-ray energies and multipolarities in the figure represent total transition intensities per 100 ^{172}Hf electron–capture decays. They are calculated from our photon intensities and from theoretical conversion coefficients of Hager and Seltzer[1] and of Dragoun et al.[2] by using the computer program developed by the Nuclear Data Project. Capture branches were deduced from intensity balance at each level.

The electron–capture decay energy for ^{172}Hf has not been determined experimentally. We used our K x-ray intensity and selected coincidence intensities to deduce the Q value. The total K x-ray intensity due to K-shell electron conversions and K captures was calculated as a function of Q. In addition, the intensities of the 114.06-, 122.92-, 125.82-, and 127.91-keV γ rays in coincidence with K x-rays were calculated, once again, for various values of the decay energy. The calculated intensities for a range of Q values were then

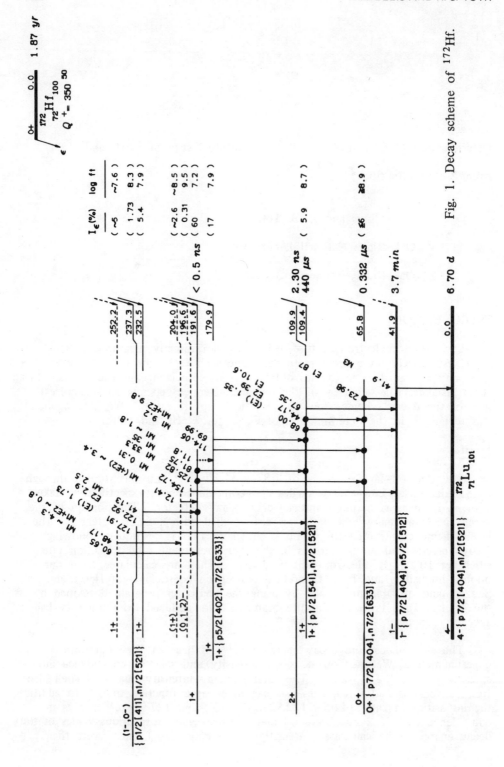

Fig. 1. Decay scheme of ^{172}Hf.

compared with the measured numbers. From this comparison, a value of 350 ± 50 keV was arrived at for the ^{172}Hf decay energy. The $Q_{E.C.}$ for ^{172}Hf has been estimated by Wapstra and Bos[3] from β-energy systematics to be 400 keV. This value is consistent with our experimental result.

In addition, the Q value can be calculated from a number of predictions of mass excesses by using semiempirical nuclear models (see Ref. 4). The calculated values vary from 0.02 MeV to 0.9 MeV, and one of them predicts ^{172}Hf to be β stable. Some selected predictions are compared with our experimental $Q_{E.C.}$ value in Table I.

Table I. Q Values (in MeV) for Electron–Capture Decay of ^{172}Hf

| | | Predictions[4] | | | |
Present work	β energy systematics[3]	Myers	Liran and Zeldes	Seeger and Howard	Bauer
0.35 ± 0.05	0.400	0.02	0.61	0.9	–0.22

^{188}Pt

The decay properties of ^{188}Pt were investigated with the use of mass-separated sources. Singles and coincidence measurements were made by using x- and γ-ray Ge(Li) detectors as well as a Si(Li) electron detector. The ^{188}Ir level scheme constructed on the bases of these experimental data is shown in Fig. 2. Total transition intensities per 100 ^{188}Pt decays are given following γ-ray energies. As in the case of ^{172}Hf, they were calculated from photon intensities and theoretical conversion coefficients.[1,2] Electron–capture intensities were deduced from intensity balances.

The ^{188}Pt decay energy has been determined by Hanson et al.[5] to be $519 \leqslant Q \leqslant 558$ keV from the upper limit on the K–to–total–capture ratio for the 478-keV level and by assuming a log ft $\geqslant 4.4$. Capture ratios for the 187.6-, 195.1-, and 478.2-keV levels, determined from our coincidence data, are listed in Table II. Theoretical values[6] and the results of Hanson et al.[5] are also included in the table. The K–to–total–capture (ϵ_K/ϵ) ratios for the first two levels were calculated from K x–rays observed in coincidence with the K conversion–electron lines of the 187.6- and 195.1-keV transitions. Specifically, they were deduced from the intensities of the K x–rays and their sum peaks. Hanson et al.[5] calculated ϵ_K/ϵ from the ratio of the K x–ray intensities seen in coincidence with K and L conversion electrons. The two sets of experimental numbers are in agreement with each other. Our ϵ_K/ϵ value for the 187.6–keV level, however, agrees somewhat better with theory.

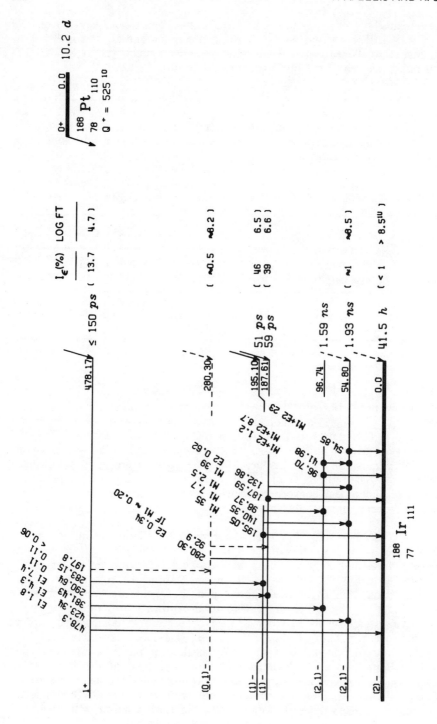

Fig. 2. Decay scheme of ^{188}Pt.

Table II. Capture Ratios for the 188-, 195-, and 478-keV Levels

	Present work	Ref. 5	Theory $Q^+ = 525 \pm 10$	$Q^+ = 550 \pm 10$
ϵ_K/ϵ (188 level)	0.747 *15*	0.766 *23*	0.748 *3*	0.755 *3*
ϵ_K/ϵ (195 level)	0.740 *15*	0.744 *20*	0.746 *3*	0.753 *3*
ϵ_K/ϵ (478 level)	<0.003	⩽0.01	0	0.00 *3*
ϵ_L/ϵ (478 level)	0.67	–	0.65 *5*	0.698 *13*

For the 478.2-keV level, ϵ_L/ϵ was determined from the γ-ray spectrum seen in coincidence with L x-rays. The 423.3-keV γ ray did not appear in the spectrum coincident with K x-rays. The limit for ϵ_K/ϵ given in Table II was deduced from a comparison of the 187.6- and 195.1-keV γ-ray intensities in that same spectrum with the upper limit which could be set for the number of 423.3-keV counts. The same method was used by Hanson et al.;[5] the two experimental limits are in agreement.

It can be seen from Table II that the experimental ϵ_K/ϵ and ϵ_L/ϵ ratios agree with theory for electron-capture decay energies in the range of 515 to 560 keV. Best overall agreement is obtained for 525 ± 10 keV. With this Q value we calculated the expected K x-ray intensity to be 99 ± 3 per hundred capture decays [$\omega_K = 0.962 \pm 0.018$ (Ref. 7) was used]. This compares well with our measured value of 96 ± 10.

For their mass adjustment, Wapstra and Bos[3] adopted a $Q_{E.C.}$ value of 535 ± 9 keV, which was obtained by using the data of Hanson et al.[5] This value agrees well with our experimental result. Calculations of the ^{188}Pt decay energy by using various mass formula range from 0.06 MeV to 1.2 MeV. The experimentally deduced $Q_{E.C.}$ values are compared with some selected predictions in Table III.

Table III. Q Values (in MeV) for Electron-Capture Decay of ^{188}Pt

Experimental		Predictions[4]			
Present work	Ref. 5	Myers	Liran and Zeldes	Seeger and Howard	Bauer
0.525 ± 0.010	⩾0.519, ⩽0.558	0.06	0.34	1.2	−0.50

References

*Research sponsored by the Division of Basic Energy Sciences, U. S. Department of Energy, under contract W-7405-eng-26 with the Union Carbide Corporation.
1. R. S. Hager and E. C. Seltzer, Nucl. Data A4, 1 (1968).
2. O. Dragoun, Z. Plajner, and F. Schmutzler, Nucl. Data Tables A9, 119 (1971).
3. A. H. Wapstra and K. Bos, At. Data Nucl. Data Tables 19, 175 (1977).
4. A. H. Wapstra and K. Bos, At. Data Nucl. Data Tables 17, 474 (1976).
5. R. J. Hanson, P. G. Hansen, H. L. Nielson, and G. Sorensen, Nucl. Phys. A115, 641 (1968).
6. N. B. Gove and M. J. Martin, Nucl. Data Tables 10, 205 (1971).
7. W. Bambynek, B. Crasemann, R. W. Fink, H. U. Freund, H. Mark, C. D. Swift, R. E. Price, and P. V. Rao, Rev. Mod. Phys. 44, 716 (1972).

COMPARISON BETWEEN DIRECT MASS MEASUREMENTS AND DETERMINATIONS OF

BETA ENDPOINT ENERGIES OF NEUTRON RICH RB AND CS ISOTOPES

H. Wollnik, F. Blönnigen, D. Rehfield
G. Jung, B. Pfeiffer, E. Koglin

II. Physikalisches Institut
der Justus Liebig-Universität
6300 Giessen, Federal Republic of Germany

Recent direct mass measurements are compared to precise Q_β-values. The corresponding beta spectra were recorded with an intrinsic Ge-detector which was calibrated with conversion electrons from the BILL spectrometer. By the use of a magnetic pre-separation recently only the high energy portion of the beta spectrum was recorded. Thus count rate limitations of the beta detector were circumvented allowing the use of stronger ß-sources and yielding beta spectra with essentially zero background.

1. Introduction

Principally two methods exist to obtain the masses of short lived neutron rich nuclei. One is to determine mass differences between neighbouring nuclei directly by employing mass spectroscopy and the other is to determine the mass difference between two nuclei of the same isobar by measuring the beta endpoint energy of the corresponding beta decay if necessary in coincidence with the γ-deexcitation of the daughter nucleus. In both cases the methods of on-line separators must be employed because of the short life times of nuclei involved.

2. The technique of direct mass measurements

Using a high resolving mass spectrometer one can directly determine the mass differences between different elements of one isobar. For neutron rich nuclei these mass differences are in the order of about 3 MeV to 10 MeV. With the fission-product-recoil-separator LOHENGRIN at the ILL in Grenoble such measurements were tried,

determining[1]) for instance the mass difference between ^{94}Rb and
^{94}Sr as 10 000 ± 500 keV.

The same types of experiments can be performed also with an ion
source on-line separator if the resolving power of the separator
can be improved sufficiently. A corresponding experiment was recently performed[2]) at the one on-line separator ISOLDE in Geneva.
Here ISOLDE provided short lived Rb or Cs ions of one specific
mass which were implanted into a hot tantalum foil. From this foil
the monoisotopic atoms diffused to the surface within some 10
milliseconds, were surface ionized, again accelerated to 10 keV
and mass analyzed in a double focusing high resolving mass spectrometer. Since here element selective surface ion sources were
used, only masses at one mass unit separation (i.e. about 1000 MeV)
were available for an investigation. Thus for the necessary peak
matching method rather larger mass jumps were required asking for
extreme precision in the determination of the quickly switched deflecting and accelerating voltages of the mass spectrometer. Since
ISOLDE supplied at one time only ions of one specific mass naturally also the magnetic flux density in the ISOLDE mass separator
had to be switched accordingly.

This method of investigation proved extremely efficient and allowed
to measure the masses of Rb$^{74-100}$ and Cs$^{117-149}$ with high precision. Some difficulties arose, however, if the isotopes under investigation were only partially produced in their ground state and
partially in some isomeric state. In this case principally a narrow
mass doublet should have been observed. Since this doublet, however, was too narrow to be resolved by the existing system, a
slightly broadened mass line was observed, the center of which
corresponds to a mass somewhere in the middle between the two
doublet masses. Since normally the isomeric state has a shorter
life time than the ground state, however, normally two experiments
were performed at two different ion source temperatures so as to
increase and decrease the hold up time in the source and thus decrease and increase, respectively, the state of shorter beta half
life.

3. The technique of determining the total decay energy from beta-spectroscopy

The classical alternative to direct mass measurements is to determine the total ß-decay energy be recording the beta endenergy
either of a beta single spectrum or of the beta spectrum obtained
in coincidence to some γ-deexcitation of the daughter nucleus.

Normally plastic scintillators are used as beta-detectors. Recently,
however, we have also used intrinsic Ge-detectors[3,4,5]) for this
purpose at the fission product on-line separator OSTIS in Grenoble[6]).

This detector was calibrated with monoenergetic electrons obtained from the conversion electron spectrometer BILL. Such Ge-detectors have much narrower response functions for monoenergetic electrons allowing to determine the beta endpoint energy with higher precision. Intrinsic Ge-detectors furthermore are sensitive to γ-radiation so that they easily can be recalibrated every day with easy to obtain γ-sources. Disadvantageous with such detectors is their limited size so that normally only 0.2 to 1 per cent of all emitted electrons were recorded. These values should be compared to the 2 to 10 percent efficiencies obtained usually with plastic detectors. The recordable beta-countrate with this detector also had to be limited to values of about 1000 per second if one wanted to avoid pile up in the neighbourhood of the endpoint of the beta spectrum. Using an electronic pile up rejector, which eliminated all electrons arriving at the detector separated by more than 200 nsec and less than 10 μsec, the beta count rate could be increased to about 10 000 per second. In order not to rely on this electronic circuit exclusively, we recorded each beta spectrum twice using beta countrates which differed by a factor of two postulating that in neither case pile up should be observed. The finally achievable errors in the determination of beta endpoint energies were optimally in the order or only a few keV with this detector. These errors have to be enlarged however, for beta energies above 6 MeV were a calibration with electrons of known energy became difficult and where no γ-rays were available for a daily recalibration.

As usual in beta spectroscopy the count rate limitation required rather long recording times for any good measurement. We used normally 20 hours for one beta spectrum. In order to have only 10^4 or 10^5 radioactive atoms per second behind the exit slit of an on-line separator one must reduce the possible ion current in most installations drastically. At our OSTIS in Grenoble about 10^7 radioactive atoms per second can be collected optimally while at the ISOLDE in Geneva even 10^9 to 10^{11} radioactive atoms per second are obtainable[2].

In order to circumvent this count rate limitation we have recently designed a stigmatic focusing magnetic sector field which finally focuses electrons of all energies onto the beta detector but which allows to intercept the trajectories of all low energy electrons in the middle of the sector field (see fig.1). Electrons with energies of ±10% around an energy which is determined by the magnet current are all transmitted without losses.

A Fermi-Kurieplot for the obtained ß-spectrum of ^{92}Rb is shown in fig.2 as an example. This spectrum was produced in about two hours with several 10^6 atoms of ^{92}Rb per second. Note the essentially zero background above the endpoint of the beta-spectrum.

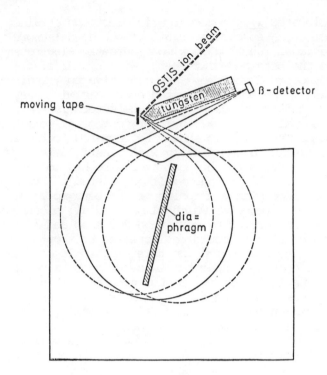

Fig.1: A magnetic sector field filter to be used together with an intrinsic Ge-detector to record ß-spectrum. The transmission of the device is 0.4% of 4π and the region of the beta spectrum transmitted corresponds to ±5% of an arbitrary chosen electron energy.

This device
  1. eliminates all pile up effects quantitatively
  2. reduces γ-radiation by many orders of magnitude because of 20 cm of tungsten shielding (see fig.1) between the radioactive source and the beta detector
  3. reduces the measuring time drastically if the on-line separator can provide strong enough beta-sources
  4. causes also all other sources of high energy background to play only a very much reduced role because of the short recording time.

In order to determine a Q_β-value it also must be established which level of the daughter nucleus corresponds to this endenergy. For this purpose, however, one must know the decay scheme of this nucleus and one must perform ß-γ-coincidences. This becomes increasingly important if one investigates nuclei very far from stability since here the ß-decay normally populates exclusively some high lying levels. To determine the lowest of these levels one can employ some γ-detector together with a plastic scintillator recording electrons with moderate accuracy in a large solid angle as has

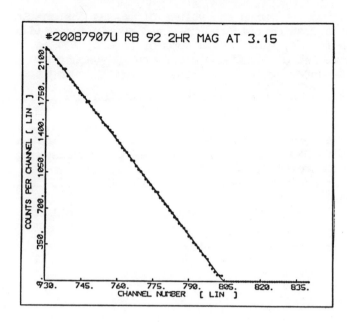

Fig.2: The last 400 keV of the ß-spectrum of ^{92}Rb as obtained with the set up of fig.1. The shown part of the beta spectrum corresponds to a region of about 400 keV.

been done in many experiments. In principle, however, one should also be able to employ a γ-detector together with the set up of fig.1. The drawback of the reduced angle of acceptance of the beta-detector here partially can be compensated by placing a thin sheet of lead before the γ-detector so that the often not interesting low energy γ-rays are drastically reduced. Up to now, however, we have measured only beta single spectra with this magnetic preselection device.

4. Results

By direct mass measurements the masses of neutron rich $^{90-99}$Rb and $^{140-147}$Cs were determined[2] relative to the masses of ^{85,88,89}Rb and ^{131,132,139}Cs. Parallely also Q_β-values were mea-

	ref.[9]	ref.[6,7]	adopted values
^{88}Rb	5309±11	5318±4	5317±3
^{89}Rb	4486±12	5410±8	4503±7
^{139}Cs	4290±70	4213±5	4213±5

Table 1: Q_β-values of ^{88,89}Rb and ^{137}Cs given in keV.

sured[3,4,6,7]) which together with the values of masses for the Sr
and Ba daughter nuclei can be compared to the directly measured
masses. First comparisons yielded deviations of several 100 keV
in all cases. Thus we tried to remeasure also the Q_β-values of
all reference and daughter nuclei involved.

The first case at which we had looked for a comparison was ^{141}Cs.
A comparison of the mass excess of 74377 ± 90 keV of ref[2]) with
the tabulated value[9]) of 75000 ± 100 keV yielded a deviation of
563 keV. Using our new Q_β-value for ^{141}Cs of 5252 ± 20 keV instead
of the old value[9]) of 5040 ± 80 keV this discrepancy decreased by
210 keV. In order to investigate this discrepancy further we remeasured the Q_β-value of ^{141}Ba finding 3208 ± 35 keV instead of[9])
3030 ± 50 keV yielding a mass excess for ^{141}Cs of 74548 ± 40 keV
so that only 173 keV were left as discrepancy to the directly determined mass value.

	mass excess in keV determined from Q_β-values				mass excess in keV determined from direct mass measurements	
mass	z = 40	z = 39	z = 38	z = 37	z = 37	devia-tion
88	–	–	−87911± 3	−82594± 5	–	–
89	–	−87695± 3	−86203± 4	−81700±11	–	–
90	−88765±3	−86481± 4	−85935± 4	−79357±16	−79330±25	−27
91	−87892±3	−86349± 3	−83652± 4	−77795±12	−77781±35	−14
92	−88456±3	−84822±16	−82892±35	−74781±56	−74818±40	+37
93	−87117±3	−84227±20	−80087±75	−72602±105	−72730±65	+128
94	−87264±3	−82350± 5	−78838±10	−68534±32	−68482±45	−52
95	−85663±3	−81218± 5	−75128±95	−65872±100	−65819±100	−53
96	−85445±4	−78325±50	−72912±70	−61242±156	−61220±70	−22
97	−82954±4	−76309±70	−68890±106	−58569±117	−58332±80	−237

Table 2: Mass excess values in keV for $^{88-97}$Zr, $^{88-97}$Y, $^{88-97}$Sr
and $^{88-97}$Rb as taken from ref[9]) and corrected according to
ref.[4,6,7]) and ref.[10,11,12,13,14]). Also listed are the
mass excess values of ref.[2]) corrected according to eq.(1b).
Note that for ^{94}Rb here the mass excess value of 86487 keV
is adopted from ref.[2]) which there was attributed to a possibly metastable state.
The beta endpoint of ^{95}Rb is changed compared to earlier
measurements[6,7]) since it was remeasured recently with the
set up of fig.1. The corresponding γ-level of 681 keV was
taken from ref.[15]). The Q_β-value of ^{96}Rb was taken from
ref.[15]). The beta endpoint energy of ^{97}Rb was taken from
ref.[7]) while the corresponding γ-level of 986 keV was
taken from ref.[15]).

As a last step we checked the Q_β-values of ^{139}Cs which was used as reference mass in ref.[2]) showing deviations to the tabulated[9]) one of 77 keV (see table 1) so that in a rough approximation the Cs masses of ref.[2]) should be increased[8]) by Δ keV where

$$\Delta^{(132+m)}Cs = m(11.3 \pm 0.7)\ keV \qquad m \overset{\geq}{=} 8 \qquad (1a)$$

Thus the mass excess for ^{141}Cs found in ref.[2]) must be increased by 9 times 11.3 = 102 keV to 74479 ± 100 keV. The finally left over discrepancy thus is only 74 keV, a value which is well inside the error bars of 40 and 100 keV.

Remeasuring also the Q_β-values of the reference masses ^{88,89}Rb used for the Rb measurements in ref.[2]) we found deviations of 9 and 24 keV (see table 1). Using a weighted average of these new and the old table values one finds[8]) that all Rb masses of ref.[2]) should be increased by

$$\Delta^{(85+n)}Rb = n(3.1 \pm 1.1)\ keV \qquad n \overset{\geq}{=} 5 \qquad (1b)$$

While the corrections of the directly measured Cs-masses by eq.(1a) are significant the corrections of the directly measured Rb-masses by eq.(1b) are almost negligible compared to the errors of the measurement.

This type of careful checking we continued for all other Rb and Cs isotopes. For this purpose we also used new Q_β-values [3,4,6,7,1o, 11,12,13,14]) obtained since the compilation of the tables of ref.[9]). It should be mentioned at this place that the values of ref.[3,4,6,7]) were determined with our intrinsic Ge-detector while the values of ref.[1o,11,12,13,14]) were determined with plastic scintillators calibrated with new Q_β-values of refs.[3,4,6]).

	mass excess in keV determined from Q_β-values				mass excess in keV determined from direct mass measurements	
mass	z = 58	z = 57	z = 56	z = 55	z = 55	devi- ation
138	–	–	–88273± 7	–82885± 25	–82908± 40	– 23
139	–	–87231± 6	–84952± 7	–80712± 7	–	–
140	–88081±6	–84320± 6	–83285± 12	–77065± 20	–77068± 95	– 3
141	–85438±6	–83008± 31	–79800± 35	–74548± 40	–74479±100	+ 69
142	–84535±6	–80018± 9	–77320±100	–70491±100	–70530±120	– 39
143	–91610±6	–78310± 80	–74051± 40	–67764± 60	–67711±130	+ 53
144	–80431±6	–74996±120	–71936±160	–63485±160	–63292±160	+193

Table 3: Mass excess values in keV for $^{138-144}$Ce, $^{138-144}$La, $^{138-144}$Ba and $^{138-144}$Cs as taken from ref.[9]) and corrected according to ref.[4,6,7]) and [11,13,14]). Also listed are the mass excess values of ref.[2]) corrected according to eq.(1a).

In tables 2 and 3 the mass excess values obtained from Q_β-measurements are listed for $^{88-97}$Rb and $^{138-144}$Cs showing excellent agreement with the mass excess values of ref.[2]) corrected according to eqs.(1). Thus one can state that if reviewed carefully both measurements can provide rather precise and reliable mass values for nuclei far from stability. This situation is illustrated graphically in fig.3 where we have compared the mass excess values for the Rb-isotopes as determined in ref.[2]) and calculated from new beta measurements.

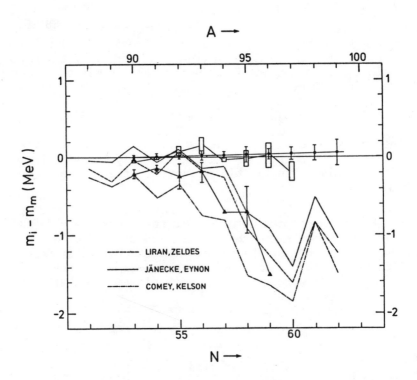

Fig.3: The masses of Rb isotopes from table 2 as determined by direct mass measurements (round dots) and Q_β-values (squares) are compared to older[9]) table values (triangles) and some theoretical mass predictions. Note the almost negligible deviation of the directly determined masses as corrected by eq.(1b) from the original values (the horizontal zero line). Note further the very good agreement between the directly measured masses and the masses as determined from Q_β-values. Note finally the drastic deviation of the theoretical mass predictions which oriented themselves on the old tabulated values apparently giving the masses of the very neutron rich isotopes a higher weight than they should have recieved in view of their large errors.

For allowing us to use a reactor neutron beam we would like to thank the Institut Laue-Langevin in Grenoble. For financial support we are grateful to the German Bundesministerium für Forschung und Technologie.

References

1.   H. Wollnik, G. Siegert, J. Greif, G. Fiedler, M. Asghar,
        J.P. Bailleul, J.P. Bocquet, M. Chavin, R. Decker,
        B. Pfeiffer and H. Schrader

        "Atomic Masses and Fundamental Constants 5" ed.
        J.H. Sanders and A.H. Wapstra, Plenum Press, New York -
        London (1975)

2.   M. Epherre, G. Audi, C. Thibault, R. Klapisch, G. Huber,
        F. Touchard and H. Wollnik
        Phys. Rev. C 19 (1979), 1504

3.   K.D. Wünsch and H. Wollnik
        Brookhaven Lab. Rep. No. 50847

4.   K.D. Wünsch, R. Decker, H. Wollnik, J. Münzel, G. Siegert,
        G. Jung and E. Koglin
        Z. Physik A 288 (1978), 105

5.   K.D. Wünsch
        Nucl. Instr. & Meth. 155 (1978), 347

6.   R. Decker
        thesis, Giessen (1979) unpublished

7.   R. Decker, K.D. Wünsch, H. Wollnik, E. Koglin, G. Siegert,
        and G. Jung
        submitted to Z. Physik A

8.   H. Wollnik
        Proc. Workshop on the spectroscopy of fission products
        in Grenoble (1979), ed. T.v. Egidy, The Institute of Physics
        London

9.   A.H. Wapstra and K. Bos
        Data Nucl. Table 19 (1977), 175

10.   R. Stippler, F. Münnich, H. Schrader, J.P. Bocquet,
        M. Asghar, R. Decker, B. Pfeiffer, H. Wollnik, E. Monnand
        and F. Schussler
        Z. Physik A 284 (1978), 95

11.   R. Stippler, F. Münnich, H. Schrader, K. Hawerkamp,
        R. Decker, B. Pfeiffer, H. Wollnik, E. Monnand and
        F. Schussler
        Z. Physik A 285 (1978), 287

12.   H. Berg, U. Keyser, F. Münnich, K. Hawerkamp, H. Schrader,
        and B. Pfeiffer
        Z. Physik A 288 (1978), 59

13.   U. Keyser, H. Berg, F. Münnich, K. Hawerkamp, H. Schrader,
        B. Pfeiffer and E. Monnand
        Z. Physik A 289 (1979), 407

14.   F.K. Wohn and W.L. Talbert jr.
        Phys. Rev. C 18 (1978), 2328

15.   U. Keyser, F. Münnich, H. Berg, B. Pahlmann, R. Decker
        and B. Pfeiffer
        contribution 6. Conf. Atomic Masses and Fundamental
        Constants, in East Lansing (1979)

MEASUREMENTS OF Q_β FOR NEUTRON-DEFICIENT NUCLEI WITH A $\sim$ 80*

P. E. Haustein, C. J. Lister, D. E. Alburger,
and J. W. Olness

Brookhaven National Laboratory
Upton, NY   11973   USA

INTRODUCTION

Measurement of the decay energies (Q_β) of isotopes which are far removed from the valley of beta stability permits checks of mass theories and provides input for the refinement of mass systematics. In particular, the nuclei which lie on or near the N = Z line with A > 60 are expected to provide some of the most stringent tests of these theories. The nuclei in this region are expected to high-light the influences of a variety of nuclear structure features which affect the nature of the mass surface. These include isospin effects, the influences of nuclear shells on binding energies, single particle versus collective phenomena, coulomb energy considerations, symmetry effects, and the role of nuclear deformation in regions of unusual N/Z ratio. In addition to the delineation of the mass surface in these regions, spectroscopic studies of nuclei which lie off of the beta stability line permit additional tests of the detailed predictions of nuclear structure theories.

Neutron deficient nuclei around ^{80}Zr constitute a rather poorly characterized region with only fragmentary decay properties reported for the isotopes in this region and minimal mass-excess or decay energy data available at the present time. Of particular interest in this region is the rapid change which occurs in the excitation energy of the first 2^+ levels in even-even nuclei. The sharp depression[1] in this quantity for Z $\sim$ 40 and N $\sim$ 40 is taken as evidence for exceptionally large oblate deformation for these nuclides.

*Research performed at Brookhaven National Laboratory under contract with the U.S. Department of Energy.

Earlier studies[2] of the nuclei in this region were performed off line with radioactive sources which had been produced by a variety of heavy-ion reactions on targets of normal isotopic distribution. These production reactions were, in general, performed at energies well above that of the coulomb barrier. Isotopic assignments were made in many cases with the assumption that the dominant reaction channel was (HI,Xn). However at the reported bombarding energies this assumption is dubious because evaporation calculations and reaction systematics indicate that proton and alpha particle emission competes fully with neutron evaporation. Conflicting reports[3,4] on the decay properties of ^{79}Sr and the reported[5] observation of anomalously long half-lives for nuclides, e.g. ^{78}Sr, with high predicted decay energies led to a decision to reinvestigate this mass region.

The present study has resulted in the observation of a new isotope, ^{80}Y, $T_{1/2}$ = 34 secs, as well as a general investigation of the decay properties of several nuclei in this region. Measurements of the total decay energy of ^{80}Y and neighboring nuclides serve to explore the interplay between the mass surface and the extent of nuclear deformation as well as providing the first definitive checks of mass theory predictions in this region. During the course of these investigations a preliminary report on a similar study was published[6] by the French group of Della Negra et al.

EXPERIMENTAL

Techniques

One of the most convenient methods for the production of neutron-deficient nuclei with N $\sim$ Z, A $\sim$ 80 is by means of heavy-ion reactions of the type ^{24}Mg + ^{58}Mg, ^{24}Mg + ^{60}Ni, or ^{40}Ca + ^{40}Ca. The excitation energy of the compound nucleus is kept low and carefully adjusted to maximize the yield of the one or two reaction products of interest which result from the evaporation of just a few nucleons from the relatively "cold" compound system. Intercomparison of "in-beam" γ-ray yields from these reactions with predictions from evaporation calculations are then used to establish relative excitation functions for various reaction channels and to select optimum bombarding energies. These results, in combination with the γ-ray spectra and half-life measurements which are obtained from the radioactive sources prepared from the same reactions, serve to establish isotopic assignments for the products of interest. Energy level schemes are then developed from the analysis of (γ,γ,t) coincidence spectra. Total decay energy measurements (Q_β) are obtained by (β^+,γ) coincidence spectroscopy. Plastic γ-ray detectors are calibrated with several radioactive sources (e.g. ^{27}Si or ^{58}Cu) whose endpoints spanned the expected endpoint energies of the

nuclides of interest. The shapes of these "standard" beta-ray
spectra are then used as input to a stretching alogorithm which
permits the determination of endpoint energies of other β-ray
spectra by comparison to expanded or contracted versions of the
"standard" spectra. In general, Q_β values for each isotope of
interest are determined from weighted averages of endpoint energies
obtained from several selected γ-ray gates.

Apparatus

The helium jet recoil transfer technique was employed as a
means of continuously preparing essentially massless radioactive
sources for β- and γ-ray spectroscopy. Heavy-ion beams from the
BNL tandem Van de Graaff facility were collimated and focussed
onto targets of natural Ni with thicknesses of 0.6-2.5 mg/cm^2.
Recoil products which emerged from the target were thermalized in
helium to which small admixtures of water vapor and isopentyl
alcohol had been added by a bubbler system to improve transport
efficiency. Using a modified version of a target and recoil
collection chamber which has been described[7] previously, the products
were collected after thermalization into a multiple capillary bundle
which consisted of six stainless steel needles. The helium flow
from the capillary bundle was combined into a single capillary of
the same size and the resulting flow was directed through a shielding
wall from the irradiation station to the counting area by means of
polyethylene surgical tubing. At the counting station[8] the helium
was skimmed off and pumped away. The recoil products were then
deposited onto an aluminized Mylar tape loop. The tape could be
advanced from the position where the recoil products were deposited
onto it to a counting area 30.5 cm away in 0.2 sec by means of a
sprocket-fed photoelectric reader system. At the counting location
detectors could be placed to within ∿6 mm of the tape: β and γ rays
emerged from the sources on the tape through a 17 mg/cm^2 Mylar
window. The time periods during which the tape was stopped for
counting of one source and collection of the next could be varied.
In this way the collection/counting system could be "tuned" to enhance
the observation of isotopes with particular half lives. The periodic
removal of the sources from the detector area in this sample shuttling
procedure effectively eliminated interference from long-lived
daughter activities. Except for the rare gas isotopes of Kr, the
helium-jet system produced high intensity sources with little or no
variation in yield from element to element. Figure 1 shows a
schematic view of the helium-jet system.

Data acquisition was accomplished with commercial Ge(Li) and
plastic detectors and standard analog electronics. Coincidence
γ-ray spectra were accumulated through "fast-slow" logic networks
and along with a TAC output for gating of prompt and delayed cascades,

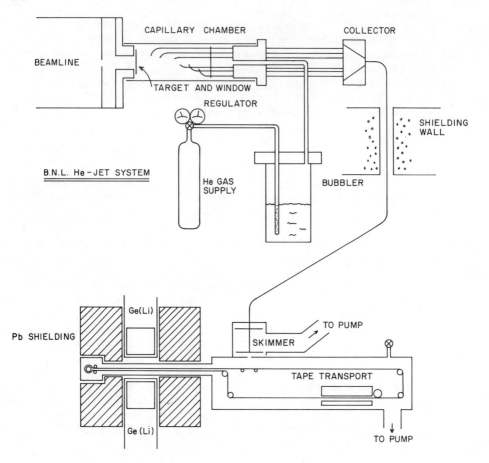

Fig. 1.   Schematic view of the helium-jet system.

the coincidence events were written serially onto magnetic tape on
a Sigma-7 computer for later off-line sorting. Singles spectra
could be recorded in a similar way with a time-from-shuttle parameter
included to permit the generation of decay curve data.

RESULTS

Gamma-ray spectra of radioactive sources which had been produced
by the ^{24}Mg + ^{58}Ni reaction (E_{lab} = 75 MeV) revealed the presence of
several products, e.g., $^{76-79}Rb$ and ^{80}Sr. In addition, several
other lines were observed. These could be attributed to either the

decay of ^{79}Sr or to the decay of a new isotope ^{80}Y, since the γ-lines in the radioactive sources were of the same energy as transitions which have been reported[2] to result from de-excitation of the ground state rotational band in ^{80}Sr. Examination of γ-ray spectra from both radioactive sources and "in-beam" studies was made as a function of incident bombarding energy in order to construct relative excitation functions for these activities and to compare these to a statistical evaporation code. In general, γ-rays were attributed to the decays of either ^{79}Sr or ^{80}Y on the basis of half-life, excitation, and/or coincidence data. These results are summarized in Table 1. A decay scheme for ^{80}Y is shown in Fig. 2. It was developed from coincidence spectra; energy level ordering and the beta-ray feeding pattern were deduced from the γ-ray population and depopulation through levels in the daughter nucleus.

Positron spectra from ^{79}Sr, ^{80}Y, and ^{82}Y, several neutron deficient Rb activities, and calibration sources such as ^{27}Si and ^{58}Cu (which had also been produced and transported in the He-jet system) were accumulated using a plastic scintillator. Two such calibration spectra are shown in Fig. 3; the insert shows the stretch factor α, (E = A + Bα) as a function of energy. It was used to compress or expand the calibration spectra and these were

Table 1.   Half-lives, Gamma-ray Energies, and Relative Intensities in the Decays of ^{79}Sr and ^{80}Y

Isotope	$T_{1/2}$	E_γ (keV)	I_γ (relative)
^{79}Sr	2.30 ± 0.10 min	104.97 ± 0.12	100
		140.98 0.14	29
		219.62 0.15	20
		507.55 0.15	16
^{80}Y	33.8 ± 0.6 sec	385.87 ± 0.10	100
		595.06 0.15	43
		690.50 0.35	3
		756.48 0.13	11
		782.81 0.16	6
		851.90 0.15	9
		1105.80 0.30	4
		1185.22 0.15	15
		1267.67 0.35	5
		1277.52 0.35	3
		1394.55 0.45	1

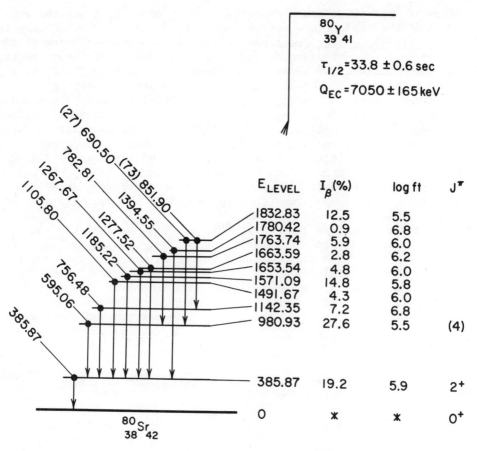

Fig. 2.   Decay scheme for ^{80}Y.

then used as standard shapes and fitted to positron spectra of ^{79}Sr, ^{80}Y, or ^{82}Y, e.g. Fig. 4. Q_{EC} values were thus determined by this method. In estimating errors, contributions from the fitting procedure, the statistical quality of both the calibration and "unknown" spectra and the linearity and reproducibility of the detector response were included. Table 2 lists the experimentally determined Q_{EC} for ^{79}Sr, ^{80}Y, and ^{82}Y along with predictions from several atomic mass theories.[9,10]

DISCUSSION

The experimentally determined value of 5.07 $\pm$ 0.07 MeV for the Q_{EC} of ^{79}Sr compares most closely with the prediction[9] of 5.14 MeV

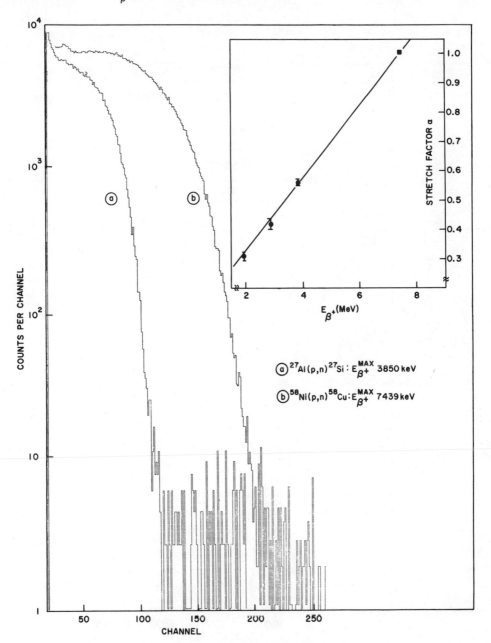

Fig. 3.   Positron calibration spectra of ^{27}Si and ^{58}Cu.   The
insert shows the energy dependence of the stretch factor
α.

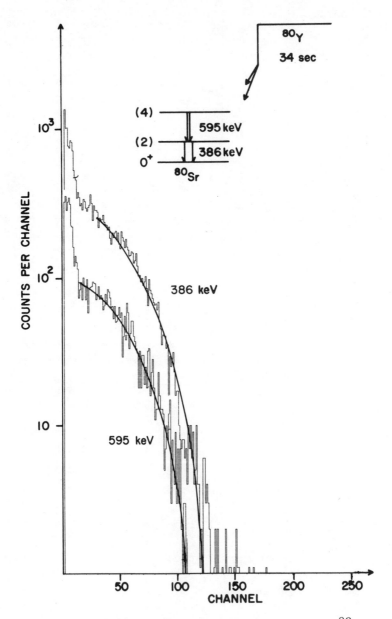

Fig. 4.   Positron spectra from the decay of [80]Y.

Table 2.   Experimental and Predicted Q_{EC} in MeV

Isotope	$^{79}_{38}$Sr	$^{80}_{39}$Y	$^{82}_{39}$Y
Experiment	5.07 ± 0.07	7.05 ± 0.17	7.75 ± 0.09
Myers[a]	5.14	8.74	7.18
Groote, Hilf, Takahashi[a]	5.41	8.80	7.22
Seeger, Howard[a]	5.5	9.1	7.6
Liran, Zeldes[a]	6.05	10.42	8.58
Beiner, Lombard, Mas[a]	6.0	---	---
Jänecke, Garvey, Kelson[a]	5.88	9.85	8.25
Comay, Kelson[a]	5.39	9.74	8.28
Jänecke, Eynon[a]	5.72	9.62	7.99
Monahan, Serduke[b]	6.09	10.10	8.25
Predicted average	5.68	9.55	7.92

[a]Ref. 9.
[b]Ref. 10.

of the Myers Semiempirical Droplet Model. It is however lower than all of the predictions by up to ∿1 MeV. Since this is of the order of the spread in the predictions, the experimentally measured Q_{EC} of ^{79}Sr, by itself, sheds little light on the predictive properties of the presently available mass models. The cases of ^{80}Y is, however, more instructive. Predictions for the ^{80}Y decay energy range from 8.74 to 10.42 MeV with an average of 9.55 MeV. Over 2.5 MeV separates the experimental result of 7.05 ± 0.17 MeV from this average with all of the predictions indicating significantly higher decay energy than that observed. Examination of Q_{EC} values determined from our data for several well characterized Rb and Sr activities which were produced along with ^{79}Sr and ^{80}Y in our experiment show general agreement with earlier work. The measured Q_{EC} for ^{82}Y ($T_{1/2}$ = 9.5 sec) of 7.85 ± 0.09 MeV, for example, is correctly predicted ($Q_{EC}{}^{ave}$ = 7.92 MeV). This tends to discount the possibility of any large systematic errors in our measurements.

It is clearly of interest to extend measurements of this type to other nuclei in this region, especially those further removed from the stability line. The region near ^{80}Zr appears to be one of rapidly changing nuclear deformation and enhanced collective effects. The low Q_{EC} of ^{80}Y may be a signature of such phenomena as they relate to sensitivity of the mass surface to nuclear deformation.

REFERENCES

1. E. Nolte, Y. Shida, W. Kutschera, R. Prestele, and H. Morinaga,
    Z. Physik, 268:267 (1974).
2. C. M. Lederer and V. S. Shirley, "Table of Isotopes," Wiley
    & Sons, New York (1978) 7th ed.
3. A. N. Bilge and G. G. J. Boswell, J. Inorg. Nucl. Chem.,
    33:2251 (1974).
4. I. M. Ladenbaner-Bellis, H. Bakhru, and B. Jones, Can. J.
    Phys., 50:3071 (1972).
5. A. N. Bilge and G. G. J. Boswell, J. Inorg. Nucl. Chem.,
    33:4001 (1971).
6. Annual Report (1978), Institut de Physique Nucléaire, Division
    de Radiochimie, Université Paris-Sud.
7. R. E. Leber, P. E. Haustein, and I.-M. Ladenbauer-Bellis,
    J. Inorg. Nucl. Chem., 38:951 (1976).
8. D. E. Alburger and T. G. Robinson, Nucl. Instr. Methods.,
    164:507 (1979).
9. At. Data Nucl. Data Tables 17:411 (1976).
10. J. E. Monahan and F. J. D. Serduke, Phys. Rev. C, 17:1196 (1977).

EXPERIMENTAL BETA-DECAY ENERGIES OF NEUTRON-RICH ISOTOPES

OF I , Xe, Ce, and Pr*

U. Keyser, H. Berg, F. Münnich, and B. Pahlman

Institut A für Physik, Technische Universität
Braunschwieg, Federal Republic of Germany

K. Hawerkamp, B. Pfeiffer, and H. Schrader

Institut von Laue-Langevin, Grenoble, France

E. Monnand

D.R.F., CEN-Grenoble, France

INTRODUCTION

The present study is part of a systematic investigation of the
decay energies of short-lived fission products, performed at the
on-line mass separators LOHENGRIN and OSTIS of the ILL in Grenoble,
France. The program was initiated to establish systematic trends of
decay energies for nuclei far from the line of β-stability and to
test the reliability of theoretical mass formulae.

The βγ-coincidence system used in these investigations consists
of a plastic scintillator telescope or an intrinsic Ge-detector for
the electrons and a big Ge(Li)-detector for the γ-rays. Details
concerning the experimental techniques applied have been given in
earlier reports[1,2,3].

The uncertainty in energy resulting from the limited calibration
accuracy of about 50 keV is included in the experimental error bars
for the measured β-endpoint energies.

*Supported in part by Bundesministerium für Forschung und Technologie
 (BMFT), Germany.

EXPERIMENTAL RESULTS

     The experimental results of the βγ-coincidence and β-singles
measurements of several heavy fission products obtained recently are
summarized in Table I. To deduce the Q_β-value from measured endpoint
energies, at least some details of the decay scheme of the nucleus
studied must be known. This information is very scarce for some of
the nuclei studied here. However, the results of the present
βγ-coincidence and β-singles measurements made it possible to derive
a reliable value of the decay energy also in these cases, as will be
shortly discussed below for these nuclides.

$^{136}_{53}$I: Western et al.[4] have observed 142 γ-rays in the decay of the
     45 s and 85 s isomers of ^{136}I; 116 of them with energies up to
6624 keV have been placed in a level scheme of ^{136}Xe. The levels
below 2869 keV are well established and have γ-transitions to the
ground-state with exception of the 2262 keV level, which is fed by
the high spin isomer of ^{136}I. This level decays only to a level of
1892 keV energy. For the β-singles spectrum, no decomposition into
components was possible due to the presence of several strong
β-transitions with energies around 5 MeV in the decay of the two
isomers of ^{136}I.

$^{137}_{53}$I: A detailed study of the decay of this nuclide has been
     performed recently by Monnand[5] also using the LOHENGRIN mass
separator. A Q_β-value of 5890±80 keV is derived from the five
βγ-coincidence measurements presented in Table I; it is in excellent
agreement with the β-endpoint energy E_β = 5880±80 keV of the singles
spectrum.

$^{138}_{53}$I: A total of 98 γ-lines have been assigned to the decay of ^{138}I;
     57 of them have been placed in a decay scheme by Hoff[6]. The
endpoint energies of β-transitions to five levels with strong
β-feeding have been evaluated; the Q_β-value deduced agrees well with
the endpoint of the β-singles spectrum measured with the intrinsic
Ge-detector.

$^{139}_{53}$I: No decay scheme of this nuclide has been published in the
     literature; only five low-energy γ-transitions have been
reported by Kratz et al.[7] and Monnand[8] with the following energies:
E_γ = 258 keV, 528 keV, 537 keV, 571 keV and 848 keV. Several
β-singles spectra have been measured both with the plastic
scintillator telescope and with the intrinsic Ge-detector.
Coincidence spectra could be evaluated for the two strongest γ-lines
of 528 keV and 848 keV. The endpoint energies of the 528 keV γ-gate
and of the β-singles spectra agree within the given errors. Therefore,
it is assumed that the ground-state is not populated by β-decay;
both spectra are attributed to the transition feeding the level of
528 keV energy which is probably the first excited state. The level
depopulated by the 848 keV γ-transition is not known. The measured

Table I.   Experimental β-endpoint energies and Q_β-values

Nucleus	$T_{1/2}$ s	γ-gate keV	Level keV	E_β keV	Q_β-this work keV	Q_β-others keV	Ref.
$^{136}_{53}I_{83}$	85.1	singles	0	6955±120	6925± 70	6980± 70	18,19
		345	2634	4240±210		6600±200	
		977	2290	4735±180			
		1313+1321	2634	4305±180			
		1536	2849	4005±240			
		2290	2290	4635±180			
		2415	2415	4570±180			
		2634	2634	4165±210			
		2869	2869	4055±210			
	44.8	370+381	1694	5850±240	7565±120	–	–
		370+381	2262	5300±240			
		1313+1321	1694	5885±150			
$^{137}_{53}I_{84}$	24.2	singles	0	5880±120	5885± 80	5500±200	18
		601	601	5350±200			
		1219	1219	4635±120			
		1303	1303	4490±200			
		1535	1535	4365±160			
		1766	1766	4180±160			
$^{138}_{53}I_{85}$	6.4	singles	0	7830±120	7820± 70	–	–
		589	589	7255±120			
		875	1464	6285±170			
		1277	1867	5925±240			
		1673	2263	5600±200			
		1809	2398	5285±240			
$^{139}_{53}I_{86}$	2.3	singles	528	6210±150	6815±100	–	–
		528	528	6345±120			
		848	(1376)	5520±150			
$^{137}_{54}Xe_{83}$	230	455	455	3570±120	4000±100	–	10–13
		849	849	3112±160			
$^{149}_{58}Ce_{91}$	3.0	58	58	4120±180	4185± 80	–	–
		87	87	3990±150			
		322	380	3760±210			
		380	380	3770±150			
		865	865	3525±250			
		893	893	3395±180			
$^{150}_{58}Ce_{92}$	3.4	110	686	2425±180	3080±100	–	–
		141	686	2475±200			
		290	686	2395±200			
		431	686	2280±200			
		1141	(1141)	1910±200			
$^{150}_{59}Pr_{91}$	6.2	130	130	5515±180	5705± 90	5680±260	17
		130	1062	4545±210			
		722	852	4900±210			
		852	852	4945±210			
		931	1062	4685±180			

endpoint energy of 5520 keV is in agreement with the assumption that this γ-transition populates the 528 keV level directly from an excited state of 1376 keV.

$^{137}_{54}$Xe: The decay scheme of this nuclide is known from literature[9].
The present Q_β-value of 4000±100 keV has been derived from the two γ-gates of 455 keV and 849 keV and is in good agreement with the values published by Onega et al.[10], and Holm[11]. The mean value of these three measurements is Q_β = 4070±50 keV. This value, however, differs outside the given errors from the value of Q_β = 4344±23 keV, obtained by Moore et al.[12] from a study of the ^{136}Xe(d,p) reaction and adopted by Wapstra and Gove[13] in their compilation. There is, however, some uncertainty in the data used for the adopted value, cf. Bunting[14] for details.

$^{149}_{58}$Ce: No decay scheme of this nuclide has been published in the literature. Preliminary results of measurements in progress[15] indicate low energy levels in ^{149}Pr with the following energies: E_γ = 58 keV, 87 keV, 380 keV, 865 keV, and 893 keV. This information has been used to derive the decay energy of ^{149}Ce from the six endpoint energies given in Table I.

$^{150}_{58}$Ce: No decay scheme of this nuclide is known, only the energy of some γ-transitions have been reported[16]. The endpoint energies of five γ-gates have been evaluated; from the results of four of them, a preliminary decay scheme can be deduced, which is corroborated by the following energy sum relationships (E_γ in keV):
$$109.9 + 145.5 = 255.4$$
$$255.2 + 141.1 = 396.3$$
$$141.1 + 289.9 = 431.0$$
$$396.3 + 289.9 = 686.2$$
$$430.7 + 255.2 = 685.9$$
The following excited levels are proposed for ^{150}Pr: 109.9 keV, 255.3 keV, 545.2 keV, and 686.0 keV.

$^{150}_{59}$Pr: The Q_β-value has been derived from five βγ-coincidence spectra; it is in a good agreement with the value of Q_β = 5680±260 keV given in the literature[16,17].

DISCUSSION

A standard way to display the systematics of beta decay energies is the Way-Wood diagram[20], where the Q_β-values are plotted as a function of the mass number A. In Figure 1, the diagram for even Z odd N nuclei in the mass region 127 ≤ A ≤ 153 is shown. The heavy dots designate some results of this study and of earlier work performed at the LOHENGRIN separator. They fit very well into the systematic pattern.

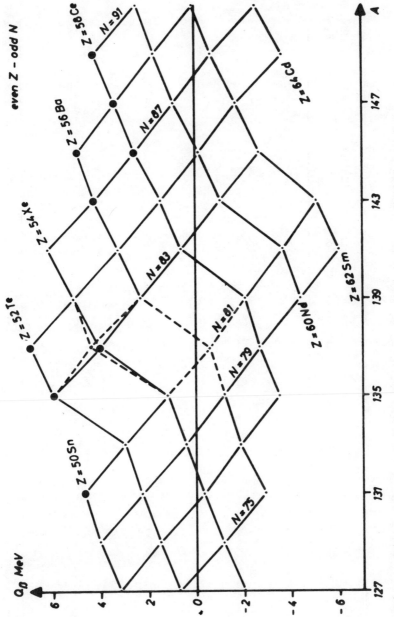

Figure 1. Way-Wood diagram for even Z – odd N nuclei with mass number $127 \leq A \leq 151$

Table II. Comparison of experimental and predicted Q_β-values (deviations in keV)

Nuclide	Exp.value	Myers[23]	Groote[24] et al.	Seeger[25] Howard	Liran[26] Zeldes	Jänecke[27]	Comay[28] Kelson	Jänecke[29] Eynon
$^{136}_{53}$I	6925± 70	+65	-155	+75	+225	+25	+75	+635
$^{137}_{53}$I	5885± 80	-385	-395	-85	-455	-255	-105	+345
$^{138}_{53}$I	7820± 70	-10	-200	+80	+500	-80	+100	+480
$^{139}_{53}$I	6815±100	-485	-485	-115	+445	-245	-85	+265
$^{137}_{54}$Xe	4000±100	+260	+210	-200	+260	-60	-220	+430
$^{149}_{58}$Ce	4185± 80	-55	+75	-285	+215	-305	-225	-335
$^{150}_{58}$Ce	3080±100	-350	-10	+20	-340	-810	-640	-810
$^{150}_{59}$Pr	5705± 90	-575	-655	-1005	+25	-305	-135	-405
$<\left|Q_\beta(exp)-Q_\beta(th)\right|>$		275	275	235	310	260	200	465

One point to be mentioned is the decay energy of ^{137}Xe. In Figure 1, the value derived from the decay studies is shown (heavy dot) together with the value adopted by Wapstra and Bos[21] (light dot). As can be seen, the adopted value seems to be somewhat too large. With the mean value of 4070±50 keV deduced from decay scheme studies, a nearly straight line for the Q_β-values of the isotones with N = 83 is obtained. The discrepancy between the two Q_β-values is possibly due to the existence of an isomeric state in ^{137}Xe, which is known for odd A Xe-isotopes of lower mass. In Table II, the experimental decay energies of this investigation are compared with the predictions of several mass calculations from the 1975 Atomic Mass Predictions [22]. For each formula, the deviation between the experimental and the predicted value is presented; the experimental errors are not taken into account. In the last row, the mean value for each mass calculation is given.

No general conclusions can be drawn from this comparison, because the number of nuclei considered is too small. This is postponed until the study of other heavy fission products in progress at the separators LOHENGRIN and OSTIS is finished.

We are indebted to the technical staff of the ILL, especially Mr. J. M. Gandit and Mr. J. Gowman for their assistance. We also thank Mr. E. Koglin for his cooperation in some measurements and Mrs. Ch. Laupheimer for her help in the preparation of the manuscript.
The financial support of the Bundesministerium für Forschung und Technologie (BMFT) is gratefully acknowledged.

REFERENCES

1. R. Stippler, F. Münnich, H. Schrader, J. P. Bocquet, M. Asghar, G. Siegert, R. Decker, B. Pfeiffer, H. Wollnik, E. Monnand, F. Schussler, Z. Physik A284,95(1978)
2. R. Stippler, F. Münnich, H. Schrader, K. Hawerkamp, R. Decker, B. Pfeiffer, H. Wollnik, E. Monnand, F. Schussler, Z. Physik A285,287(1978)
3. H. Berg, U. Keyser, F. Münnich, K. Hawerkamp, H. Schrader, B. Pfeiffer, Z. Physik A288,59(1978)
4. W. R. Western, J. C. Hill, W. L. Talbert jr., W. C. Schick jr., Phys. Rev. C15,1822(1977)
5. E. Monnand, priv. comm.
6. P. Hoff, J. Inorg. Nucl. Chem., in print
7. K.-L. Kratz, W. Lauppe, and G. Hermann, Inorg. Nucl. Chem. Letters 11,331(1975)

8. E. Monnand, priv. comm.
9. W. R. Western, J. C. Hill, W. L. Talbert jr., and W. Schick jr.,
   Phys. Rev. C15,1024(1977)
10. R. J. Onega, and W. W. Pratt, Phys. Rev. B136,368(1964)
11. G. Holm, Ark. Phys. 37,1(1967)
12. P. A. Moore, P. J. Riley, C. M. Jones, M. D. Mancusi,
    J. L. Foster jr., Phys. Rev. 175,1516(1968)
13. A. H. Wapstra, and N. B. Gove, Nucl. Data Tables 9(1971)
14. R.L. Bunting, Compiler, Nucl. Data Sheets 15,335(1975)
15. E. Monnand, priv. comm.
16. C. M. Baglin, Compiler, Nucl. Data Sheets 18,223(1976)
17. G. Sharnemark, Thesis
18. R. L. Bunting, and J. J. Kraushaar, Compilers, Nucl. Data
    Sheets 13,191(1974)
19. K. Aleklett, E. Lund, G. Nyman, and G. Rudstam,
    CERN-Report 76-13,113(1976)
20. K. Way, and M. Wood, Phys. Rev. 94,119(1954)
21. A. H. Wapstra, and K. Bos, The 1977 Atomic Mass Evaluation,
    Atomic Data and Nuclear Data Tables 19, Number 3(1977)
22. The 1975 Atomic Mass Predictions, Atomic Data and Nuclear
    Data Tables 17, Numbers 5-6(1976)
23. W. D. Myers, 22., p. 411
24. H. V. Groote, E. R. Hilf, and K. Takahashi, 22., p. 418
25. P. A. Seeger, and W. M. Howard, 22., p. 428
26. S. Liran, and N. Zeldes, 22., p. 431
27. J. Jänecke, 22., p. 455
28. E. Comay, and J. Kelson, 22., p. 463
29. J. Jänecke, and E. P. Eynon, 22., p. 467

^{151}Pr, ^{153}Nd AND ^{153}Pm AS NEW ELEMENTS OBTAINED AT SIRIUS

J.A. Pinston[†], F. Schussler[†], E. Monnand[†]
J.P. Zirnheld[*], V. Raut[*], G.J. Costa[*], A. Hanni[*],
R. Seltz[*]

† Institut Laue-Langevin, 156 X Centre de Tri,
38042 Grenoble Cedex, France
* Centre de Recherches Nucléaires et Université Louis
Pasteur, 67037 Strasbourg Cedex, France

INTRODUCTION

During preliminary works on rare-earth elements produced as fission fragments of ^{235}U at the SIRIUS on-line mass separator at the C.R.N. Strasbourg reactor, we have been able to observe two new nuclides, ^{151}Pr and ^{153}Nd and complete the β-decay scheme of ^{153}Pm[1]).

DETECTION AND MEASUREMENTS APPARATUS

The reaction chamber is a cylinder slipped in an horizontal channel of the reactor, so that the uranium target ($\sim$ 27 mg UO$_2$) is submitted to a thermal neutron flux of about 5.10^{11} n/cm^2.s. Inside the reaction chamber, the fission fragments recoil in a gas mixture consisting of helium charged in graphite clusters by the pyrolise of C$_6$H$_6$ vapors in an electric oven. Both the size and concentration of the clusters are adjustable. The fission fragments and the gas mixture are forced to flow through a capillary tube to the ion source, where the fission products are ionized. The ions are then accelerated to about 50 keV and focused near the entry of the analysing magnet. The mass resolving power is of the order of 420 for mass 132. After passing through the spectrometer slits, the ion beam is focused and collected on an aluminium tape transport controlled by a computer M-20 system (Intertechnique).

For these preliminary works we used a 15 % efficiency Ge(Li) with a 2.7 keV resolution for the 1.33 MeV ^{60}Co γ-ray. The M-20 Intertechnique acquisition system was used in the "multiscaling mode".

Experimental results

   a) ^{151}Pr

   Until now, the β-decay of this nucleus was completely unknown;
nevertheless, the ^{151}Nd level scheme has been well established
through (n,γ) and (d,p) reactions[2,3]. From these results, we could
assign unambiguously the γ transitions following the β-decay of
^{151}Pr and consequently, deduce the period $T_{1/2}$.

   The following timing procedure was adopted :
- Irradiation : 20 s.
- "Cooling down" : 4 s.
- Acquisition of eight successive spectra of 6 s. counting time
  each.

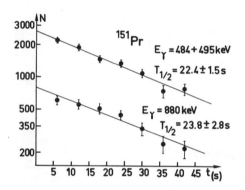

Fig. 1   Decay curve of the 484 + 495 keV and the 880 keV
         γ-rays.

   Fig. 1 shows the intensity decrease of the sum of the 484 +
495 keV and the 880 keV γ-rays. The period of ^{151}Pr found by this
method is : $T_{1/2}$ = 22.4 ± 1.5 s. In these experiments, we were
unable to confirm the 4 s. period previously attributed to this
nucleus[4].

   Knowing the excited states of ^{151}Nd (above mentioned references) we could establish a β-decay scheme of ^{151}Pr (Fig. 2) and
find a new level at 685 keV in ^{151}Nd

   b) ^{153}Nd and ^{153}Pm

   In order to deduce the period of ^{153}Nd, we used a different
timing procedure i.e. 90 s. irradiation and 8 γ-spectra of 8 s.
each. The final spectrum corresponds to a collection of about 30
cycles. Table 1 gives the list of the γ-transitions attributed to
the β-decay of ^{153}Nd, and Fig. 3 shows the decrease of the strongest

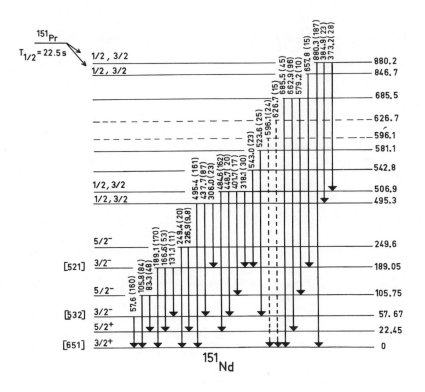

Fig. 2 Decay scheme of levels in ^{151}Pr

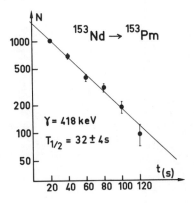

Fig. 3 Decay curve of the 418.3 keV γ-ray in ^{153}Nd

Table 1   γ-rays observed in the β-decay of ^{153}Nd

E_γ	F_γ
32.2 ± 0.3	a)
83.0 ± 0.3	27 ± 11
105.4 ± 0.2	36 ± 5
255.2 ± 0.3	14 ± 4
345.2 ± 0.2	18 ± 4
418.3 ± 0.2	100
475.2 ± 0.2	33 ± 5

a) No significant value due to uncertainties

Table 2   γ-rays observed in the β-decay of ^{153}Pm.

E_γ	I_γ
28.3 ± 0.3	a)
35.8 ± 0.3	a)
83.3 ± 0.3	11.2 ± 2.0
90.9 ± 0.3	15.5 ± 3.0
91.5 ± 0.3	11.7 ± 2.0
119.8 ± 0.2	41.5 ± 5.0
127.3 ± 0.2	100
129.4 ± 0.3	
147.1 ± 0.3	3.0 ± 0.5
166.6 ± 0.3	2.6 ± 0.5
175.4 ± 0.2	11.5 ± 2.0
182.9 ± 0.2	14.0 ± 2.0
196.9 ± 0.3	0.8 ± 0.3
254.8 ± 0.3	1.9 ± 0.5
269.2 ± 0.3	1.5 ± 0.4
276.7 ± 0.3	2.0 ± 0.5
321.1 ± 0.3	1.3 ± 0.4
397.9 ± 0.3	1.5 ± 0.4
442.5 ± 0.3	1.8 ± 0.4
630.2 ± 0.3	2.2 ± 0.5

a)   No significant values, due to uncertainties.

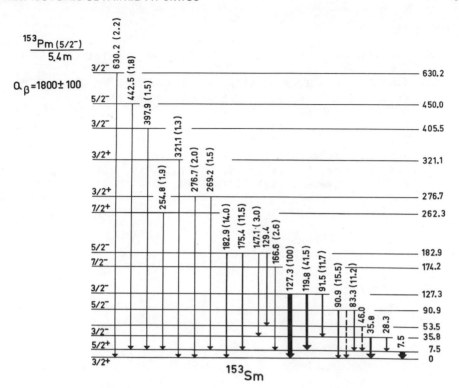

Fig. 4   Decay scheme of levels in ^{153}Pm.

γ-ray (418.3 keV) belonging to this nucleus. The period obtained is : $T_{1/2}$ = 32 ± 4 s.

For the study of ^{153}Pm, whose period was known[5] ($T_{1/2}$ = 5.4 mn) the following timing was chosen :
- Irradiation : 10 mn.
- Cooling : 20 s.
- Acquisition : 4 γ-spectra of 5 mn. each.
The final spectrum was the result of 4 successive cycles.

The list of the γ transitions attributed to ^{153}Pm is shown in Table 2. The spectrum of the γ-ray following the β-decay of ^{153}Pm has been previously obtained by Smither et al.[6], from the ^{154}Sm(γ,p)^{153}Pm reaction, but only γ-ray transitions with energy below 183 keV were observed. Due to better experimental conditions we were able to place in the level scheme of ^{153}Sm, γ-rays of energy as high as 630 keV. All the γ-ray lines observed in our experiment are reported in the preliminary decay scheme (Fig. 4) and confirm the results based on (n,γ) and (n,e⁻) reactions on ^{152}Sm[6].

J. A. PINSTON ET AL.

References

1. J.P. Zirnheld, L. Schutz and K.K. Wohn, Nucl.Inst. <u>158</u> (1979) 409
2. J.A. Pinston, R. Roussille, H. Börner, W.F. Davidson, P. Jeuch, H.R. Koch, K. Schrekenbach and D. Heck, Nucl.Phys. <u>A270</u>(1976) 61
3. Nucl.Data, <u>19</u> (1976) 33
4. U.C.R.L. 19530 (1970) 178
5. Nucl.Data Sheets 10 (1973) 429
6. R.K. Smither, E. Bieber, T. Von Egidy, W. Kaiser and K. Wien, Phys.Rev. 187 (1969) 1632

GAMMA-RAY ENERGY CALIBRATION STANDARDS

C. van der Leun*, R.G. Helmer** and P.H.M. Van Assche***

Task group of the IUPAP Commission on Atomic Masses and
Fundamental Constants

INTRODUCTION

The task group on gamma-ray calibration standards was established in 1972 by the IUPAP Commission on Atomic Masses and Fundamental Constants. Its task is the production, recommendation and
publication of a consistent set of calibration standards for use
in gamma-ray spectroscopy.
    The present paper is the first official report of the task
group. It presents a consistent set of gamma-ray energies, all with
uncertainties of at most 10 ppm. The energies, in the range E_γ =
60 - 6100 keV, are based on the value[1] E_γ = 411 804.4 $\pm$ 1.1 eV for
the gamma-ray from the decay of ^{198}Au. The highest gamma-ray energy
listed, that of the 6129 keV line in ^{16}O, can be compared with an
energy measurement[2] based on a mass-doublet scale.
    A more detailed account of the activities of the task group
will be published in the near future[3].

GAMMA-RAY ENERGY SCALES

Shortly after the establishment of the task group it became
apparent that two completely different experiments were underway,
that would both provide a definition of the gamma-ray energy scale

*   Fysisch Laboratorium, Rijksuniversiteit, Princetonplein 5,
     3508 TA Utrecht, The Netherlands; task group chairman.
**   EG&G Idaho, Idaho Falls, Idaho, USA; work performed under the
     auspices of the U.S. Department of Energy.
*** SCK-CEN, Nuclear Energy Centre, Mol, Belgium.

with a precision that is about an order of magnitude better than
those available previously. The task group therefore decided to
wait for the outcome of these experiments before making any recom-
mendations.

The following two experiments lead to two basic scales:
1. The double flat crystal spectrometer measurements by Kessler
et al.[1] of the absolute wavelength of the 412 keV gamma-ray from
[198]Au. The result of this experiment, which essentially relates the
gamma-ray wavelength to optical wavelength standards, is a wavelength of 3 010 788.8 $\pm$ 1.1 am for the [198]Au line. The corresponding gamma-ray energy, calculated with the voltage-wavelength conversion factor from the 1973-adjustment of the fundamental constants[4]
is 411 804.4 $\pm$ 1.1 eV. In the final uncertainty the error in the
wavelength measurement (0.37 ppm) is considerably smaller than that
in the conversion factor (2.6 ppm).
2. A second scale is equally important especially for high-energy
gamma-rays. It is based on mass spectroscopic measurements of isotopic mass differences with the mass spectrometer of Lincoln Smith[5].
The deduction of gamma-ray energies from these measurements, with a
quoted precision of about 6 ppm[6], involves the deuteron binding
energy. Since a discrepancy of about 50 eV exists among the two
most recent precision determinations[7],[8] of this binding energy, the
mass doublet scale for gamma-ray energies is currently not in a satisfactory position.

On this basis it was decided to restrict this 1979-recommenda-
tion to gamma-ray energies based on the wavelength scale only.

SELECTION CRITERIA

Source selection

The present list of recommended gamma-ray calibration energies
does not include all the precise gamma-ray energies that have been
published. A selection was made with the general user in mind.
The following criteria were applied:
- Preference is given to lines from long-lived sources, i.e. with a
  half-life of at least 30 d. As an exception lines are included
  from [24]Na, since this short-lived source can be easily produced.
- Sources should be commercially available.
- Sources with spectra too complex for calibration of Ge(Li) and Ge
  gamma-ray spectrometers are not listed.

Data selection

Three criteria have been used in the selection of data:
- They should be of good quality and high precision (error <10 ppm).
- They should be based on the 412 keV [198]Au scale, or easily conver-
  table to this scale.
- The total uncertainty should be separated into a statistical and
  a systematic error.

Ge-spectrometer data

Additional requirements for data from Ge-spectrometers are:
- At least two different spectrometers have been used.
- The analysis is performed without relying on the 511 keV distances between photopeaks and single- and double-escape peaks.
- The lines studied and the calibration lines are measured simultaneously and uni-directionally.

HIGH-ENERGY GAMMA RAYS

The ^{198}Au-standard and the low-energy (E_γ < 3.5 MeV) gamma-rays based on this standard, have been discussed in some detail at this conference by Deslattes and Greenwood, respectively. This discussion might be complemented by a few short remarks on high-energy gamma-rays.

The number of long-lived sources providing high-energy gamma-rays is very limited. One of the best known sources in this category consists of a mixture of enriched ^{13}C and an alpha-particle emitter. The reaction ^{13}C(α,n)^{16}O excites the 6.13 MeV level of ^{16}O which has a long lifetime (τ_m = 27 ps). The decay gamma-ray of 6.13 MeV is therefore not Doppler-broadened.

The cascade-crossover technique cannot be directly applied to measure the energy of this line, since the 6.13 MeV level decays exclusively to the ground state. Cascade-crossover relations were generated in two low-energy resonances (E_p = 319 and 391 keV) in the reaction ^{25}Mg(p,γ)^{26}Al, thus providing very precise energies for two transitions in ^{26}Al with energies $E_\gamma \simeq$ 6.2 MeV.

The latter then were used as calibration standard for the 6.13 MeV line. The dispersion at $E_\gamma \simeq$ 6 MeV has been deduced from the peak-positions of several primary transitions in the ^{25}Mg(p,γ)^{26}Al resonances. The energy differences between these primary transitions are calculated from precision measurements of ten of the lowest bound states of ^{26}Al.

Even at the low proton energies mentioned above, special precautions have to be taken to reduce Doppler-shift effects. Precision geometric alignment and alternating measurements at the detection angles θ = +90° and -90°, reduce this error to a negligible effect.

The final result[9] is
$$E_\gamma = 6\ 129\ 266 \pm 54\ \text{eV}.$$
This value, exclusively based on the ^{198}Au-scale can be compared to the value published by Shera[2], which is based primarily on the mass-doublet scale
$$E_\gamma = 6\ 129\ 170 \pm 43\ \text{eV}.$$
The tempting conclusion that the wavelength scale and the mass-doublet scale thus are practically in agreement, however, might be premature. Gamma-ray energies based on the mass doublets, depend on the deuteron binding energy, on which conflicting recent data exist[7,8]. The energies based on the wavelength scale, on the other hand, might change due to a possibly sizable change in the wavelength-energy con-

C. VAN DER LEUN ET AL.

TABLE I

Recommended gamma-ray calibration energies arranged by energy [a]

Source	E_γ (eV)	Source	E_γ (eV)	Source	E_γ (eV)
^{182}Ta	67 750.0 ± 0.2	^{108}Ag^m	614 281 ± 4	^{182}Ta	1 231 016 ± 5
^{153}Gd	69 673.4 ± 0.2	^{110}Ag^m	620 360 ± 3	^{56}Co	1 238 287 ± 6
^{170}Tm	84 255.1 ± 0.3	^{124}Sb	645 855 ± 2	^{182}Ta	1 257 418 ± 5
^{182}Ta	84 680.8 ± 0.3	^{110}Ag^m	657 762 ± 2	^{182}Ta	1 273 730 ± 5
^{153}Gd	97 431.6 ± 0.3	^{137}Cs	661 660 ± 3	^{22}Na	1 274 542 ± 7
^{182}Ta	100 106.5 ± 0.3	^{198}Au	675 887.5 ± 1.9	^{182}Ta	1 289 156 ± 5
^{153}Gd	103 180.7 ± 0.3	^{110}Ag^m	677 623 ± 2	^{59}Fe	1 291 596 ± 7
^{182}Ta	113 672.3 ± 0.4	^{110}Ag^m	687 015 ± 3	^{124}Sb	1 325 512 ± 6
^{182}Ta	116 418.6 ± 0.7	^{144}Ce	696 510 ± 3	^{60}Co	1 332 502 ± 5
^{152}Eu	121 782.4 ± 0.4	^{94}Nb	702 645 ± 6	^{56}Co	1 360 206 ± 6
^{57}Co	122 061.4 ± 0.3	^{110}Ag^m	706 682 ± 3	^{124}Sb	1 368 164 ± 7
^{192}Ir	136 343.4 ± 0.5	^{124}Sb	713 781 ± 5	^{24}Na	1 368 633 ± 6
^{57}Co	136 474.3 ± 0.5	^{124}Sb	722 786 ± 4	^{182}Ta	1 373 836 ± 5
^{182}Ta	152 430.8 ± 0.5	^{108}Ag^m	722 929 ± 4	^{110}Ag^m	1 384 300 ± 4
^{182}Ta	156 387.4 ± 0.5	^{95}Zr	724 199 ± 5	^{182}Ta	1 387 402 ± 5
^{182}Ta	179 394.8 ± 0.5	^{110}Ag^m	744 277 ± 3	^{124}Sb	1 436 563 ± 7
^{182}Ta	198 353.0 ± 0.6	^{110}Ag^m	763 944 ± 3	^{110}Ag^m	1 475 788 ± 6
^{192}Ir	205 795:5 ± 0.5	^{124}Sb	790 712 ± 7	^{144}Ce	1 489 160 ± 5
^{182}Ta	222 109.9 ± 0.6	^{110}Ag^m	818 031 ± 4	^{110}Ag^m	1 505 040 ± 5
^{182}Ta	229 322.0 ± 0.9	^{54}Mn	834 843 ± 6	^{110}Ag^m	1 562 302 ± 5
^{228}Th	238 632 ± 2	^{56}Co	846 764 ± 6	^{228}Th	1 620 735 ± 10
^{152}Eu	244 698.9 ± 1.0	^{228}Th	860 564 ± 5	^{124}Sb	1 690 980 ± 6
^{182}Ta	264 075.5 ± 0.8	^{94}Nb	871 119 ± 4	^{207}Bi	1 770 237 ± 10
^{203}Hg	279 196.7 ± 1.2	^{192}Ir	884 542 ± 2	^{56}Co	1 771 350 ± 15
^{192}Ir	295 958.2 ± 0.8	^{110}Ag^m	884 685 ± 3	^{56}Co	1 810 722 ± 17
^{192}Ir	308 456.9 ± 0.8	^{46}Sc	889 277 ± 3	^{88}Y	1 836 063 ± 13
^{192}Ir	316 508.0 ± 0.8	^{228}Th	893 408 ± 5	^{56}Co	1 963 714 ± 12
^{51}Cr	320 084.2 ± 0.9	^{88}Y	898 042 ± 4	^{56}Co	2 015 179 ± 11
^{152}Eu	344 281.1 ± 1.9	^{110}Ag^m	937 493 ± 4	^{56}Co	2 034 759 ± 11
^{198}Au	411 804.4 ± 1.1	^{124}Sb	968 201 ± 4	^{124}Sb	2 090 942 ± 8
^{192}Ir	416 471.9 ± 1.2	^{56}Co	1 037 844 ± 4	^{56}Co	2 113 107 ± 12
^{108}Ag^m	433 936 ± 4	^{124}Sb	1 045 131 ± 4	^{144}Ce	2 185 662 ± 7
^{110}Ag^m	446 811 ± 3	^{207}Bi	1 063 662 ± 4	^{56}Co	2 212 921 ± 10
^{192}Ir	468 071.5 ± 1.2	^{198}Au	1 087 691 ± 3	^{56}Co	2 598 460 ± 10
^{7}Be	477 605 ± 3	^{59}Fe	1 099 251 ± 4	^{228}Th	2 614 533 ± 13
^{192}Ir	484 577.9 ± 1.3	^{65}Zn	1 115 546 ± 4	^{24}Na	2 754 030 ± 14
^{207}Bi	569 702 ± 2	^{46}Sc	1 120 545 ± 4	^{56}Co	3 009 596 ± 17
^{228}Th	583 191 ± 2	^{182}Ta	1 121 301 ± 5	^{56}Co	3 201 954 ± 14
^{192}Ir	588 585.1 ± 1.6	^{60}Co	1 173 238 ± 4	^{56}Co	3 253 417 ± 14
^{124}Sb	602 730 ± 3	^{56}Co	1 175 099 ± 8	^{56}Co	3 272 998 ± 14
^{192}Ir	604 414.6 ± 1.6	^{182}Ta	1 189 050 ± 5	^{56}Co	3 451 154 ± 13
^{192}Ir	612 465.7 + 1.6	^{182}Ta	1 221 408 ± 5	"^{16}O"	6 129 270 ± 50

[a] For details and references, see ref.3.

version factor resulting from the next evaluation of the fundamental constants[10].

Although the consistency of the two fundamental scales thus remains a topic to be studied in the years to come, it is gratifying to see that the actual values for the energy of the 6.13 MeV line of ^{16}O based on the two different scales are in reasonable agreement.

SUMMARY AND PROGRAM

The recommended calibration standards for gamma-ray energies are listed in table 1. In cases where more than one measurement is available, the weighted average is listed. In general the different values are in reasonably good agreement. A notable exception is the ^{170}Tm line. For details and references to the original literature the reader is referred to a full article to appear in Atomic Data and Nuclear Data Tables[3].

The about 120 energies listed in the table cover the range E_γ = 60 - 6100 keV, of which the low-energy part of course is covered much better than the high-energy part. The first future aim of the task group obviously is the extension of the present evaluation to the higher gamma-ray energies.

REFERENCES

1. E.G. Kessler, R.D. Deslattes, A. Henins, and W.S. Sauder, Phys. Rev. Lett. 40:171 (1978).
2. E.B. Shera, Phys. Rev. C12:1003 (1975).
3. R.G. Helmer, P. Van Assche, and C. van der Leun, Atomic Data and Nuclear Data Tables, to be published.
4. E.R. Cohen and B.N. Taylor, J. Phys. Chem. Ref. Data 2:663 (1973).
5. L.G. Smith and A.H. Wapstra, Phys. Rev. C11:1392 (1975).
6. A.H. Wapstra, Proc. 2nd Int. Conf. on Neutron Capture Gamma-Ray Spectroscopy, p. 686; R.C.N., Petten, The Netherlands (1975).
7. Ts. Vylov et al., Yad. Fiz. 28:1137 (1978); translation p. 585.
8. R.C. Greenwood and R.E. Chrien, Proc. 3rd Int. Conf. on Neutron Capture Gamma-Ray Spectroscopy, p. 618; Plenum, New York, U.S.A. (1979).
9. P.F.A. Alkemade, C. Alderliesten, P. de Wit, and C. van der Leun, Nuclear Instruments (to be published).
10. E.R. Cohen, communication to AMCO6 (1979).

COMPARISON OF PRECISION MASS MEASUREMENTS OF LIGHT ISOTOPES

Jerry A. Nolen, Jr.

Physics Department and Heavy Ion Laboratory
Michigan State University
East Lansing, MI   48824

This talk is a survey of the present status of precision mass measurements of isotopes below ^{19}F. The need for new precise mass measurements of certain stable as well as unstable light isotopes is presented. Three different techniques for precise mass measurements of unstable isotopes are briefly described and compared. These techniques are the ion source calibration technique currently being developed in Auckland, the rf time-of-flight technique used in Munich, and the kinematic calibration technique developed at Michigan State University. Four selected examples of recent measurements done here at MSU are discussed and compared with the results of other methods.

Despite the tremendous progress made in atomic mass measure-ments in the past 30 to 40 years there are still important reasons for improved measurements of several stable light isotopes. The important work done at Princeton with the precision rf mass spectrometer of Lincoln Smith has provided the masses of 11 light isotopes[1,2] (^{1}H, ^{2}H, ^{3}H, ^{3}He, ^{4}He, ^{13}C, ^{14}C, ^{14}N, ^{15}N, ^{16}O, and ^{19}F) relative to ^{12}C with precisions of better than 100 eV. Unfortu-nately, the remaining stable light isotopes (^{6}Li, ^{7}Li, ^{9}Be, ^{10}B, ^{11}B, ^{17}O, and ^{18}O) still have mass uncertainties of about 1 keV, and in several cases these are the average of inconsistent experi-mental results. Fortunately, Smith's spectrometer has been moved to Delft by Koets and Wapstra and will soon be ready for new mass measurements.[3]

Mass measurements of ^{6}Li and ^{10}B with uncertainties of less than 100 eV are currently needed. The ^{6}Li mass is essential for a search for a parity and isospin violating component of

the strong interaction, an experiment currently in progress by
Robertson, et al.[4]   Plans are underway to attempt this mass
measurement at Delft in the very near future. An error in the
presently accepted value for the ^{10}B mass could explain the discrepancy between the Auckland and MSU determinations of the
^{10}C mass discussed in more detail below.

In addition to the needs for precise masses of stable isotopes
there are also a variety of reasons for very accurate mass measurements of unstable isotopes. In this context, precise excitation
energy measurements in stable or unstable isotopes are often
physically as important as ground state masses. One specific
example is the 0^+, T=1 excited state of ^{6}Li at 3.56 MeV. A precise
measure of the mass of this state relative to the mass of ^{2}H + ^{4}He
is crucial for the success of the parity-violating resonance
experiment mentioned above.   Similarly, the uncertainties in
the masses of the lowest lying T=3/2 levels in ^{9}Be and ^{9}B are
currently the dominant terms in the test of the isobaric mass
multiplet equation discussed by Kashy at this conference.[5]   Of
course, the ground state masses of ^{9}Li and ^{9}C are also important
ingredients of this test. Another important example is the ground
state mass of ^{14}O, which must be determined reliably with a precision of about 500 eV or better. The ^{14}O to ^{14}N, 0^+ to 0^+ superallowed beta decay transition is the best known of the low-Z
elements, and it, therefore, has an important role in establishing
the charge dependency of these ft-values.[6,7] A change of 2 keV
in the mass of ^{14}O is enough to completely eliminate the charge
dependent trend in the ft-values discussed by Wilkinson.[6]

Mass determinations of unstable isotopes are generally carried
out via reaction Q-value measurements. The most precise have
usually been (n,γ) reaction energies which can generally be determined with uncertainties of between 100 and 500 eV depending
on the gamma energies involved. Similar precision has been
obtained in some charged-particle threshold measurements such
as (p,γ) when the beam energy was low enough and special techniques
such as electrostatic analyzers were utilized to determine the
beam energy.

A new fairly general method for very accurate charged-particle
threshold measurements and magnetic spectrograph calibration
is currently being developed in Auckland, New Zealand. A progress
report on this work was given at this conference by Barker.[8]
This method permits the direct comparison of the magnetic rigidities
of protons with energies of several MeV with singly charged heavy
ion beams at less than 100 keV. This method of magnetic spectrograph calibration is potentially more accurate and flexible than
techniques involving the use of alpha sources as standards.

A similar ion-source apparatus for direct determination of accelerator beam energies is being developed at MSU by Dyer and Robertson.[9]

Many charged-particle reaction Q-values have also been determined in magnetic spectrographs with beam energies in the 10-
100 MeV range. However, when alpha sources were used for the
spectrograph calibration there were accuracy limitations at the
1 to 10 keV level imposed by the extrapolations to higher magnetic
rigidities and/or target thickness problems. Recently, however,
two techniques have been developed which should permit the determination of a large number of reaction energies with uncertainties
of 0.1 to 1. keV.

The first of these methods involves the rf time-of-flight
apparatus in Munich. This apparatus was described at AMCO-V[10]
with some important $(^3He,t)$ reaction Q-values being published
in 1977.[11] The Q-values were determined with uncertainties of
a few tenths of a keV even though beam energies of over 20 MeV
were used. These measurements greatly reduced the uncertainties
of several of the super-allowed beta decay ft-values as discussed in references 6 and 7. With this apparatus the beam energy
from the Munich MP-tandem can be measured quickly and very accurately with the simultaneous observation of the elastically
scattered beam and/or reaction products in the Q3D spectrograph.
Unfortunately, no new results have been recently obtained with
this apparatus because of problems with the accelerator in Munich.

Another fairly general method for reaction Q-value measurements with sub-keV uncertainties has been developed at MSU.
This kinematic technique for spectrograph calibration was described[12] in 1974 and some preliminary results of Q-value and
excitation energy measurements were presented[13] at AMCO-V. This
technique combines momentum matching[14] and kinematics for spectrograph calibration with dispersion matching[15] to obtain good energy
resolution. The most precise measurements are done on nuclear
emulsion plates so that all calibration lines and unknown lines
are simultaneously recorded. Excitation energies up to about
7 MeV have been measured with uncertainties of 100-200 eV and
Q-values for reaction such as (p,d) and (p,t) have been measured
with uncertainties of 200-500 eV. Measurements on light isotopes
have included excitation energy measurements in $^6Li, ^{10}B, ^{12}C, ^{14}N, ^{15}N,$
and ^{16}O as well as reaction energies leading to the isotopes
$^6He, ^{10}C, ^{11}C, ^{12}N, ^{13}N, ^{14}O,$ and ^{15}O. Not all of the analysis of
these data is complete, but some examples are discussed below.
In most of this work the energy standards have been the $^{15}N(p,d)^{14}N$
Q-value as determined by the $^{14}N-^{15}N$ mass doublet of Smith and
Wapstra[2] and the related excitation energies in ^{15}N determined
via the $^{14}N(n,\gamma)$ work of Greenwood and Helmer.[16]

The mass of ^{6}He was recently measured at MSU by Robertson, et al,[17] because of its importance as a calibration mass in many other reaction energy measurements such as the (^{4}He,^{8}He) reactions reported at this conference by Tribble. It is also of interest because of the importance of the ^{6}He-^{6}Li ft-value in the physics of beta decay. This measurement is presented here to emphasize the need for more precise mass measurements of certain stable light isotopes. The experiment consisted of a simultaneous comparison of the particle energies from the ^{7}Li(d,^{3}He)^{6}He and ^{19}F(d,^{3}He)^{18}O* (1.982 MeV) reactions on a LiF target. A spectrum of ^{3}He ions recorded by a focal plane detector at 0° scattering angle is shown in figure 1. The use of the uniformly mixed (LiF) target, the 0° scattering angle, and the near degeneracy of the two ^{3}He energies shown in figure 1 made possible the very small total measurement error of 410 eV for this experiment. The final total uncertainty of 1.1 keV for the mass of ^{6}He, however, is dominated by the present uncertainty of 0.9 keV in the mass of the stable isotope ^{7}Li,[18] with the uncertainty of 0.3 keV in the ^{18}O mass[18] also contributing. Hence, better measurements of the masses of these two stable isotopes could reduce the total uncertainty in the mass of ^{6}He to less than 500 eV.

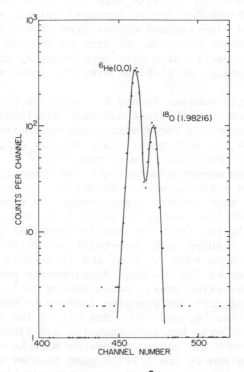

Fig. 1.   Spectrum recorded at 0° of ^{3}He particles resulting from 20.8-MeV deuteron bombardment of a LiF target.

The other examples of recent measurements at MSU, to be discussed here, are the masses of ^{10}C and ^{14}O. These measurements were initially carried out[19] in collaboration with Paul Barker during his visit from Auckland, New Zealand and have since been checked[20] because of discrepancies with other work. These measurements utilized the kinematic technique of focal plane calibration[12,13] mentioned above. In each case (p,p) elastic scattering and (p,p') lines to well known excited states were used to determine the focal plane momentum calibration, while (p,d) reactions were used to determine the beam energy. The ^{16}O(p,t)^{14}O and ^{12}C(p,t)^{10}C reactions were used to measure the ^{14}O and ^{10}C masses, respectively. As mentioned above, the ^{15}N(p,d)^{14}N Q-value and the energy levels of ^{15}N were used as energy standards for this work. This Q-value and several of the ^{15}N energy levels are now known with uncertainties of 200 eV or less permitting measurements of (p,t) Q-values with total uncertainties of about 500 eV or less.

A typical spectrum used in the determination of the ^{10}C mass is shown in figure 2. Here the separation of the ^{12}C(p,d)^{11}C ground state peak from the ^{24}Mg(p,p')^{24}Mg(1.4 MeV) energy level determines the beam energy and the separation of the ^{12}C(p,t)^{10}C ground state peak from the ^{16}O(p,p')^{16}O(6.1 MeV) energy level determines the ^{10}C mass. It was necessary to independently measure the ^{12}C(p,d)^{11}C Q-value relative to that of ^{15}N(p,d)^{14}N since the ^{11}C mass was previously only known to ±1.1 keV.[18] The ^{14}O mass was determined by a similar procedure, also using ^{11}C as an intermediate reference mass.

The result obtained for the ^{10}C mass is 2.7 keV or about 3 standard deviations less than that obtained previously in Auckland[21] using an alpha source to calibrate the beam energy in a (p,n) threshold measurement. It is possible that an error in the mass of the stable isotope ^{10}B could explain the discrepancy. In fact, the discrepancy has already decreased from 3.3 to 2.7 keV due to a change in the ^{10}B mass reported in the 1971 and 1977 mass adjustments.

The present result for the mass of ^{14}O is 0.4±0.5 keV larger than the value reported in the 1977 Mass Table,[18] which primarily represents the weighted average of 3 sub-keV precision ^{12}C(^{3}He,n) threshold measurements. However, the present result is about 2.5 standard deviations larger than the recent value from a ^{14}N(^{3}He,t) measurement at the Munich[21] rf-TOF facility and about 2 standard deviations above the value from a ^{14}N(p,n) threshold measurement at Auckland[22] with an alpha source calibration.

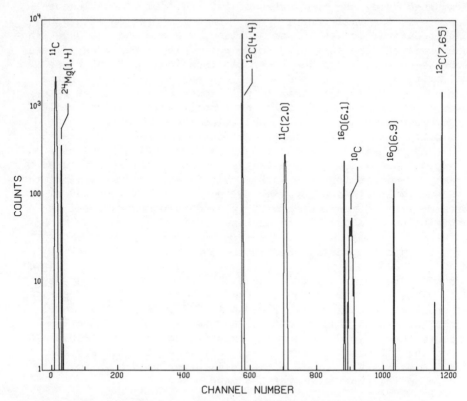

Fig. 2.   Spectrum of protons, deuterons, and tritons recorded
          on nuclear emulsion resulting from a proton beam bom-
          barding a mixed target containing ^{12}C, ^{16}O, and ^{24}Mg.

At this conference Barker reported[8] a very preliminary,
but potentially much more precise measurement of the $^{14}N(p,n)$
threshold using their new ion-source calibration method.   Indications are that their new ^{14}O mass may be slightly above, but
consistent with, the older Auckland value.[22]

The Munich measurement may be checked[23] in the near future
when the upgraded MP Tandem becomes operational. With the higher
beam energies expected from the upgraded accelerator it may be
possible to check the ^{14}O via the $^{16}O(p,t)$ Q-value as well as
with the $^{14}N(^3He,t)$ reaction which was used previously.

Since the MSU value for the ^{10}C mass was less than the Auckland
value, whereas the ^{14}O mass was larger than the other two recent
measurements a direct measure of the $^{10}C-^{14}O$ mass difference

was carried out[20] as a consistency check. In this check the
$^{12}C(p,t)^{10}C$ and $^{16}O(p,t)^{14}O$ Q-values were directly compared as
indicated in figure 3. This difference measurement is more
straight-forward and precise than the separate Q-value measurements,
because several systematic uncertainties are eliminated. Using
the 1977 Mass Table values as references, the ^{10}C-^{14}O difference
as determined by this direct comparison[20] was 3.44±0.27 keV,
whereas the difference obtained by subtraction of our earlier
results[19] was 3.06±0.42 keV. Hence the MSU values are internally
consistent with the direct measurement indicating a slight tendancy
towards a larger discrepancy rather than a smaller one.

   Finally, because of the importance of the mass of ^{6}Li mentioned
above, Robertson and Nolen have recently measured the $^{6}Li(p,^{4}He)^{3}He$
reaction energy by techniques similar to those discussed above.
In this case, it is the mass of the target which is of interest
rather than the reaction products. The preliminary results are
that this Q-value is less than that derived from the 1977 Mass

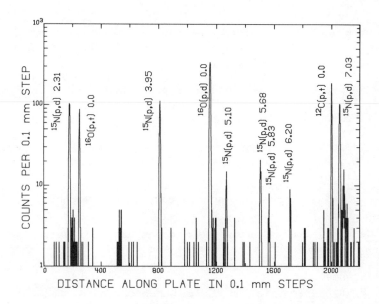

Fig. 3. Spectrum of deuterons and tritons recorded on a nuclear
         emulsion to determine the difference between the
         $^{16}O(p,t)^{14}O$ and $^{12}C(p,t)^{10}C$ Q-values. The target was
         a mixture of ^{15}N, ^{12}C, and ^{16}O.

Table 18. Although this analysis is not yet final it does indicate the need for the potentially much more precise measurement to be carried out at Delft.

In summary, there remain several specific reasons for more precise mass measurements of both stable and unstable light isotopes. It is hoped that by the time of AMCO-VII several new measurements will have been carried out and some of the existing discrepancies clarified.

This research was supported by the National Science Foundation under Grant No. Phy 78-22696.

1. L.G. Smith, Phys. Rev. C4 (1971) 22.
2. L.G. Smith and A.H. Wapstra, Phys. Rev. C11 (1975) 1392.
3. E. Koets, Progress report presented at this conference.
4. R.G.H. Robertson and P. Dyer, private communication.
5. E. Kashy, talk presented at this conference.
6. D.H. Wilkinson, Phys. Lett. 67B (1977) 13.
7. I.S. Towner and J.C. Hardy, Phys. Lett. 73B (1978) 20.
8. P.H. Barker, talk presented at this conference.
9. P. Dyer and R.G.H. Robertson, Michigan State University Heavy Ion Lab. Progress Report, 1979.
10. P. Glässel, E. Huenges, P. Maier-Komor, H.-Rösler, H.J. Scheerer, H. Vonach and D. Semrad, Atomic Masses and Fundamental Constants 5, J.H. Sanders and A.H. Wapstra (Eds.), Plenum Press, New York (1976) 110.
11. H. Vonach, P. Glässel, E. Huenges, P. Maier-Komor, H. Rösler, H.J. Scheerer, H. Paul and D. Semrad, Nucl. Phys. A278 (1977) 189.
12. J.A. Nolen, Jr., G. Hamilton, E. Kashy, and I. Proctor, Nucl. Instr. and Methods 115 (1974) 189.
13. J.A. Nolen, Jr., Atomic Masses and Fundamental Constants 5, J.H. Sanders and A.H. Wapstra (Eds.), Plenum Press, New York (1976) 140.
14. G.F. Trentelman and E. Kashy, Nucl. Instr. and Methods 82 (1970) 304.
15. H.G. Blosser, G.M. Crawley, R. de Forest, E. Kashy and B.H. Wildenthal, Nucl. Instr. and Methods 91 (1971) 61.
16. R.C. Greenwood and R.G. Helmer, Nucl. Instr. and Methods 121 (1974) 385, also R.C. Greenwood, talk presented at this conference.
17. R.G.H. Robertson, E. Kashy, W. Benenson and A. Ledebuhr, Phys. Rev. C17 (1978) 4.
18. A.H. Wapstra and K. Bos, Atomic Data and Nuclear Data Tables 19 (1977) 175.
19. P.H. Barker and J.A. Nolen, Jr., Proceedings of the Inter. Conf. on Nucl. Structure, Tokyo (1977) 155.

20. J.A. Nolen, Jr., P.H. Barker and M.S. Curtin, Bull. Amer.
     Phys. Soc. 24 (1979) 63.
21. D.C. Robinson and P.H. Barker, Nucl. Phys. A225 (1974) 109.
22. R.E. White and H. Naylor, Nucl. Phys. A276 (1977) 333.
23. H. Vonach, private communication.

PROGRESS AND FUTURE OF MASS MEASUREMENTS

AMONG NUCLEI FAR FROM β-STABILITY

J.C. Hardy

Atomic Energy of Canada Limited
Chalk River Nuclear Laboratories
Chalk River, Ontario, Canada

The past decade or so has seen an enormous increase in our knowledge of nuclei far from the region of β-stability. A comparison of the 1977 nuclear chart with that of 9 years earlier shows 300 new isotopes lighter than uranium discovered in that time - an increase of about 20% in the total number known. As large as this increase is, it still cannot do justice to the enormous expansion in our experimental knowledge of all β-unstable nuclei. And this expansion is still going on, as you have heard from many talks during the past few days.

Indeed, when I agreed to summarize the conference talks in this field, I imagined that they would comprise only a small fraction of the total material given. Obviously I was wrong. Even people like Barber and Johnson, whom I might reasonably have expected to stick to other things, have let me down. As a result, I cannot possibly make a complete summary in the time alotted, and shall instead concentrate on a few results that seem particularly interesting, couching my remarks in a broader context suggested by the question: why should we measure the masses of nuclei far from β-stability?

In attempting to answer the question, let us first discount the reply: because they are there. Not that that is invalid as a justification, simply that it does not lead to a very profound intellectual development. At the moment there are three broad reasons for measuring the masses of β-unstable nuclei:

1) To test specific theories:
    - Conserved Vector Current (CVC) hypothesis (superallowed β-decay),

       - Isobaric Multiplet Mass Equation (IMME), and charge dependence,
       - Coulomb Energies

2) To observe systematic effects:
     - evidence for magic numbers,
     - regions of deformation.

3) To test mass calculations:
     - refine parameters for further predictions,
     - probe fundamental nuclear properties(?).

The first category, though very important, is one that concerns relatively few nuclei and involves very specialized techniques. The previous speaker has already dwelt on this subject so I shall not emphasize it in my own remarks. However, I cannot pass on without some comment on the precision threshold measurements reported preliminarily by P.H. Barker. If his final results live up to expectations, their small uncertainties ($\sim$ 0.1 keV) will have profound effects on the study of superallowed β-decay, particularly in defining the role of charge-dependent mixing. This could be an important development.

It is in the study of systematic effects and the testing of mass calculations, the second and third categories, where we really move into the territory of exotic nuclei. It is here where we learn about new effects rather than honing our understanding of familiar ones. But we must make the distinction between observed effects that tell us something independent of theory - or at least independent of controversial theory - and those that help to refine or shape some theory. The first type of effect can be observed in the plot of two-neutron separation energies that appears in Fig. 1. It is taken directly from the 1977 mass tables[1] and simply illustrates what we have seen in a number of contributions to this meeting - that systematic studies of relative masses can unequivocally signal a closed shell (the discontinuity above N=82 in the figure) or a region of deformation (the bump near N=91).

I should like to mention two examples of this kind of approach. The first is the contribution of K.-H. Schmidt and D. Vermeulen, in which they demonstrate the importance of <u>proton</u> number in determining the shell strength at <u>neutron</u> magic numbers, and vice versa. Their evidence is presented in Fig. 2. The difference between two-neutron (two-proton) separation energies just above and just below a closed neutron (proton) shell is plotted as a function of proton (neutron) number, the energies having been corrected first for the macroscopic features of the binding energy surface as predicted by the droplet model[2] without shell corrections. The sharp increase in the

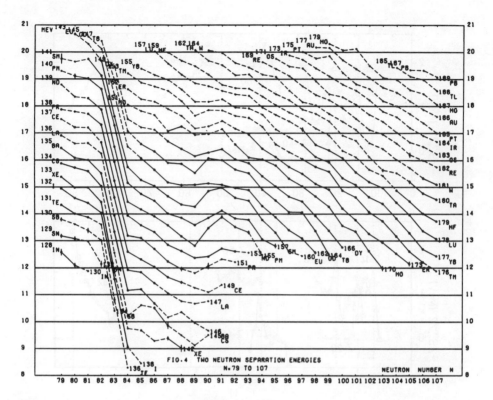

Fig. 1   Two neutron separation energies plotted against neutron
number for a variety of elements (from ref. 1).

separation energy of one type of nucleon, as the other type approaches a magic number, must arise from a mutual support of shell
closures by both neutrons and protons. This conclusion, though its
illustration was aided by the droplet model, does not depend essentially upon any model of masses. It is the product only of
experimental systematics.

A second example of systematic observation comes from the talk
by E. Roeckl, in which he presented his results for Q_α values in
the decays of light tellurium isotopes: $^A_{52}\text{Te} \xrightarrow{\alpha} {}^{A-4}_{50}\text{Sn}$. Since the
lightest case he has studied is the decay of ^{107}Te, it is still not
possible to draw final conclusions about the shell closure expected
at ^{100}Sn. Nevertheless Roeckl demonstrated a significant increase
in Q_α as he went from ^{111}Te down to ^{107}Te, an effect that strongly
points to the approach of a doubly-magic nucleus.

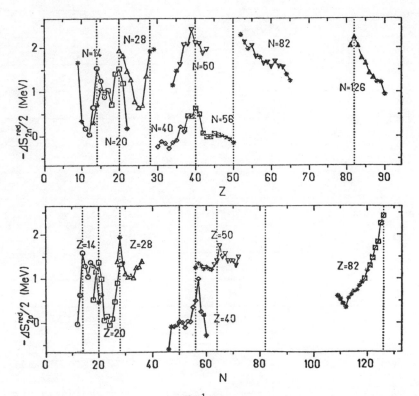

Fig. 2 Top: A function ΔS_{2n}^{red} of two neutron separation energies
        plotted for various neutron numbers as a function of proton
        number. Here $\Delta S_{2n} = \Delta S_{2n} - \Delta S_{2n}^{DM}$, where $\Delta S_{2n} = M(Z,N+2) -$
        $2M(Z,N) + M(Z,N-2)$ and ΔS_{2n}^{DM} is the corresponding quantity
        as calculated in the droplet model[2]. Bottom: Equivalent
        plot for two-proton separation energies. The figure is
        taken from the contribution of K.-H. Schmidt and D.
        Vermeulen.

        "Model-independent" conclusions such as these do not account
for the bulk of what we have heard in the last few days concerning
nuclei far from β-stability. Most experiments just produce the
masses of a few nuclei - sometimes connected systematically, sometimes not - and often little can be said about these masses except
to compare them with the predictions of various mass formulas.
After listening to many contributions of this sort, as well as to

various descriptions of mass calculations, I am impressed by the
discrepancy that seems to exist between the experimenters' and the
theorists' expectations for a set of mass predictions. Kelson has
already alluded to this problem, but I think it bears repetition
and emphasis here.

A schematic view of the situation appears in Fig. 3. Presumably we are seeking some ideal untruncated microscopic model, which
on the one hand accurately calculates experimental masses and on the
other relates to fundamental nuclear interactions. Whether we ever
will reach this goal is debatable, since for the moment we seem to
be a long way off. However, though we may be far from the ideal,
still there are many calculations that successfully reproduce the
general trends of experimental masses. To do so, most subdivide
the problem into two parts: one that takes account of the bulk
nuclear properties through some macroscopic technique - e.g. the
droplet model; and one that describes some microscopic features
through shell corrections. Being only an approximation to the ideal
full microscopic model, this approach can at best approximate the
experimental masses. Even if all the parameters of a given approximation are optimized, there must remain a discrepancy with experiment
that can best be described as "irrelevant noise". The magnitude of
the noise depends upon the sophistication of the approximation used.

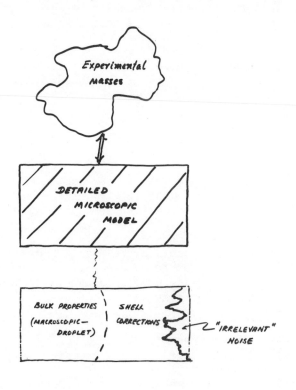

Fig. 3 Schematic repre-
        sentation of the
        comparison between
        experimental masses
        and theoretical
        calculations.

Most important, we can gain very little insight into the RMS value
of this noise from one or even a few mass measurements. It takes
many.

     Some qualitative feeling for the accuracy we might expect from
global mass calculations can be derived from considering conventional
spectroscopic calculations. In attempting to calculate with the
shell model, for example, the level scheme of a particular nucleus,
we employ a high degree of sophistication, incorporating many active
nucleons in a variety of well-defined orbitals. Yet we rarely ask
for better than a few hundred keV agreement with experimental energy
levels (only a relative mass measurement), and are often overjoyed
if simply the ordering of states is correct. How can we expect a
mass formula, which does not even incorporate nuclear spin let
alone further complications, to do nearly as well in predicting
absolute masses?

     It is natural for the experimenter to tend to overinterpret
his mass measurement in terms of its usefulness for assigning merit
to mass predictions. Without this apparent impact, individual mass
measurements seem unrewarding. Ultimately, though, it is only when
many data are accumulated - as at this conference - that the real
rewards start to emerge. One example that comes immediately to mind
is the realization that nuclei with a neutron excess seem in general
to be less tightly bound than predicted. This is most obvious from
the 40 or so Q_β-values reported by the Studsvik group (contribution
by K. Aleklett, E. Lund and G. Rudstam), and from the series of
direct mass measurements on caesium and rubidium isotopes that we
heard about from M. Ephere et al. The results of the latter group
are shown in Fig. 4, where the measured rubidium masses are compared with four sets of predictions[3]. They typify the perceived
trend. It is a trend that will no doubt be soon absorbed into more
refined and more accurate mass predictions.

     From the foregoing discussion of the reasons for measuring
masses, as illustrated by contributions to this conference, a set
of goals can be derived for experiments in the future. In a few
cases - tests of CVC or the IMME for instance - extreme precision
is still called for. But in general for nuclei far from stability
a precision of perhaps ±100 keV is more than adequate to test any
forseeable mass formula or to contribute to a systematic trend. Of
far greater importance than high precision to an experimental technique is its applicability to many nuclei; and of greater importance
still is its reliability. After all, nothing mimics a systematic
mass trend better than a systematic error.

     With these criteria, it is hardly possible to give any experi-
mental technique a completely clean bill of health. Reaction Q-
value measurements are probably among the least susceptible to
significant (at the ±100 keV level) systematic errors, although they

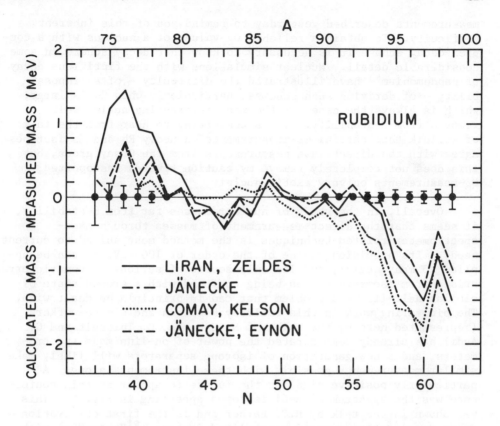

Fig. 4 Comparison of experimental and predicted masses for rubidium isotopes. The figure is taken from the contribution
of M. Epherre, G. Audi, C. Thibault, R. Klapisch, G. Huber,
F. Touchard and H. Wollnick.

are of somewhat limited applicability. Despite considerable ingenuity, such as that exemplified by the beautiful new results on pion
induced double charge exchange reactions (reported by H. Nann), they
usually apply to nuclei relatively near stability where most mass
formulas already do quite well. The more exotic reactions are also
expected to apply only to light nuclei; here there may be some hope
for expansion, since R.C. Pardo has found that cross sections for
(^{3}He, ^{6}He) reactions do not continue to drop above A=100.

Radioactivity measurements, both with and without isotope
separators, have a much broader applicability, but are frequently
plagued by a lack of reliability. Without attempting to denigrate
all measurements of this type, I am prompted by the spate of Q_β

measurements described yesterday to remind you of this inherent
difficulty. To obtain a reliable Q_β-value for a nucleus with a complex decay scheme it is necessary to know that decay scheme in some
considerable detail. Nuclear simulations with the fictitious decay
of pandemonium[4]) have illustrated the difficulty - often impossi-
bility - of deriving such schemes, particularly when Q_β is large,
which is always the case for the most interesting nuclei well-
removed from β-stability. It is comforting to hear from the talk
of Wollnik that careful Q_β-measurements on heavy Rb and Cs isotopes
agree with the direct mass measurements from the Orsay group, but
this does not completely remove my caution towards unsupported
Q_β-measurements on very exotic nuclei.

        Overall, in the study of nuclear masses far from β-stability,
it seems that the direct measurement of masses through mass-
spectrometer related techniques is the method most suited to current
needs. Its precision can be of the order of 100 keV, it can be applied to many nuclei, and its only source of serious systematic error
arises from isomeric states being confused with a ground state of
similar half-life - a problem that can in principle be dealt with.
The pioneering work in this field by Klapisch and his co-workers
(represented here by the talks of M. Epherre, C. Thibault and G.
Audi) has already demonstrated the power of on-line mass spectrometers, and a new generation of isotope separators will likely make
similar measurements possible with these instruments alone. A
particularly positive sign for the future to appear at this conference was the spectrum of A=61 isotopes appearing in Fig. 5. This
was shown in the talk by R.C. Barber and is the first observation
of an accelerator-produced short-lived isotope (^{61}Zn;$t_{\frac{1}{2}}$=90s) with
"Manitoba II" - a high performance conventional mass spectrometer.
It will not surprise me if AMCO VII is overwhelmed by data from the
descendents of these direct measurements.

        As risky as that prediction may be, there is little risk in
assuming that there will indeed be an AMCO VII meeting. Atomic
masses continue to fascinate both the theorist and the experimenter,
and work in the field will not stop growing. Perhaps it is because
an atomic mass is a simple uncomplicated number that contains every-
thing we know - and much that we do not know - about the nuclear
Hamiltonian. Its simplicity is deceptive and forever tantalizing.

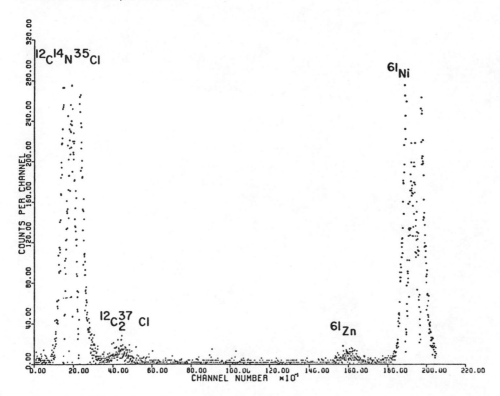

Fig. 5   Mass spectrum at A=61 obtained with the "Manitoba II" mass
          spectrometer on-line with a cyclotron. The figure is taken
          from the contribution of R.C. Barber, K.S. Kozier, K.S.
          Sharma, V.P. Derenchuk, R.J. Ellis, V.S. Venkatasubramanian
          and H.E. Duckworth.

REFERENCES

1) A.H. Wapstra and K. Bos, Atomic Data and Nucl. Data Tables
     19 (1977) 177.
2) W.D. Myers, Droplet Model of Atomic Nuclei (Plenum, New York,
     1977).
3) 1975 Mass Predictions, S. Maripuu editor, Atomic Data and Nucl.
     Data Tables 17 (1976) 411.
4) J.C. Hardy, L.C. Carraz, B. Jonson and P.G. Hansen, Phys. Lett.
     71B (1977) 307.

STATUS OF THE FUNDAMENTAL CONSTANTS*

E. Richard Cohen

Rockwell International Science Center
1049 Camino Dos Rios
Thousand Oaks, California 91360

INTRODUCTION

In 1973 Cohen and Taylor[1] published an analysis and adjustment of the fundamental physical constants which was subsequently adopted and recommended for general use by CODATA.[2] As is the usual fate of such publications, the paper was rapidly challenged by new measurements of the velocity of light, c; the Rydberg constant, R_∞; the ampere realization, K; the Faraday constant, F; the proton gyromagnetic ratio, γ'_p; the Avogadro constant, N_A; and the gas constant, R_0; as well as improved theoretical calculations of the anomalous moment of the electron and of the higher order corrections to fine structure and hyperfine structure separations in positronium, muonium and helium. The entire situation was reviewed by Taylor and Cohen at AMCO-V in 1975, where much of this new data was also reported.[3]

The situation has not changed greatly in the past four years. Improvements on previously reported measurements have in general led only to a sharpening of certain discrepancies which were indicated then. Only one disagreement — in the gas constant measured by acoustic thermometry — has been resolved in the interim. In this paper I shall not attempt to discuss in detail all of the pertinent data on fundamental constants; I shall instead concentrate primarily on those areas where there are disagreements and inconsistencies. Unfortunately, there are disagreements at the level of 5 to 10 parts per million, while the new measurements are

*This work was supported in part by the National Science Foundation under grant #PHYS78-26467.

reported to achieve precision at the 1 ppm level or better.
Because of this I shall invoke only approximate conversions of
laboratory standards of electrical units — the ampere and the
ohm — to SI, since that question is a difficult and extensive one
if it is to be carried out to the full precision which may be
possible (of the order of 2×10^{-7}). We need not explore in
detail corrections and renormalizations at the sub-ppm level in
order to demonstrate the primary issue.

NEW DATA

Velocity of Light

In 1973 the Consultative Committee on the Definition of the
Meter (CCDM) adopted the provisional value,[4] $c = 299792458$ ms^{-1}
for the velocity of light, based on Evenson, *et al.*'s measurement
of the frequency of a methane-stabilized laser[5] and a consensus of
several consistent measurements of the wavelength. Independent
confirmation of the adequacy of this adopted value comes from at
least three other determinations using different transitions
and/or different frequency comparison chains to relate the wavelength to the Cs frequency standard. These experimental data,
corrected to the convention that the Kr wavelength applies to the
mid-point between the peak and the center of gravity of the Kr
line shape, are

Evenson, *et al.*, NBS[5]	299792457.4 ±1.1 ms^{-1}
Blaney, *et al.*, NPL[6]	459.0 ±0.6
Woods, *et al.*, NPL[7]	458.8 ±0.2
Baird, *et al.*, NRC[8]	458.1 ±1.9
Javan, *et al.*, MIT[9]	457.6 ±2.2

These measurements indicate that the wavelength of the Kr standard
may be defined with an accuracy of the order of 2×10^{-9} and that
within those limits measurements are consistent with the adopted
interim value of c.

Rydberg Constant

In 1973 the Rydberg constant had been given with a precision
of 0.08 ppm. This is probably the limit of what could be achieved
with incoherent light. In 1974, Hansch, *et al.*, used laser
saturation spectroscopy to measure the fully resolved fine structure of Hα and Dα and were thereby able to obtain an increase in
precision of a factor of ten. More recently Goldsmith, Weber and
Hansch[10] have used polarization spectroscopy of Hα to improve the
resolution of the Hα fine structure and have achieved an improvement by another factor of three. Their value,
$R_{\infty} = 10973731.476(32)$ m^{-1}, is based on assigning a wavelength

$\lambda = 632991.3998$ pm to the $^{127}I_2(i)$ absorption peak.[11] On the other hand they also quote an alternative value in frequency units based on the CO_2 (R12) frequency of Evenson[5] and the iodine to CO_2 wavelength ratio of Layer, Deslattes and Schweitzer[12] which corresponds to $\lambda = 632991.3981$ pm. Since the CCDM recommendation lies nicely between the two we prefer to give the unweighted mean of Goldsmith, *et al.*'s two alternatives

$$R_\infty = 10973731.492 \text{ m}^{-1} . \tag{1}$$

An appropriate uncertainty may be ±0.050 m^{-1} (0.0045 ppm).

A similar measurement has been carried out at NPL by Petley and Morris;[13] they find a slightly higher value, but one which is consistent within the quoted errors:

$$R_\infty = 10973731.513(85) \text{ m}^{-1} \ (0.0077 \text{ ppm}) . \tag{2}$$

The resolution of the spectrum itself in both these measurements is such that it is realistic to anticipate an increase in accuracy by perhaps another factor of 10 when systematic errors (corrections for uncertainties in wavelength standards, Stark shifts, buffer gas density shifts, resonance asymmetries, etc.), which are the major contributors to the error budget, can be better controlled.

Gas Constant

At AMCO-V Quinn, Colclough and Chandler reported a new determination of the gas constant,[14] $R_O = 8.31600(17)$ J mol^{-1} K^{-1}, that was higher by 191(40) ppm than the 1973 recommended value (this was reduced slightly to 8.31573(17) J mol^{-1} K^{-1} in a later publication[15]). This experiment measured the velocity of sound in argon using an acoustic interferometer at the triple point of water. The acoustic isotherm was measured as a function of pressure from 30 to 200 kPa (0.3 to 2 atmospheres) and extrapolated to zero pressure after correction for boundary layer effects. The corrected isotherm was then fitted to a quadratic to give the velocity of sound in the perfect gas limit. This procedure was criticized by Rowlinson and Tildesley[16] who point out that although the experimental data on virial coefficients of argon may not be adequate to allow an *ab initio* calculation of the acoustic isotherm, such data must still be considered. The available information on the compressibility of argon and calculations based on models of the argon-argon intermolecular potential indicate that Quinn, *et al.*'s quadratic extrapolation is inappropriate. Using an independent estimate of the second acoustic virial coefficient they find $R = 8.31485(35)$ J mol^{-1} K^{-1} to be compared with $R = 8.31478(10)$ J mol^{-1} K^{-1} from a linear fit to Quinn's original data.

Furthermore Colclough,[17] in an analysis of low temperature acoustic thermometry (4.2 − 20 K) interpreted a systematic deviation from absolute gas thermometry in terms of a correction to the accepted value of the gas constant. From this he deduced a value $R = 8.31462(29)$ J mol^{-1} K^{-1}. In a similar analysis of the velocity of sound in helium in the temperature range $-175°C$ to $150°C$, B.E. Gammon[18] obtained a value $R = 8.31479(35)$ J mol^{-1} K^{-1}.

Colclough and Quinn[19] have recently reconsidered their original measurements and have been able to identify a nonlinearity in the diaphragm of the transducer in the acoustic interferometer. In addition, in order to evaluate the contribution due to the third virial coefficient, their measurements were extended to 1.3 MPa (13 atmospheres). The quadratic dependence observed between 2 and 13 atmospheres was then used to correct the lower pressure data. With the nonlinearity correction applied their corrected isotherm is then linear with pressure below 200 kPa and the extrapolation to zero pressure gives

$$R = 8.31449(18) \text{ J mol}^{-1} \text{ K}^{-1} \qquad\qquad (21 \text{ ppm}) \qquad (3)$$

where the quoted error includes 11 ppm from the statistics of the fitting (random error) and 18 ppm for an estimate of systematic errors. This value is now only 10 ppm higher than the 1973 recommendation and consistent with it and with the other recent determinations. Thus the apparent discrepancy in the gas constant has been essentially resolved. It would be pleasant to be able to say the same for the discrepancies to be discussed below.

Ampere Realization

The SI electrical units are defined coherently from Ampere's Law and the requirement that electrical and electromagnetic energy are measured in the same units as mechanical energy. The realization of this — the establishment of standards and units against which electrical quantities may be measured — is in terms of a "standard" or "international" volt and the "standard" ohm. Thus we must know the relationship $K = A_{BIPM}/A$ and $\bar{R} = \Omega_{BIPM}/\Omega$ between the standards and the units they are intended to represent. (From time to time the standards may be redefined in an attempt to set K and $\bar{R}$ closer to unity.) The ohm is now maintained in terms of the calculable capacitor and the volt in terms of the Josephson effect. The definitions,

$$2e/h = 483594 \text{ GHz/V}_{BIPM} = 483593.420 \text{ GHz/V}_{NBS}$$

establish the corresponding laboratory standards of the volt.

However, it is the ampere, using a form of current balance, that can be measured. There are no new results on the quantity $K = A_{BI69}/A$ since AMCO-V;[3] we shall list them here for completeness:

$$
\begin{aligned}
(K-1) \times 10^6 &= -2.3 \pm 6.0 \quad \text{(VNIIM, 1966–69)} && (4.1) \\
& -1.2 \pm 7.7 \quad \text{(NBS, 1958)} && (4.2) \\
& 0.0 \pm 5.5 \quad \text{(NPL, 1965–70)} && (4.3) \\
& 1.7 \pm 8.0 \quad \text{(ASMW, 1974)} && (4.4) \\
& 1.8 \pm 9.7 \quad \text{(NBS, 1968)} . && (4.5)
\end{aligned}
$$

The mean of these data is -0.4 ± 3.1 (σ_i), or ± 0.8 (σ_e) so that they are much more consistent than the assigned uncertainties might lead one to expect (probability of $\chi^2 \geqslant 0.25$ for 4 degrees of freedom is approximately 0.99).

Gyromagnetic Ratio

The gyromagnetic ratio of the proton, $\gamma_p' = 2\mu_p'/h$ = $(\mu_p'/\mu_N)(e/m_p)$, in which the prime indicates an effective value due to diamagnetic shielding (standardized to protons in a spherical sample of water) may be measured in two different ways. The "low-field" measurement determines the proton resonance frequency in a field whose magnitude is determined by the dimensions and geometry of a precision solenoid carrying a measured current. The measured gyromagnetic ratio, ν/B, is then inversely proportional to the maintained unit of current. The "high-field" measurement determines the field by measuring the forces on a current-carrying conductor placed in it. In this case the deduced magnetic field is inversely proportional to the maintained unit of current and γ_p' (high) is proportional to that unit. Following the suggestion of Huntoon and McNish[20] it is useful to emphasize this distinction by expressing γ_p' (low) in units, $s^{-1}T^{-1}$, and γ_p' (high) in units $A \cdot s \cdot kg^{-1}$. Thus if the high and low field measurements can be expressed in terms of the same arbitrary unit of current, the geometric mean of the two measurements gives the true gyromagnetic ratio, independent of the arbitrary current unit and the square root of the ratio gives the calibration factor for the arbitrary unit in terms of absolute (SI) amperes. The two measurements taken together are thus the equivalent of an ampere balance experiment in which the proton resonance frequency serves as a means of transferring a low field, in which the solenoid geometry may be calculated, to a high field in which the magnetic forces may be more accurately measured.

Two measurements of γ_p' whose preliminary results were discussed at AMCO-V are the high field measurement of Kibble and Hunt[21] at NPL

$$\gamma'_p \text{ (high)} = 267517010(270) \text{ A}_{BI69} \cdot \text{s} \cdot \text{kg}^{-1} \tag{5}$$

and the low field measurement of Williams and Olsen at NBS[22,23]

$$\gamma'_p \text{ (low)} = 267513229(57) \text{ s}^{-1}\text{T}^{-1}_{NBS}$$

$$= 267513542(57) \text{ s}^{-1}\text{T}^{-1}_{BI69} . \tag{6}$$

The NPL high-field measurement has a quoted precision of 1 ppm and the NBS low-field measurement a precision of 0.21 ppm. These measurements represent sufficiently large increases in precision that for the present discussion no other data on γ'_p need be considered.

Electron Magnetic Moment and the Fine-Structure Constant

The anomalous magnetic moment of the electron has been evaluated theoretically using Feynman-diagram perturbation expansions in powers of the fine-structure constant:[24]

$$\mu_e/\mu_o = 1 + \frac{\alpha}{2\pi} + C_4\left(\frac{\alpha}{\pi}\right)^2 + C_6\left(\frac{\alpha}{\pi}\right)^3 + \cdots \tag{7}$$

where $C_4 = -0.3284784458$

$C_6 = +1.184 \pm 0.007$.

The fourth order perturbation term C_4 has been calculated analytically while for C_6 some of the integrals involved have only been evaluated numerically — hence the associated numerical uncertainty.[25]

The most accurate experimental measurement of the anomalous moment is the one recently reported by van Dyck, Schwinberg and Dehmelt[26]

$$a_e = \mu_e/\mu_o - 1 = 1159652200(40) \times 10^{-12} .$$

From this one could calculate a value of the fine structure constant:

$$\alpha^{-1} = 137.036007 . \tag{8}$$

An uncertainty of ±5 in the final digit arises directly from the quoted uncertainty in the measured value of a_e; the uncertainty in C_6 contributes an additional uncertainty of ±10. Furthermore the uncalculated eighth order term is important; the sensitivity, $\partial(1/\alpha)/\partial C_8$, is 3.4×10^{-6} so that, if C_8 is estimated to be $C_8 = -10$, the reciprocal fine structure constant becomes

$\alpha^{-1} = 137.035973$. At the level of parts per billion in α (or parts in 10^{12} in μ_e/μ_0) one can not ignore contributions from hadronic interactions and other non-electromagnetic forces.[23] We note here that C_4 given above includes a term, $(m_e/m_\mu)^2/45 = 5.198 \times 10^{-7}$, which arises from virtual-muon pair production.

We compare this value of α^{-1} with that which may be calculated from the low-field γ_p measurement. It follows from the definition of γ_p given above and the definition of the Rydberg constant that we may write

$$\alpha^{-1} = \left[\frac{(\mu'_p/\mu_B)(2e/h)c}{4R_\infty \gamma'_p} \right]^{\frac{1}{2}} . \tag{9}$$

Because $2e/h$ is used to define the maintained volt realization and γ'_p (low) is proportional to the unit of current with respect to which it is measured, if we express both these quantities in consistent units we need introduce only a relatively well-known factor for the ratio of the maintained laboratory ohm to the SI ohm.[1] Williams and Olsen give[23]

$$\alpha^{-1} = 137.035963(15) . \tag{10}$$

This value for the fine-structure constant is consistent with the value deduced from the anomalous moment; the difference between them is only 44×10^{-6} (0.32 ppm). It is more difficult to assign a meaningful uncertainty to this number however. Using only the uncertainties in the two experiments and the uncertainty assigned to C_6 would give $\pm 19 \times 10^{-6}$ but that assumes $C_8 = 0$ and no other higher order terms. A more fruitful approach is to use the anomalous moment and the value of α deduced from γ'_p (low) to calculate

$$C_6 + C_8 \left(\frac{\alpha}{\pi} \right) + \cdots = 1.154 \pm 0.10 \pm 0.003 \tag{11.1}$$

where the first component of the uncertainty is ascribable to α and the second to a_e. Alternatively this result may be expressed as

$$C_8 + \cdots = -12.9 \pm 5.5 . \tag{11.2}$$

No calculation of C_8 exists but this "experimental" evaluation may not be unreasonable. It is clear however that one can not use the anomalous moment of the electron to evaluate the fine structure constant; it can only be used to check the validity of QED.

This is also the case with all of the other atomic spectroscopic data (fine structure measurements in H, D and He$^+$, as well
as hyperfine structure measurements in H, muonium and positronium);
the so-called "non-QED" value of α is so accurately determined
that none of the QED data can compete with it — either because
of the inherent experimental uncertainty in the data or because
of the inadequacy of the theoretical calculations.

Faraday

Two measurements of the Faraday are reported at this conference. R.S. Davis has given the final results of the NBS silver
coulommeter program:

$$F = 96486.33 \pm 0.24 \ A_{NBS} \cdot s \cdot mol^{-1} \ (2.5 \ ppm)$$

$$= 96486.22 \pm 0.24 \ A_{BI69} \cdot s \cdot mol^{-1} \tag{12.1}$$

and W.F. Koch reported the results of the 4-amino-pyridine
($C_5H_6N_2$) work

$$F = 96484.52 \pm 1.08 \ A_{NBS} \cdot s \cdot mol^{-1} \ (11 \ ppm)$$

$$= 96484.41 \pm 1.08 \ A_{BI69} \cdot s \cdot mol^{-1} . \tag{12.2}$$

The large error here is dominated by a 10 ppm allowance for
possible impurities. If improved chemical analysis could reduce
the uncertainty from this cause to the 2-3 ppm level the measurement would be capable of yielding a 5 ppm uncertainty in F.

These two measurements are somewhat inconsistent (the differ-
ence between them is $(1.81\pm1.11) \ A \cdot s \cdot mol^{-1}$, $(19\pm12 \ ppm)$),
indicating that a 10-15 ppm impurity contribution in defining the
effective electrochemical equivalent to $C_2H_6N_2$ may indeed be
present.

The Avogadro Number

R.D. Deslattes and his collaborators have exploited the x-ray
optical interferometer (XROI) to measure the atomic repeat distance
in a single-crystal of Si directly in terms of optical wavelengths,
thus completely bypassing the need to introduce the x-unit as a
"local" unit of measurement. This was coupled with an accurate
mass-spectroscopic determination of the isotopic composition of
the boule from which the silicon crystal was cut, and an accurate
hydrostatic weighing procedure to determine the density of similar
silicon crystal samples. The most recent result from this work
is[27]

$$N_A = 6.0220978(63) \times 10^{23} \text{ mol}^{-1} \tag{13}$$

with an astonishing uncertainty of only 1.04 ppm. There are however some unresolved sources of possible systematic error in the determination of the mean diameter of stainless steel spheres which were used as density standards. Two different methods of cleaning the spheres led to measured diameters which differed by 0.55 ppm (1.64 ppm in volume) although no evidence could be found of any surface films or other differences in the surface of the spheres which could explain the difference, which was highly consistent for two different spheres. Because of this, the value of N_A given above is based on a density standard defined by the average of the two diameter measurements with an additional systematic error component of 0.82 ppm, pending a resolution of the discrepancy.

CONSISTENCY

It is premature to attempt a full least squares analysis of the presently available data on the fundamental constants but the data mentioned above already shows important discrepancies. First we compare the ampere realization factor determined directly with the value deduced by combining the high-field and low-field γ_p measurements. We find

$$K = [\gamma'_p \text{ (low)}/\gamma'_p \text{ (high)}]^{\frac{1}{2}} = 1 - (6.5 \pm 0.5) \times 10^{-6} \tag{14}$$

from eqs. (5) and (6), to be compared with $K - 1 = -(0.4 \pm 3.1) \times 10^{-6}$ from the mean of the direct measurements, eqs. (4.1 − 4.5). The discrepancy inherent in this is shown in Fig. 1. Instead of forming a triangle, the three lines in the figure should intersect at a point. This disagreement is less serious if one looks at the individual measurements of K, whose errors range from 5.5 ppm to 9.7 ppm and four of the five determinations have error bars which overlap with (14). What is disturbing here however is the clustering of the direct data around K = 1 which may be an indication of the effect known as "intellectual phase-locking"* — the tendency to obtain the answer that one thinks is correct.

A more serious discrepancy becomes apparent if one looks at the Faraday data. A value of the Faraday can be calculated from γ'_p (high):

$$F = \frac{M_p \gamma'_p \text{ (high)}}{(\mu'_p/\mu_N)} \times (10^{-3} \text{ kg/mol})$$

---

*I believe this term was first introduced into experimental physics by William W. Hansen.

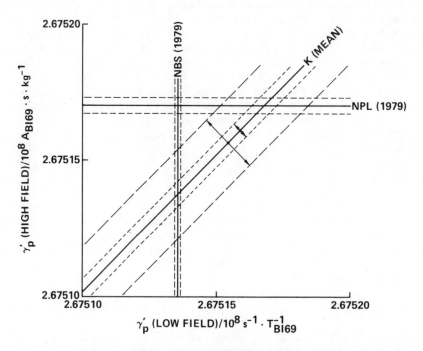

Fig. 1.   Gyromagnetic Ratio and the Ampere Realization.
          The locus K is the mean of five consistent
          measurements with uncertainties of 5 to 10 ppm.
          The larger error band (3 ppm) is the *à priori*
          (internal consistency) estimate; the smaller
          error band (0.8 ppm) is based on the mean
          square deviation of the data (external
          consistency).

$$F = 96486.00 \pm 0.10 \ A_{BI69} \cdot s \cdot mol^{-1} \tag{15}$$

where M_p is the relative atomic mass of the proton and μ'_p/μ_N is
the proton moment expressed in nuclear magnetons. This is shown

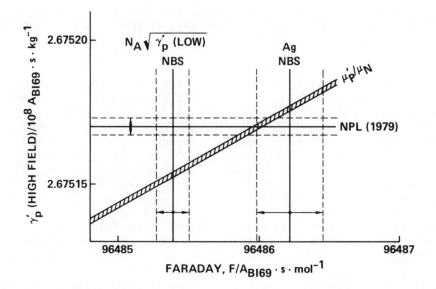

Fig. 2.   Faraday Constant, Gyromagnetic Ratio and the Avogadro
          Constant

in Fig. 2 as the intersection of the horizontal line representing
the Kibble and Hunt measurement with the cross-hatched diagonal
labeled μ_p'/μ_N. (The width of the cross-hatching represents the
one-sigma error range.)

A second independent indirect determination of the Faraday[28]
involves the Avogadro constant and the square root of γ_p' (low):

$$F = N_A e = 4N_A \alpha / \mu_o c (2e/h)$$

$$= \frac{8}{\mu_o} \left(\frac{\overline{R}}{cE} \right)^{3/2} N_A \left(\frac{R_\infty \gamma_p' \ (low)}{\mu_p'/\mu_B} \right)^{\frac{1}{2}}$$

$$= 96485.39 \pm 0.10 \ A_{BI69} \cdot s \cdot mol^{-1} \qquad (16)$$

where $E \equiv 483594$ GHz/V_{BI69} and $\bar{R}$ is the ratio of the maintained ohm to the SI ohm, $\bar{R} = \Omega_{BI69}/\Omega$.

This value of F is 6.3 ppm lower than the value given in eq. (15), yet each has an indicated uncertainty of approximately 1 ppm.

It would be intriguing to ascribe the discrepancies in both Fig. 1 and Fig. 2 to γ_p' (low). An increase of 12 ppm in this datum would be needed to increase K and F by the required 6 ppm. Such a large change however would be hard to understand in view of the careful attention to detail that went into that result and the consequent 0.21 ppm uncertainty assigned to it. It would also lead to a decrease of 6 ppm in α^{-1}, [eq. (9)], an equally unacceptable result. If we hold to the conclusion that Fig. 1 represents an acceptable discrepancy in view of the large uncertainties assigned to the individual direct measurements of K, we are still faced with the need to increase the Avogadro constant by approximately 6 ppm. Since the stainless steel density standards show a variation of 1.6 ppm in measured volume dependent on the details of surface treatment and cleaning, it may not be unreasonable to expect that at least part of this elusive 6 ppm discrepancy may be related to a possible systematic error here.

THE MASS-ENERGY CONVERSION CONSTANT

One of the close ties between atomic masses and fundamental constants is the mass-energy conversion factor that relates gamma-ray energies to atomic mass units. We speak of this as if the quantity of importance is the conversion from mass units to electron volts but it is extremely difficult to attach a voltmeter to a gamma ray, and the quantity actually measured is usually the wavelength. If gamma-ray energies are calibrated in terms of the annihilation radiation the significant quantity is the electron-proton mass ratio, m_e/m_p. Thus, although the 1973 recommended value for the energy equivalent of the electron mass (511.0034 keV) has an assigned uncertainty of 2.7 ppm, the atomic mass of the electron (548.58026 μu) has an uncertainty of only 0.38 ppm. If all gamma rays could be accurately related to the 511 keV standard it is only the latter uncertainty that would enter into an atomic masses adjustment.

Deslattes, *et al.*[27,29] have recently determined the wavelength of the [198]Au 411 keV line directly in terms of the wavelength of an I_2-stabilized He-Ne laser. This value, $\lambda = 3.0107788(11)$ pm, has been converted to energy units as 411.80441(15) keV using a conventional value for the wavelength-energy conversion factor. This has been used by Helmer, van Assche and van der Leun[30] as the

calibration value for a set of recommended standard gamma-ray
energies. The appropriate conversion factor is then the Compton
wavelength of one atomic mass unit:

$$\lambda \cdot \Delta M = hN_A(10^3 \text{ mol kg}^{-1})/c$$

$$= 2F(10^3 \text{ mol kg}^{-1})/(2e/h)c$$

$$= \frac{2F_{BI69}K^2\overline{R}(10^3 \text{ mol kg}^{-1})}{c(2e/h)_{BI69}} \tag{17}$$

where λ is the gamma-ray wavelength and ΔM the mass equivalent in
mass units, and the factor $K^2\overline{R}$ may be recognized as the ratio
of the joule as maintained in electrical units to the SI joule.

The uncertainty in the voltage-wavelength conversion factor
is of no direct concern as long as an adopted value is used
consistently. This requires that the calibration chain of
gamma-ray energies is traceable and consistent; the recommended
standards of Helmer, et al. satisfy this condition.

The proton Compton wavelength (and hence the mass-unit
Compton wavelength) in the 1973 adjustment of the physical constants carried an error of 1.7 ppm. However we are presently
faced with the 6 ppm discrepancy of Fig. 2 in the "best" value
for the Faraday constant, or equivalently, the high-field γ_p,
which brings one back again to the Avogadro constant and the
Ampere realization.

In terms of the low-field γ_p measurement, eq. (17) would be
written

$$\lambda \cdot \Delta M = \frac{2M_p\gamma_p'(\text{low})_{NBS}(\Omega_{NBS}/\Omega)}{(\mu_p'/\mu_N)c(2e/h)_{NBS}} \ .$$

If we use the results of Williams and Olsen for γ_p' (low) we
would have,

$$\lambda \cdot \Delta M = 1.3310650(58) \text{ fm} \cdot u \ .$$

Adopting the voltage-wavelength product $V\lambda = 1.239852 \times 10^{-6}$ eV·m,
this corresponds to

$$1 \ \mu u = 931.50061(41) \text{ eV}$$

with a precision of 0.5 ppm. This is 1 ppm lower than the 1973 recommended value, 931.5016(26) eV/μu. If we use the Kibble and Hunt value of γ'_p (high) and $K - 1 = (-0.4 \pm 3.1) \cdot 10^{-6}$ we obtain

$$1 \ \mu u = 931.4886(60) \ eV \ .$$

A proposed direct measurement of the wavelength of a nuclear γ-ray, or of a cascade of γ-rays, associated with a known mass difference has been discussed here by Deslattes. Some useful information may in fact already exist in the data and calculations of Wapstra and Bos[31] if the mass-energy factor is taken as an adjustable parameter in their least squares fitting. This would require that all of the β and γ decay energies that enter into that fitting are on a consistent energy scale.[27]

It may not be necessary to carry out a least-squares adjustment if one can identify equivalent mass-doublet-reaction pairs. The most obvious of these would be mass doublets of the type $^AZH - {}^{A-1}ZD$ and the equivalent reaction, $^{A-1}Z(d,p)^AZ$. The comparison of mass differences and reaction energies would give the mass energy conversion constant. The advantage of a least squares analysis is that it allows us in essence to construct these mass doublets in those cases where they have not been observed.

The important question is whether such fittings could achieve a 1 ppm accuracy level for the conversion constant. If so it would be a useful and important input for the next adjustment of the fundamental constants. On the other hand, if it can not yield an accuracy better than 10 ppm the result may be even more interesting for the next adjustment of the atomic masses. It would imply that the mass adjustment is not sensitive to the value of the mass-energy conversion factor at that level. Hence the inconsistencies that have been discussed in this paper would be of minor concern and continued use of the current value would be adequate — at least until there would be a significant improvement in the precision of the atomic mass and reaction energy data.

CONCLUSIONS

The precision of the data available for a new adjustment of the numerical values of the physical constants has increased significantly since 1973. The question of the adequacy of QED which was dominant in the 1969 adjustment may not yet be fully resolved but it no longer enters directly into any new adjustment. The spectroscopic fine-structure and hyperfine-structure data are now limited by the accuracy of the theoretical expressions so that they can not give significant inputs on the value of the fine structure constant but can only be used with a totally "non-QED" value as a check on the validity of the QED calculus. This of course is not a position that is absolute in time. With improved calculations those experiments should in the future provide quantitative values of α comparable to the value derivable from the gyromagnetic ratio data, and perhaps then reopen an old question on a new level of precision.

REFERENCES

1. E. Richard Cohen and B.N. Taylor, J. Phys. and Chem. Reference Data, *2*, 663–734 (1973).
2. E. Richard Cohen, CODATA Task Group on Fundamental Constants, CODATA Bulletin 11, December 1973.
3. B.N. Taylor and E. Richard Cohen, in *"Atomic Masses and Fundamental Constants 5,"* J.H. Sanders and A.H. Wapstra, editors, Plenum Press (New York; London) 1976; pp. 663–673. See also, in particular, pp. 48–53, 450–614, for papers on specific experiments.
4. J. Terrien, Metrologia, *10*, 75 (1974).
5. K.M. Evenson, J.S. Wells, F.R. Petersen, B.L. Danielson and G.W. Day, Appl. Phys. Lett. *22*, 192 (1973).
6. T.G. Blaney, C.C. Bradley, G.J. Edwards, B.W. Joliffe, D.J.E. Knight, W.R.C. Rowley, K.C. Shotten and P.T. Woods, Proc. Roy. Soc. (London) *A355*, 89 (1977).
7. P.T. Woods, K.C. Shotten and W.R.C. Rowley, Applied Optics *17*, 1048 (1978).
8. H. Preston-Thomas (private communication, 13 September 1979).
9. J.P. Monchalin, M.J. Kelly, J.E. Thomas, N.A. Kurnit, A. Szoke, A. Javan, F. Zernicke and P.H. Lee, Optics Letters *1*, 5 (1977).
10. J.E.M. Goldsmith, E.W. Weber and T.W. Hansch, Phys. Rev. Lett. *41*, 1525 (1978).
11. W.G. Schweitzer, E.G. Kessler, R.D. Deslattes, H.P. Layer and J.R. Whetstone, Applied Optics *12*, 2927 (1973).
12. H.P. Layer, R.D. Deslattes and W.G. Schweitzer, Applied Optics *15*, 734 (1976).
13. B.W. Petley and K. Morris, Nature *279*, 141 (1979).
14. T.J. Quinn, A.R. Colclough and T.R.D. Chandler, *"Atomic Masses and Fundamental Constants 5,"* J.H. Sanders and A.H. Wapstra (editors), Plenum Press (New York, London) 1976, p. 608.
15. T.J. Quinn, A.R. Colclough and T.R.D. Chandler, Phil. Trans. Roy. Soc. London *283*, 367 (1976).
16. J.S. Rowlinson and D.J. Tildesley, Proc. Roy. Soc. London *A358*, 281 (1977).
17. A.R. Colclough, Proc. Roy. Soc. London *A365*, 349 (1979).
18. B.E. Gammon, J. Chem. Phys. *64*, 2556 (1976).
19. A.R. Colclough and T.J. Quinn (private communication, to be published); A.R. Colclough, Acustica *42*, 28 (1978).

20. R.D. Huntoon and A.G. McNish, Nuovo Cimento, Suppl. *6*, 146 (1957).

21. B.P. Kibble, in *"Atomic Masses and Fundamental Constants 5,"* J.H. Sanders and A.H. Wapstra, editors, Plenum Press, New York–London (1976) p. 545; B.P. Kibble and G.J. Hunt, Metrologia *15*, 5 (1979).

22. E.R. Williams and P.T. Olsen, in *"Atomic Masses and Fundamental Constants 5,"* J.H. Sanders and A.H. Wapstra, editors, Plenum Press, New York–London (1976) p. 538.

23. E.R. Williams and P.T. Olsen, Phys. Rev. Lett. *42*, 1575 (1979).

24. J. Calmet, S. Narison, M. Perrottet and E. de Rafael, Rev. Mod. Phys. *49*, 21 (1977).

25. P. Cvitanovic and T. Kinoshita, Phys. Rev. *D10*, 4007 (1974).

26. R.S. van Dyck, P.B. Schwinberg and H.G. Dehmelt, Phys. Rev. Lett. *38*, 310 (1977); Bull. Am. Phys. Soc. *24*(5), 758 (1979) Abstract D5.

27. R.D. Deslattes, in "Metrology and Fundamental Constants," — Course LXVII, Enrico Fermi Summer School, Varenna (1976) — to be published; see also contribution to this conference.

28. Ref. 1, eq. (33.1).

29. E.G. Kessler, R.D. Deslattes, A. Henins, and W.C. Sauder, Phys. Rev. Lett. *40*, 171 (1978).

30. R.G. Helmer, P.H.M. van Assche and C. van der Leun, Atomic Data and Nuclear Data Tables (to be published); also C. van der Leun, this conference.

31. A.H. Wapstra and K. Bos, Atomic Data and Nuclear Data Tables *19*, 177 (1977).

CITATION COUNTS FOR TABLES

Katharine Way

Triangle Universities Nuclear Laboratory
Duke University
Durham, N. C. 27706

The Science Citation Index[1] has been employed to estimate the numbers of users of mass tables and other data compilations in the atomic and nuclear fields. Although the number of citations of a paper is not a true measure of the number of users (especially in the case of tables), nevertheless, it is the only objective, international criterion of use which is presently available.

Fig. 1 shows citation numbers for the mass tables of Mattauch and colleagues[2] and of Wapstra and coworkers[3] who continued this mass adjustment by essentially the methods of the Mattauch group. The main difference is that the Wapstra work was published in Nuclear Data Tables and the Mattauch work in Nuclear Physics. The graph shows the two publications are nearly equal in citations, so the conclusion seems justified that these two journals are equally good publication mediums for mass tables.

When is the best time to present a new or revised table? Fig. 1 shows that if the period is two or three years as in the case of Everling, (1960)[2a]; König, (1962)[2b]; and Mattauch (1965)[2c] the citation situation becomes rather chaotic. Apparently users don't learn about the new table (or don't want to hear) and persist with the familiar edition. But when the revision delay becomes some six years (as in the case of Wapstra) the new table becomes established in people's minds. Also users seem to abandon the old tables more rapidly as the reputation of the new table (and its authors) becomes more firmly established.

The mass-table counts show a rise time of about two years, a leveling off, and a decline when new tables are introduced.

The tables on internal conversion coefficients of Sliv and Band[4], Rose[5], and Hager and Seltzer[6] in Fig. 2 show a similar behavior.

The stopping-power and gamma-attenuation curves of Fig. 3 show a different behavior, one of constant rise. Northcliffe and Schilling[7] has had no real competition. Its increased citations seem due to the increasing needs for stopping-power values.

The gamma-attenuation tables of McMaster et al.[8], Storm and Israel[9], and Hubbell[10] all started at about the same time and so did not cut each other off. Since they are all products of reliable groups, one wonders if the different results do not reflect the different modes of presentation: report, NBS pamphlet, and journal. However, the citation-number differences are so small that they may not be significant.

The magnetic-moment results shown in Fig. 4, make a strong case for publication in a regular journal rather than in appendices to conference proceedings. None of the frequent Shirley revisions[11] appearing in rather specialized volumes has established itself as a widely consulted source in the same way the Fuller and Cohen[12] work has done.

Finally, a table is presented comparing citation counts for research papers [13-16] and a data compilation[17] all having the same author. The moral is easily seen to be that if you want your name to make a splash in the Citation Index, publish a table.

Citations of Research Papers and Compilation by Same Senior Author

	Research Papers				Compilation
	1961[13]	1962[14]	1964[15]	1965[16]	1965[17]
61-64	25	12			
65	7	11	0	2	
66	4	8	1	13	0
67	8	12	3	17	44
68	7	12	3	11	55
69	8	8	3	12	37
70	6	8	0	14	77
71	3	9	1	3	49
72	0	4	1	6	58
73-78					221
Totals	68	84	12	78	541

13-17. See References

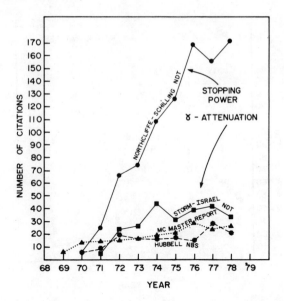

Fig. 3   Citation numbers for stopping-power and gamma-attenuation tables.   See Refs. 7-10.

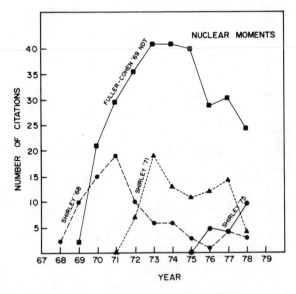

Fig. 4.   Citation numbers for nuclear-moment tables.   See Refs. 11 and 12.

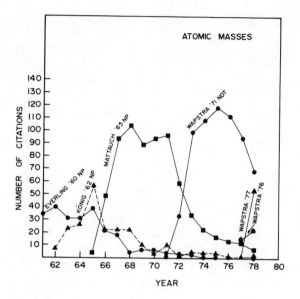

Fig. 1.   Citation numbers for mass tables (not Q-values).   See
Refs. 2 and 3.

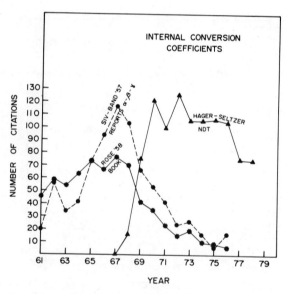

Fig. 2.   Citation numbers for tables of internal-conversion
coefficients.   See Refs. 4,5, and 6.

1. Science Citation Index, Inst. for Scientific Information, Inc. Philadelphia. (1961-   ).

2. F. Everling, L. A. Konig, J. H. E. Mattauch, and A. H. Wapstra, Relative Nuclidic Masses, Nucl. Phys. 15:342 (1960).

b. L. A. Konig, J. H. E. Mattauch, and A. H. Wapstra, 1961 Nuclidic Mass Table, Nucl. Phys. 31:18 (1962).

c. J. H. E. Mattauch, W. Thiele, and A. H. Wapstra, 1964 Atomic Mass Table, Nucl. Phys. 67:1 (1965).

3. A. H. Wapstra and N. B. Gove, At. Data Nucl. Data Tables 9:267 (1971).

b. A. H. Wapstra and K. Bos, 1975 Midstream Atomic Mass Evaluation, At. Data Nucl. Data Tables 17:474 (1976).

c. A. H. Wapstra and K. Bos, Atomic Mass Table, At. Data Nucl. Data Tables 19:177 (1977).

4. L. A. Sliv and I. M. Band, Reports 57 ICC K, 58 ICC L, Univ. of Illinois (1957, 1958).

b. L. A. Sliv and I. M. Band, Tables of Internal Conversion, p. 1639 in:"Alpha-, Beta-, and Gamma-Ray Spectroscopy". K. Siegbahn, ed., North-Holland Publishing Co. Amsterdam (1965).

5. M. E. Rose, "Internal Conversion Coefficients", North-Holland Publishing Co., Amsterdam (1958).

6. R. S. Hager and E. C . Seltzer, Internal Conversion Tables, Nucl. Data Tables 4:1 (1967).

7. L. C. Northcliffe and R. F. Schilling, Range and Stopping-power Tables for Heavy Ions, Nucl. Data Tables 7:233 (1970).

8. W. H. McMaster, et al., Compilation of X-Ray Cross Sections, 1 kev to 1 MeV, UCRL-50174 (1969, 1970).

9. E. Storm and H. I. Israel,Photon Cross Sections from 1 keV to 1 MeV for Elements z=1 to z=100, Nucl. Data Tables 7:565 (1970).

10. J. H. Hubbell, Photon Cross Sections, Attenuation Coefficients, and Energy Absorption Coefficients from 10 keV to 100 GeV, NSRDS-NBS 29, U. S. Government Printing Office, Washington, D.C. (1969).

11. V. S. Shirley, Table of Nuclear Moments, p. 985 in: "Hyperfine Structure and Nuclear Radiation", E. Matthias and D. A. Shirley, eds., North-Holland Publishing Co., Amsterdam (1968).

b. V. S. Shirley, Table of Nuclear Moments, p. 1255 in: "Hyperfine Interaction in Excited Nuclei". G. Goedring and R. Kalish, eds., Gordon and Breach Science Publishers, New York (1971).

c. V. S. Shirley, and C. M. Ledrer, Table of Nuclear Moments, pp. I-XXII in: "Hyperfine Interactions Studied in Nuclear Reactions and Decay", E. Karlsson and R. Wappling, eds., Almqvist and Wiksell International, Stockholm (1975).

12. G. H. Fuller and V. W. Cohen, Nuclear Spins and Moments, Nucl. Data Tables 5:433 (1969).

13. P. H. Stelson and F. K. McGowan, Coulomb Excitation of Second
     2^+ States in Even-Even Medium-Weight Nuclei, Phys. Rev.
     121:209 (1961).

14. P.H. Stelson and F. K. McGowan, Coulomb Excitation of the
     First 2^+ State of Even-Even Nuclei with 58 A 82, Nucl.
     Phys.32:652 (1962).

15. P. H. Stelson and F. K. McGowan, Cross Sections for (,n)
     Reactions in Medium-Weight Nuclei, Phys. Rev. 133:B911
     (1964).

16. P. H. Stelson, R. L. Robinson, H. J. Kim, J. Rapaport, and
     G. R. Satchler, Excitation of Collective States in the
     Inelastic Scattering of 14-MeV Neutrons, Nucl. Phys. 68:97
     (1964).

17. P. H. Stelson, Nuclear Transition Probability B (E2), Nucl.
     Data Tables 1:21 (1965).

PROGRESS IN ATOMIC MASS DETERMINATIONS SINCE 1977

Aaldert H. Wapstra
Instituut voor Kernphysisch Onderzoek, Amsterdam
University of Technology, Delft, The Netherlands
and K. Bos
Laboratory for High Energy Physics, ETH-Zürich
CERN, Geneva, Switzerland

INTRODUCTION

The present paper summarizes the situation, after adding the new data reported at this conference, in the field of the experimental determination of masses of atoms, that is, essentially, of total nuclear binding energies. Our starting point will be our 1977 evaluation[1], and we will add several points not yet mentioned at this conference. A general overview is as follows.

Along the line of beta-stability, the quantity of precise experimental data formed already in 1977 a "backbone" of considerable rigidy. We use this fact now to disregard a large collection of old and somewhat questionable mass spectroscopic data. Some consequences, and a few remaining problems along this backbone will be discussed.

In the region of far beta-unstable nuclides, many new data have become available. Foremost among them are the on-line mass spectroscopic measurements reported in the talks of Thibault, of Epherre and of Audi on sometimes very short-lived Na, Rb, Cs and Fr isotopes. The many new determinations of alpha- and beta-decay energies of far unstable nuclei reported in several other talks also yield much information. The agreement between these new data, and with older data, though not yet always perfect, is improving. This will be discussed below, among others hoping to encourage further experimental work.

DATA ON NEAR BETA-STABLE NUCLEI

A very important part of the backbone is now formed by (n,γ)
reactions, with often a precision of a few tenths of a keV.
Greenwood gave us examples in his report. An important development
in this respect are the recommended standard gamma ray energies
presented by Van der Leun, Helmer and Van Assche, based on the
calibration work discussed by Deslattes. Even the number of known
cases where the bombarded isotope is unstable is increasing. These
reactions connect only isotopes of one element. Connections between
different elements are given by ^{A+2}Z ^{35}Cl – $^{A}Z'$ ^{37}Cl type mass
doublets such as the ones reported by Barber and by Johnson. Others
follow from a variety of nuclear reactions (among them, (p,γ) ones)
and by beta decay energies. Unfortunately, the art of very precise
measurements of beta endpoints seems to have fallen into decay,
except for a number of quite usefull measurements of Reddy and
collaborators[2]. A further exception is Tretyakov et al.'s[3] measurement on ^3He which, after a correction for atomic effects[4], yields

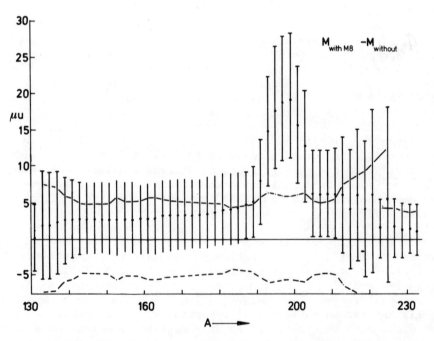

Fig. 1   Differences between atomic masses (for odd mass numbers,
         most stable isobars) calculated with and without use of
         pre-1962 mass spectrometric results (among them, the Hg-
         results marked M8). Error bars indicate statistical errors
         in the "without" values. The errors in the "with" values
         are indicated by the two broken lines.

a ^{3}He - ^{3}He mass difference of 18616 ± 13 eV. Unfortunately, this
value does not quite agree with Bergquist's[4] earlier value
18651 ± 16 eV. It does agree somewhat better with a value derived
from mass spectroscopic work of Smith[5], but I feel that the latter
measurement is not quite without objection and should be repeated.

This result has a bearing on the neutron mass, as has the
neutron capture energy in H(n,γ)D. Here the new measurement
reported at this conference by Greenwood yields a value
2224564 ± 17 eV that unfortunately does not agree with Vylov's
recent value 2224631 ± 16 eV[6]. A measurement by Deslattes[7] may
solve this matter.

A few other problems in mass values at low A have been
mentioned by Nolen. We hope that at least one of these problems
will soon be solved by a mass spectroscopic measurement on ^{6}Li in
Delft.

A few problems like these do not detract from the fact that
the backbone is so rigid that we now judge advisable no longer to
use mass doublets still measured in the ^{16}O atomic mass scale -
that is before 1962 - or with low precision. This had a somewhat
unexpected consequence in the Au-Hg region. In fig. 1 we show the
differences in mass values along the line of stability in two
adjustments made with and without these doublets. A sharp bump is
seen at A = 191 - 203. Fig. 2 shows the differences with the
omitted mass doublets. Differences with the remaining mass doublets,
among them the new values reported here by Johnson, are shown in
fig. 3. The reader is also reminded that the way of assigning
errors to mass doublets causes then to be regularly underestimated
by a factor[1] between 1.5 and 5; this factor has not been applied
to the error bars in fig. 3. Apparently, the 1977 mass excess
values for ^{197}Au and $^{198-202}$Hg are about 20 keV high. It was therefore interesting that Barber reported at this conference differences
of some 20 keV with the 1977 masses. Unfortunately, the differences
are in the wrong direction, as shown in fig. 3. It is planned to
check the situation with the Delft mass spectrometer. Also it
appears to be desirable that results calculated from the backbone
are checked every 10-20 mass numbers by mass spectrometers on what,
mixing metaphors, may be called anchor points. The work depicted
in fig. 3 is a valuable step in this direction, but one would like
an approximately five-fold increase in precision.

Finally, I want to stress that the backbone still has weak
spots, and to ask people with the right tools to improve them. The
worst spots are the following:
a) The nuclides $^{81-82}$Br + ^{82}Kr are connected to higher mass
   nuclides by a doublet with an error of 30 keV (including the
   general correction for consistency) and to lighter ones + ^{81}Se

by links with a combined error of 20 keV.
b) The group nuclides $^{104-105}$Ru + ^{105}Rh + $^{104-105-106}$Pd + ^{106}Ag is
   connected to lighter isotopes with a combined error of 15 keV,
   with one of 10 keV to the group ^{107}Pd + $^{107-108}$Ag which last
   group is in its turn connected to heavier nuclides with a
   combined error of 15 keV. The connection between Cd and Sn could
   be improved somewhat too.
c) A complicated group of connection between ^{133}Cs + $^{132-133}$Ba and
   lighter nuclides has a combined error of 12 keV.
We analyze here the data in the latter region (see fig. 4), in view
of their importance in next paragraph. The closure error in the
cycle ^{129}Xe – ^{136}Xe – ^{130}Ba – ^{129}Ba – ^{129}Xe is 112 ± 32 keV. A
pointer to what may be wrong can be found in a mass spectroscopic
measurement of Epherre et al. (see below) of the $^{128-129-131}$Cs mass
triplet of which the authors derive M – A = –85792 ± 60 keV for

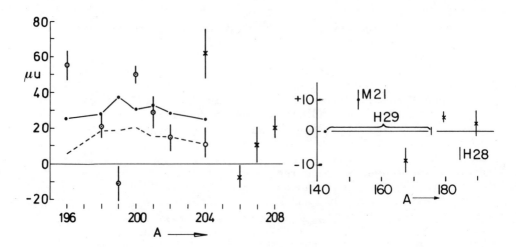

Fig. 2   Differences between the M8 mass doublets and a mass
         evaluation made without these doublets. Circles refer to
         Hg isotopes, crosses to Pb ones. The dots connected by line
         segments represent the 1979 Winnipeg doublets. The broken
         line corresponds to the 1977 mass adjustment. Also in
         fig. 3, positive values correspond to doublets giving
         higher atomic masses than the adjustment.

Fig. 3   Differences between recent mass doublets and a mass
         evaluation made with them. H-values have been measured in
         Winnipeg; H29 is a Nd-Lu comparison. M-values have been
         measured in Minneapólis, the crosses are the measurements
         reported at this conference. Several points represent
         averages for measurements at two nearby mass numbers.

^{128}Cs, which they compare with the 1977 value −85935 ± 6 as a check
on their method. If reinterpreted as a measurement of ^{129}Cs it
yields M − A = −87665 ± 32 keV to be compared with the 1977 value
−87563 ± 24keV. Together with the consideration of fig. 4, this
makes us doubt the correctness of the reported decay energy
^{129}Cs(β$^+$)^{129}Xe and perhaps too of the reaction and mass spectroscopic links of ^{130}Ba with higher mass numbers. Omitting them would
decrease the ^{130}Cs (and ^{130}Ba) mass by 30 keV, the ^{129}Cs one by
60 keV. The situation requires further study. A great help would be
measurement of the ^{133}Cs(p,α)^{130}Xe reaction energy with a precision
of 5 keV or better.

THE PARIS-ISOLDE MASS SPECTROSCOPIC RESULTS.
COMPARISON WITH BETA DECAY DATA.

The measurements reported by Thibault confirm and extend the
data indicating deformation at and beyond N = 20, for Z ≃ 11. Those
by Audi, combined with series of known α-decay energies[1], considerably extend our knowledge of masses in the region of neutron
deficient Tl, Bi, At, Fr and Pa isotopes. Even more valuable is

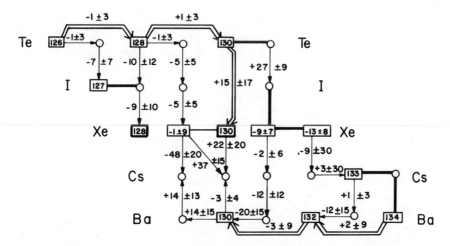

Fig. 4   Connections between data with a bearing on the masses of
         ^{129}Cs and ^{130}Cs. The numbers give the differences between
         the experimental values, with their errors, and the 1979
         mass adjustment. The signs are positive if the experiment
         indicates greater stability for the nuclide at the arrow-
         head. Double lines represent differences measured by mass
         spectroscopy. Rectangles with content B ± b give other mass
         spectroscopic data indicating B keV more stability. The
         measurements at ^{128}Xe and ^{130}Xe, and reaction links drawn
         with thick lines, are so precise (error < 2 keV) that their
         results are only very little changed in the adjustment.

the expansion of the body of known masses in the region of Rb and
Cs masses. The discussion of these data is not only based on
Epherre's report to the conference but also on a paper published
very recently[8]. The authors measure triplets of Rb and Cs isotopes.
They represent each measurement by the resulting mass of one of
these isotopes calculated using adopted values for the two others.
These mass values are strongly correlated[9] and cannot be used as
input values in a least squares mass evaluation. We therefore
replaced them by calculated values for linear combinations of the
three masses selected as discussed elsewhere[9]. The choice of the
combinations is not unambiguous and it is hoped that the Paris-
Isolde group themselves will reformulate their results in this way.

We combined the linear combinations in a least squares
calculation with nuclear reaction and decay data as in our 1977
adjustment[1]. Among these data were provisional values of the beta
decay energies in the neutron-rich Rb and Cs isotopes reported
at this conference by Keyser, Lund, Munnich and Wollnik, and
revised values of Wohn and Talbert[10]. The resulting partial
consistency factor[1] 1.8 for Epherre et al.'s results belongs to the
best ones among mass spectroscopic data (see above). This value is
probably still somewhat high for three reasons. The newer beta decay
energy values agree better with the mass spectroscopic ones than
those used in the adjustment. The linear combinations that we used
to represent the mass spectroscopic data were not yet completely
adequate. And, finally, we did not yet treat the problem of the
influence of isomerism on the mass spectroscopic Rb or Cs lines
at some mass number correctly. Yet, it is interesting to use this
calculation as basis for a comparison (fig. 5).

For the neutron rich Rb isotopes, substantial agreement
exists. Remarkable is the strong correlation causes by the mass
spectroscopic data which makes the adjusted masses at A = 96 - 98
systematically higher not only than the values due to the decay
energies but also than the original mass values of Epherre et al.
This feature requires more study. The neutron rich Cs data are
influenced in an understandable way by Epherre et al.'s choice of
a ^{139}Cs mass value which later appeared to be somewhat high. The
neutron deficient Cs data also show the systematic and progressive
influence of variations in the selected standard masses, in this
case the isotopes ^{129}Cs and ^{130}Cs discussed above.

We are inclined to think that the real masses for neutron
deficient Cs isotopes may differ even somewhat more from the
original values given by Epherre et al. than suggested by fig. 4.
The reason for this is a comparison with other data, among them
the alpha decay measurements reported here by Roeckl et al.
Other new data concern decay energies in beta-delayed proton-[11,12]
and alpha decay[13,14].

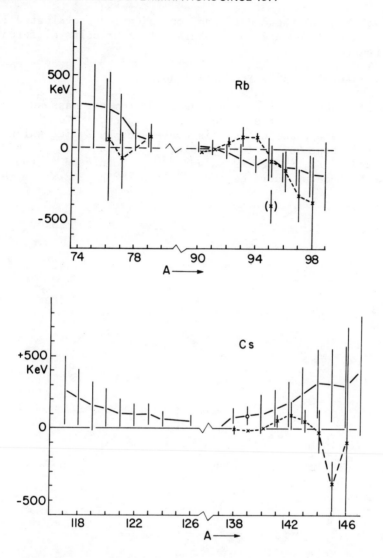

Fig. 5   Differences between mass values as given by Epherre et al.
for Rb and Cs isotopes and such as derived
from available beta-decay energies respectively, with
values calculated in a least squares adjustment in which
both groups of data have been used as input data. The
errors in the mass spectroscopic data (bars connected by a
full line) are the original ones multiplied by 1.5. Of the
values derived from decay energies (crosses) those at mass
numbers 96-96-98 have not been used, and only the (older)
value in parenthesis for A = 95. The value at A = 139
marked with a circle was used as input value by Epherre.

The impact of these data can conveniently be evaluated in a graph of mass differences of isobars with four units different charge number

$$M(Z,A)-M(Z + 4,A) = \sum_{0}^{3} \bar{Q} (Z + i,A).$$

The values plotted in fig. 6 are derived from our mass adjustment of July 1979. The input data do not contain the ^{104}Cd, ^{106}Sn and ^{108}Sn decay data reported at this conference by Zylics which make the first isotope 96 keV less stable, ^{106}Sn and ^{110}Te 475 keV more stable, and ^{108}Sn, ^{109}Te, ^{105}Sn and ^{113}Xe (connected to the first of these by α-decay energies and ^{109}Te(εp)^{108}Sn) 252 keV more

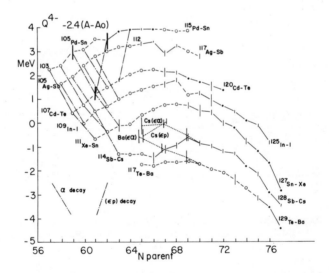

Fig. 6   Difference in mass between isobaric nuclides with ΔZ = 4. Linear functions of A were subtracted in order to show details more clearly. Circles represented values obtained by multiple extrapolation. Dash-dot lines connect points calculated with help of known α-decay or β-delayed p-decay energies. Such lines have not been drawn (in order to avoid confusion) to displayed values for Cs and Ba isotopes calculated from (εp) and (εα) decay energy values not accepted for use in the mass adjustments.

stable. In fact, consideration of the diagrams shown below will
show that these results do not agree nicely with extrapolation from
and interpolation between other data; even worse is the disagreement
caused in α-decay energy systematics (see e.g. ref. 9) by an 348
keV lower ^{108}Sn α-energy as would follow from Zylics data. Some more
confirmation would be useful.

The data on Cs, marked Sn - Cs, are most interesting. The Cs
values plotted correspond up to 200 keV more stability (see fig. 5)
than the values given by Epherre et al. Fig. 6 indicates, though,
that the lightest isotopes measured by them should still be some
200 keV more stable. This conclusion depends strongly on the values
in fig. 6 calculated with help of the new α-decay chains. A systematics of the α-decay energies themselves (not shown here) also
shows a break between ^{116}Cs and ^{117}Cs pointing in the same direction.
Even more stability would be required to improve the fit with
experimental delayed-proton and delayed-alpha decay energies. The
values derived from them deviate, in fig. 6, in the opposite
direction, not only for Cs but also for Ba. Systematics of these
quantities (fig. 7, 8) indicate, though, that these experimental
results underestimate the real values, even if it is assumed (as
done in this adjustment) that the experimental ^{115,117}Xe(εp) decay
energies are somewhat high.

We are incline to evaluate the situation as follows. Systematic
and progressive deviations from the mass values of Epherre et al.

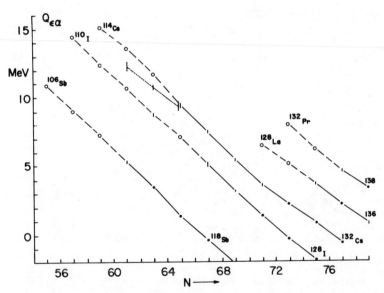

Fig. 7   Beta delayed alpha particle decay energies for odd-odd
         nuclides. Explanation see fig. 8.

up to a few hundred keV for the furthest unstable isotopes do not seriously spoil the consistency with their primary experimental data (see fig. 4-5). Other experimental data indicate that such a shift is necessary at least for the neutron deficient Cs isotopes. The great value of the work of Epherre et al. demands that an effort be made to make an absolute measurement (something like C_9H_{10} - ^{118}Cs) for some of the most unstable of these isotopes.

CONNECTING α-CHAINS IN THE REGIONS Z = 66 - 78, N = 82 - 105.

New alpha particle energies in this region have been measured by several groups[25-29] and reviewed here by Roeckl and Toth. They give a very useful extension to our knowledge of the mass data. As mentioned earlier[1], most experiments do not prove that the α-branch measured feeds a ground state. Fig. 9 shows that this effect is not very important, but the effects might accumulate except for even-even nuclides where the ground state transitions are always by far the most intensive ones. A better check could therefore be obtained by making a systematics of sums of alpha decay energies in chains

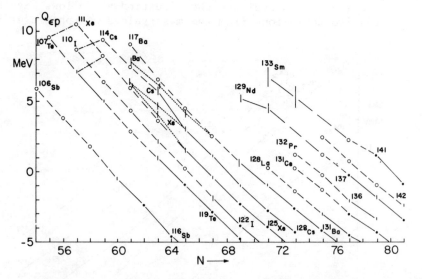

Fig. 8   Delayed proton decay energies for odd-N nuclides. Circles represent points obtained by interpolation, dash-dot lines mark values connected through known α-decay energies. Dotted lines connect experimental values for Xe, Cs and Ba not accepted for use in our mass adjustment.

of three or four successive decays as given in fig. 10. It does
not appear here that the points for even-odd nuclides are
systematically low compared with those for neighbouring even-even
nuclides. We therefore use these data with confidence though we
increase their errors (most < 10 keV) to 50 keV. Most of these
very neutron deficient chains do not contain a member for which
other data are known with a bearing on its mass. Pardo et al.[19]
recently made a break-through by measuring the reaction energy
in ^{144}Sm(^{12}C, ^{10}Be)^{147}Gd. Together with the data shown in fig. 9,
including an estimated value for ^{162}Hf(α) with an error of
decidedly better than 200 keV, this allows a determination with
essentially this error of the mass of ^{178}Hg, 18 neutrons away
from stability.

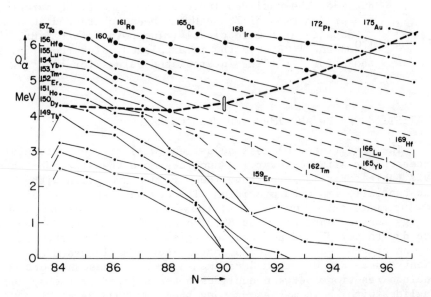

Fig. 9   Systematics of alpha decay energies as a function of the
neutron number of the parent. New values (since 1977) are
plotted as thick dots. The alpha decay chain marked with
a thick broken line allows to calculate the mass of ^{178}Hg
from experimental data and only one interpolation, for the
alpha decay energy of ^{162}Hf (shown).

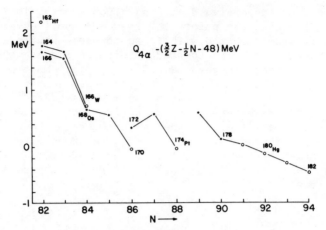

Fig. 10   Systematics of sums of decay energies in four successive
          α-decays, as a function of the neutron number of the final
          nuclide. The nuclides indicated are the first parents. A
          linear function of Z and N has been subtracted in order
          to show details more clearly. The points for ^{162}Hf, ^{166}W,
          ^{170}Os and ^{174}Pt use the interpolated value from fig. 1,
          those for $^{179-182}$Hg interpolated values for $^{167-170}$W.

STATISTICS OF NEW AND IMPROVED MASS EXCESSES

     The next (interim) mass adjustment will give experimental mass
excesses for slightly over 130 nuclides more than the published
1977 table[1]. For 130 other nuclides, values are given with
substantially improved precision. The new mass spectroscopic
measurements take care of 50 of these 230 values, and new reaction
energy measurements of about 80. The majority, about 130 cases,
are due to new β⁻ and β⁺ decay energies, about 50 of them in the
region of the neutron-rich Rb and Cs isotopes discussed above. Some
twenty cases are due to α-decay measurements. These numbers do not
include reactions between nuclides not connected to the system of
nuclides with known masses – among them numerous α-decays mentioned
above between far β-unstable nuclides – but also not many new or
improved links along the "backbone".

     In some forty cases, the precision found in 1979 is less than
that reported in the 1977 tables. About half of them, mostly in
the mass number region 20-40, are caused by no longer using
pre-1962 doublets. In four cases, the cause was the emergence of
new reaction or decay data somewhat at variance with earlier
result, in three cases we discarded results in view of disagreement
with systematics. Twelve cases result from a different way of
treating nuclides for which two reaction links give discrepant

values for the mass excesses. In such cases, a direct least squares adjustment gives the right average result but a too small error estimate. These cases have now been treated in the way discussed in the description of the 1977 mass adjustment (p. 14 in[1]) but there not yet carried hrough.

Among the reaction energy measurements some new types occur' that deserve mention (see table 2). The most exciting are the pion double charge exchange reactions discussed at this conference by Nann. Their reaction energies give directly the mass difference between two nuclides with $\Delta Z = 2$ (double beta decay energies). The first value ^{16}Ne $- ^{16}$O $= 29140 \pm 500$ keV determined this way[20] was not more accurate than the value 28830 ± 140 keV in the 1977 mass table derived, though, with an isobaric mass equation. Both are outdated by the value 28669 ± 80 keV following from the ^{20}Ne$(\alpha, ^{8}$He$)^{16}$Ne reaction[21]. The new pion charge exchange reactions mentioned in table 2 are considerably more precise and give very valuable information. We hope that this type of measurement will be continued.

If the stream of valuable new data continues at the present rate, the next adjustment masses tables – which we expect to publish in 1982 – will be more different from its predecessors than many scientists might have expected.

Table 2. Nuclides for which mass excesses are newly determined (underlined) or much improved from Q-value measurements on the previously often uncommon type reactions mentioned.

$(\pi^{\mp}, \pi^{\pm})$	^{9}He	^{12}O	^{16}Ne	^{18}C	^{24}Si	^{26}Ne	^{32}Ar
$(^{3}$H$, ^{3}$He$)$	^{54}V	^{58}Mn	^{70}Cu	^{80}As	^{82}As	^{100}Nb	
$(d, ^{3}$He$)$	^{69}Cu	^{73}Ga	^{75}Ga	^{147}Pr	^{149}Pr	^{153}Pm	
$(^{3}$He$, d)$	^{121}I	^{123}I	^{130}Cs				
$(^{3}$He$, ^{6}$He$)$	^{61}Zn	^{87}Zr	^{103}Cd	^{109}Sn	^{141}Sm		
$(^{3}$He$, ^{8}$Li$)$	^{15}F	^{27}P					
$(\alpha, ^{8}$He$)$	^{12}O	^{16}Ne	^{36}Ca	^{50}Fe	^{54}Ni		
$(\alpha, ^{7}$Be$)$	^{61}Fe						
$(^{7}$Li$, ^{7}$Be$)$	^{48}K						
$(^{12}$C$, ^{9}$Be$)$	^{147}Gd						
$(^{12}$C$, ^{10}$Be$)$	^{146}Gd						
$(^{16}$O$, ^{14}$C$)$	^{98}Pd						
$(^{18}$O$, ^{15}$O$)$	^{21}O						
$(^{18}$O$, ^{17}$F$)$	^{19}N	^{27}Na					
$(^{18}$O$, ^{20}$Ne$)$	^{62}Fe	^{68}Ni					
$(^{48}$Ca$, ^{51}$V$)$	^{15}B						

Acknowledgements. The calculations mentioned in the text have been performed on the IBM .370-68 of CERN. We thank CERN's DD-division for allowing this work.

REFERENCES

1. A.H. Wapstra and K. Bos, Atomic Data and Nuclear Data Tables 19 (1977) 175; 20 (1977) 1.
2. T.S. Reddy et al., Zs. Physik 284 (1978) 403; J. Phys. G3 (1977) 633, 637 and others.
3. E.F. Tretyakov et al., Izw. Ak. Nauk SSSR 40 (1976) 2026.
4. K.E. Bergquist, Nucl. Phys. 39B (1972) 371.
5. L.G. Smith and A.H. Wapstra, Phys. Rev. 11C (1975) 1392.
6. T. Vylov et al., Yadern. Fiz 28 (1978) 1137.
7. R.D. Deslattes, priv. comm. 1979.
8. M. Epherre et al., Phys. Rev. 19C (1979) 1504.
9. A.H. Wapstra, Proceedings Conference Far from Stability, Nashville 1979, North Holland Publ. Cy.
10. E.F. Wohn and W.L. Talbert Jr., Phys. Rev. 18C (1978) 2328.
11. J.M. d'Auria et al., Nucl. Phys. 301A (1078) 397.
12. D.D. Bogdanov et al., Nucl. Phys. 303A (1978) 145.
13. B. Jonson and P.G. Hansen, priv. comm. 1976.
14. D.D. Bogdanov et al., Phys. Letters 71B (1977) 67.
15. C. Cabot et al., Zs. Physik 283A (1977) 221, 287A (1978) 71.
16. U.J. Schrewe et al., Zs. Physik 288A (1978) 189;
    S. Hoffman et al., Zs. Physik 291A (1979) 53.
17. E. Hagberg et al., Nucl. Phys. 318A (1979) 29.
18. G.D. Alkhazov et al., Zs. Physik 291A (1979) 397.
19. R.C. Pardo et al., Bull. Am. Phys. Soc. 24 (1979) 666.
20. R.J. Holt et al., Phys. Letters 69B (1977) 55.
21. G.J. Kekelis et al., Phys. Rev. 17 (1978) 1929.

Participants

Alburger, David	Brookhaven National Lab.
Arnould, Marcel	Université Libre de Bruxelles
Audi, Georges G.	Laboratoire René Bernas Orsay
Barber, R.C.	University of Manitoba
Barker, Paul	Auckland University
Bauer, Mariano	Universidad N.A. de Mexico
Becchetti, Fredrick	University of Michigan
Behrens, Heinrich	Fachinformationszentrum, Leopoldshafen
Benenson, Walter	Michigan State University
Berg, Valter	Institut de Physique Nucleaire, Orsay
Blosser, Henry	Michigan State University
Borchert, Gunther	CERN

Brenner, Daeg	Clark University
Browne, Cornelius	University of Notre Dame
Cerny, Joseph	University of California
Cohen, E. Richard	Rockwell International Science Center
Crawley, Gerard	Michigan State University
Davids, Cary	Argonne National Lab.
Davis, Richard	National Bureau of Standards
Derenchuck, Vladimir	University of Manitoba
Deslattes, Richard	National Bureau of Standards
Duckworth, H.E.	University of Winnipeg
Ellis, Y.A.	Oak Ridge Nat'l Lab.
Epherre, Marcelle	Laboratoire René Bernas, Orsay
Galonsky, A.	Michigan State University
Gelbke, K.	Michigan State University
Gierlik, E.	Joint Institute for Nuclear Research, Dubna
Greenwood, R.C.	E G & G Idaho
Hardy, J.C.	Atomic Energy of Canada
Haustein, Peter	Brookhaven National Lab.
Harwood, L.	Michigan State University
Helmer, Richard	E G & G Idaho
Herrera, Julio	Universidad N.A. de Mexico
Hilf, Eberhard	University of Darmstadt

Jacobs, Ludo	K.U. Leuven
Janecke, Joachim	University of Michigan
Johnson, Walter	University of Michigan
Kasagi, J.	Michigan State University
Kashy, Edwin	Michigan State University
Kelson, I.	Tel Aviv University
Keyser, Uwe	Techn. University, Braunschweig
Koch, William	National Bureau of Standards
Koets, E.	Delft University of Technology
Koslowski, Vernon	University of Toronto
Kouzes, Richard	Princeton University
Kummel, Hermann	Ruhr-Universitat Bochum
Kundig, Walter	University of Zurich
Lizurej, H.	Joint Institute for Nuclear Research, Dubna
Lund, Eva	Studsvik Science Research Laboratory
McHarris, William	Michigan State University
Monahan, James	Argonne National Lab.
Morimura, Masanao	Nat'l Res. Lab of Metrology, Tokyo
Munnich, Fritz	Techn. Universitat Braunschweig
Myers, William	Lawrence Berkeley Lab.

Nann, Hermann	Los Alamos Scientific Lab
Naulin, Francois	Institut de Physique Nucleaire, Orsay
Newmann, David	University of Michigan
Nolen, Jerry A.	Michigan State University
Pardo, Richard	Michigan State University
Pearson, J.M.	Universite de Montreal
Petley, Brian	National Physical Lab Teddington
Quentin, Philippe	Institut Laue Langevin, Grenoble
Ragnarsson, Ingemar	Lund T.H.
Resmini, F.	Michigan State University
Robertson, H.	Michigan State University
Roeckl, Ernst	GSI Darmstadt
Rytz, A.	Bureau International des Poids et Mesures
Schmidt, Karl	GSI Darmstadt
Schult, Otto	Institut fur Kernphysik Julich
Sharma, K.S.	University of Manitoba
Sierk, Arnold	Los Alamos Scientific Lab.
Sood, P.C.	Banaras Hindu University
Symons, T.J.M.	Lawrence Berkeley Lab
Takahasi, Kohji	Technischehochschule, Darmstadt
Thibault, Catherine	Laboratoire Rene Bernas Orsay

Theis, Wolf-Gerolf	Universitat Karlsruhe
Tondeur, F.	Universite Libre de Bruxelles
Toth, K.S.	Oak Ridge National Lab.
Tribble, Robert	Texas A & M University
Uno, Masahiro	Waseda University
Van Assche, Pieter	Nuclear Energy Centre & Leuven University
Van der Leun, C.	Rijksuniversiteit Utrecht
Vandeplassche, Dick	K.U. Leuven
VanDyck Jr., Robert	University of Washington
VanWormer, Marc	Michigan State University
Verplancke, Jan	K.U. Leuven
Vieira, David	Los Alamos Scientific Lab.
Wapstra, A.H.	I.K.O., Amsterdam
Way, Katharine	Duke University
Weisser, D.C.	Australian National University
Wildenthal, Bryan	Michigan State University
Wohn, Fred	Iowa State University
Wollnik, H.	Justus Liebig Universitat, Giessen
Zeldes, Nissan	Racah Institute of Physics
Zylicz, Jan	Warsaw University and G.S.I. Darmstadt